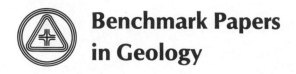 **Benchmark Papers
in Geology**

Series Editor: Rhodes W. Fairbridge
Columbia University

A selection from the published volumes in this series

Related titles of interest

A complete listing of volumes published in this series begins on p. 383.

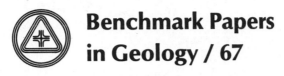

**Benchmark Papers
in Geology / 67**

A BENCHMARK® Books Series

ECONOMIC EVALUATION OF MINERAL PROPERTY

Edited by

SAM L. VanLANDINGHAM
Consulting Geologist/Environmentalist
Cincinnati, Ohio

Hutchinson Ross Publishing Company

Stroudsburg, Pennsylvania

LIBRARY OF CONGRESS CATALOGING IN PUBLICATION DATA
Main entry under title:
Economic evaluation of mineral property.
   (Benchmark papers in geology; 67)
   Bibliography: p.
   Includes indexes.
   1. Mine valuation—Addresses, essays, lectures.
I. VanLandingham, Sam L.   II. Series.
TN272.E32   333.33'9   82-1025
ISBN 0-87933-423-1   AACR2

Distributed worldwide by Van Nostrand Reinhold Company Inc.,
135 W. 50th St., New York, NY 10020.

CONTENTS

Contents

SERIES EDITOR'S FOREWORD

The philosophy behind the Benchmark Papers in Geology is one of collection, sifting, and rediffusion. Scientific literature today is so vast, so dispersed, and, in the case of old papers, so inaccessible for readers not in the immediate neighborhood of major libraries that much valuable information has been ignored by default. It has become just so difficult, or so time consuming, to search out the key papers in any basic area of research that one can hardly blame a busy person for skimping on some of his or her "homework."

This series of volumes has been devised, therefore, as a practical solution to this critical problem. The geologist, perhaps even more than any other scientist, often suffers from twin difficulties—isolation from central library resources and immensely diffused sources of material. New colleges and industrial libraries simply cannot afford to purchase complete runs of all the world's earth science literature. Specialists simply cannot locate reprints or copies of all their principal reference materials. So it is that we are now making a concerted effort to gather into single volumes the critical materials needed to reconstruct the background of any and every major topic of our discipline.

We are interpreting "geology" in its broadest sense: the fundamental science of the planet Earth, its materials, its history, and its dynamics. Because of training in "earthy" materials, we also take in astrogeology, the corresponding aspect of the planetary sciences. Besides the classical core disciplines such as mineralogy, petrology, structure, geomorphology, paleontology, and stratigraphy, we embrace the newer fields of geophysics and geochemistry, applied also to oceanography, geochronology, and paleoecology. We recognize the work of the mining geologists, the petroleum geologists, the hydrologists, and the engineering and environmental geologists. Each specialist needs a working library. We are endeavoring to make the task of compiling such a library a little easier.

Each volume in the series contains an introduction prepared by a specialist (the volume editor)—a "state of the art" opening or a summary of the object and content of the volume. The articles, usually some twenty to fifty reproduced either in their entirety or in significant extracts, are selected in an attempt to cover the field, from the key papers of the last century to fairly recent work. Where the original works are in foreign languages, we have endeavored to locate or commission translations. Geologists, because of their global subject, are often acutely aware of the oneness of our world. The selections cannot therefore be restricted to any one country, and whenever possible an attempt is made to scan the world literature.

To each article, or group of kindred articles, some sort of "highlight commentary" is usually supplied by the volume editor. This commentary should serve to bring that article into historical perspective and to emphasize its particular role in the growth of the field. References, or citations, wherever possible, will be reproduced in their entirety—for by this means the observant reader can assess the background material available to that particular author, or, if desired, he or she too can double-check the earlier sources.

A "benchmark," in surveyor's terminology, is an established point on the ground that is recorded on our maps. It is usually anything that is a vantage point, from a modest hill to a mountain peak. From the historical viewpoint, these benchmarks are the bricks of our scientific edifice.

RHODES W. FAIRBRIDGE

PREFACE

Mineral property evaluation has such vast dimensions that it is almost impossible to adequately survey in one book. This task becomes considerably more difficult if one is to cover mineral property evaluation by choosing the most noteworthy of benchmark papers from the more than 30,000 titles on the subject. Although it probably is most closely related to geology and engineering, economic evaluation of mineral property in the loose sense bears directly on the disciplines of mathematics, risk analysis, chemistry, physics, geophysics, biology, ecology, surveying, management, finance, marketing, accounting, real estate, mineralogy, petrology, stratigraphy, electronics, computer technology, cybernetics, law, taxation, government, conservation, planning, transportation, and even aesthetics and philosophy. In the broad sense it is difficult to define what is not included in mineral property evaluation; practically every phase of such fields as mining engineering, mineralogy, and geology has at least an indirect bearing on the subject.

The importance of mineral property evaluation was recognized long before the time of Agricola (1490–1555) and probably had its beginning before recorded history. Nevertheless, it was not until the latter part of the nineteenth century that mineral property evaluation began to emerge as the sophisticated but widely inclusive science we recognize today. Although there are many historically significant papers on this science, the emphasis in this volume is on modern concepts, particularly geostatistics and computer technology, and their rapid development in the last fifty years.

SAM L. VanLANDINGHAM

CONTENTS BY AUTHOR

ECONOMIC EVALUATION OF
MINERAL PROPERTY

INTRODUCTION

One of the principal reasons there is so much literature on economic mineral-property evaluation is it has been appreciated since medieval times that the accurate assessment of the reserves of a mineral deposit represents the most important and difficult task to the economic mineral industry. Because errors made in assessment of reserves will be transferred through all subsequent phases of a mining operation, accuracy in estimation has always been heavily stressed. Recent advances in geostatistics and computer technology have made significant contributions to increasing the accuracy of reserve assessment, sometimes at the cost of simplicity. This emphasis on mathematics and computers makes it all too easy to form the illusion that this sophisticated technology will in the end finally enable us to arrive at an accurate estimate of the value of a property. Experience, however, has left us with the admonition that the value of a property is determined in the market place and not in the computer room, and that fortunes are as likely to be lost in poor management and financing as in poor prospecting and evaluation.

Due to space limitations it is impossible to give as much attention to many important subjects as would be desired such as formation and log evaluation, basin analysis, hypothetical reserves, reservoir simulation, risk analysis, geochemical prospecting, mine examination, and surveying. *Economic Evaluation of Mineral Property* emphasizes those works that deal specifically with methodology and theoretical applications of reserve estimation as opposed to the abundance of works concerned primarily with descriptions and tabulation figures for specific deposits. Many excellent publications on mineral prospecting and exploration, especially those dealing with extensive geographical areas, are excluded because they do not bear directly on economic evaluation of mineral property or reserve calculations. Many outstanding works could not be included as benchmark papers because of difficulties in obtaining reprints, locating authors, length of publication, and other problems.

With the exception of a few papers of obvious importance such as those by Hoskold, Krige, and Matheron, the items reproduced here are intended only as excellent examples of the abundance of fine publications in economic mineral-property evaluation. In order that many of the noteworthy publications will not be neglected, a review of the literature on economic mineral-property evaluation is presented in Part III of this book. The works on such subjects as exploration, prospecting, and oil reserves are very extensive and a representative sampling of them is included in the review.

Part I

VALUATION AND APPRAISAL

Editor's Comments
on Papers 1 Through 9

Paper 1 is a refinement and condensation of portions of Hoskold's treatise of 1877, which was the most significant treatment of mine valuation at that time and explained the time-tested Hoskold formula still in limited use today. This paper reviews financial elements of mine valuation, summarizes the derivation of formulas for determining present and deferred money values, and discusses the need of a redemption fund.

Papers 2, 4, and 6 are good examples of the abundance of information on metallic ore-deposit appraisal. Paper 2 by Johnson and Bennett is an excellent work on computer applications to ore evaluation. Accompanied by printouts, this article is noteworthy for its continuity, conciseness, and clarity. Various factors that determine the viability of an iron deposit are reduced to a monetary value by Ohle (Paper 4), enabling all factors to be totalled economically; this paper is representative of many good papers explaining special problems that geologists and engineers face in appraising a particular ore type. Paper 6 is concerned with such topics as relationships of operating costs, capital costs, and coordinating formulas for metal prices, smelting schedules, and core samples. Olle and Kneller (Paper 3) have done a brief comparative study of some methods of mineral property evaluation with simplified formulas, using sand deposits as an example. One of the best general treatments of modern aspects of valuation is Raymond's article (Paper 5).

Papers 7, 8, and 9 display some of the great variety of topics to be found in evaluation from the financial, accounting, and legal standpoints. Grant's article (Paper 7) is outstanding in that it deals directly with the complexities of negotiations by use of computers for obtaining information on financial analysis by sensitivity of profit indicators, sharing, or risk and profit. Grant uses hypothetical examples to clarify this financial evaluation. Krige (Paper 8) has given us an excellent view of the relations between mine taxation and mine economics by comparisons of mining tax rates, bases of depreciation of capital assets, and special allowances that occur in different countries, all of which can affect the cutoff grade; the ideal type of tax structure for mining is discussed. The field of forensic mineral-property evaluation often has been neglected and it is for this reason that papers such as Wing's (Paper 9) are of value. Paper 9 is related to discussions concerning jury, expert witness, client, and attorney. Wing suggests that a geologist in court keep the answers to the following fundamental questions in mind at all times. What are your qualifications as an appraiser? What is the basis for your assignment of value? What method did you use? What is the value of the property?

1

Reprinted from *Am. Inst. Mining Engineers Trans.* **33**:777–789 (1903)

The Valuation of Mines of Definite Average Income.

BY H. D. HOSKOLD, INSPECTOR-GENERAL OF MINES OF THE ARGENTINE
REPUBLIC, AND CHIEF OF THE NATIONAL GOVERNMENT OFFICES
OF MINES AND GEOLOGY, BUENOS AIRES, S. A.

(New Haven Meeting, October, 1902.)

As the theory and the practice of valuing mines have never
been discussed in the *Transactions*, a paper on the subject may
be acceptable, even though not exhaustive. The method here
indicated is set forth, not as a model of perfection to be followed absolutely, to the exclusion of any other system based
on sound scientific, commercial and equitable principles, but
merely as an aid to mining engineers and financiers in estimating the value of mining-property.

This paper is intended merely to point out certain financial
features of the question, and not at all to go into the theories
of the formation of mineral veins, extremely valuable as it
must, of course, be to understand the laws of the earth's formation and of its various changes, which are the basis of such
important sciences as geology and mineralogy, and highly useful as guides in determining the probable existence and extension of mineral deposits.

Financial Elements of Mine-Valuation.

The practice of basing the stock-exchange price of mining
shares upon reports, periodically received from the mine-cap-
tain or manager, stating that in his opinion the mine is "improving" in this or that level, and is worth so much "per
fathom," is, of course, absurd. Everybody experienced in
metalliferous mining is aware that a mineral vein varies in
thickness and also in percentage of metal in different parts of
the same mine. According to the practice just indicated, a
mine must have had as many values "per fathom" as there
have been given reports upon it at intervals during the year,

and the shares bought and sold according to such representations must have varied similarly in price. Hence, if all the shares of the mine had been sold for cash according to any one of such stock-exchange valuations, the purchase-money may have been in excess or defect, according to the local variations in richness indicated by the periodical reports. It would be safer to assess the value of a mine upon the annual net yield, as determined in a proper financial manner.

When a mine is not fully developed, and its capacity of product and profit awaits future determination, the estimate of its present value is difficult. If attempted at all, it must be based upon the quantity and the quality of the mineral already opened up to view, and available for immediate extraction; and, in addition, upon the estimated future extension and yield under probable future conditions—a problem of "deferred benefit," the present value of which is to be treated in a different, special manner.

The estimate of probable future benefit from veins on which little or no work has been done is, of course, still more difficult and doubtful, because there is nothing to rest upon, except the general conclusions of science and practice, and also, perhaps, the analogies furnished by similar mines under exploitation in the same district. Frequently, however, such veins, called "mines," are to be inspected for vendors or intending purchasers. If any estimate of value can be given, it must be that of a probable future benefit, not only uncertain in itself, but also deferred for as many years as would be required to bring the mine to the anticipated productive capacity.

In the valuation of collieries there is greater certainty of a constant yield of mineral. For a fully-developed and producing concern, the present annual yield in tons of coal can be known, and the only difficulty in assessing the present cash-value of the mine consists in estimating the cost and market-price of a ton of coal for a future period of years. In the case of a proposed colliery, in land of which the valuable contents are, through geological investigations, local borings, actual mining operations in the vicinity, etc., more or less definitely known, all the conditions of development, markets, etc., still remain to be estimated. When this has been done as accurately as possible, the present cash-value, as already explained, is to be calculated as the present worth of a probable future value.

When a mine has been laid out, and coal or other mineral has been extracted through a considerable period, it is often assumed, especially in the case of a colliery, that it will continue a given annual output for a certain number of years to come; and on this series of deferred benefits the present value is calculated. But circumstances may permit or require an augmented annual yield in the future; and this may shorten the duration of exploitation, and probable increase to that extent the series of annual benefits. Of course, future variations in cost of labor and supplies, or in market-price of product, may likewise affect, for better or worse, the estimate of present value. Some of these elements of cost may be foretold—such as the augmented cost of extraction due to greater depth attained in mining, or greater distance of underground transportation, as the levels are extended.

Evidently, in the face of such uncertainties, a precise determination of present value is not to be expected. Nevertheless, much can be done in the way of proper and prudent calculations; and estimates based on the scientific use of even partial data are much better than mere speculative guesses.

First, the quality of the available mineral, the cost of installation, the time required for preliminary development, the total period of productiveness and the annual profit have to be determined as correctly as the case permits. Next, the vendor and purchaser must agree upon the rate per cent. expected to be received by the latter for his capital or purchase-money; and also upon the rate per cent. for its eventual redemption. The present value of a unit of the yearly income (or the year's purchase) and the total value, including proper allowance for the redemption-fund, may then be determined by the technical expert in accordance with such agreement, or, in its absence, fixed by him as an arbiter.

In England, where the valuation of mines has long been practiced, it has been customary to allow the purchaser of mining-property a high annual interest. Upon collieries, for instance, the rate is from 14 to 20 per cent. per annum; and upon metalliferous mines still higher, because the risk is greater. For foreign mines, the details of management, economy and profit are further removed from control, and consequently, as the risk is proportionally increased, the purchaser should reckon upon the allowance of a far higher rate, depending upon the class and character of the mine, and probably from 25 to 35 per cent. The lower the percentage allowed upon the capital to be invested, the higher the present value of "*unity*," or the year's purchase, and consequently of the total purchase-money,—and *vice versâ*.

The recouping, by the time the mine is exhausted, of the capital originally invested, is an element that should enter into every valuation; but in practical mining, unfortunately, it is too often neglected. Generally, the only thing done with relation to it is to "write off" the books of accounts a certain sum every year, but without depositing the sum in any bank as a redemption-fund proper. Such a fund should be made up by annual deposits or secure investments which will amount, on the exhaustion of the mine, to the original purchase-money. The need of such a fund will be more fully discussed further on.

Before 1877, valuations of mines were effected in England upon an erroneous basis. Such tables of values as were employed to aid financial transactions were calculated upon the blind assumption that the annual rate of interest for a redemption-fund would be the same as the rate expected or to be paid annually upon the capital invested, whatever that rate might be. In other words, it was assumed that the high rate of profit justly allowed to hazardous investment could be also obtained on deposits for the redemption of capital.

When the two rates did not differ much from 3 or 4 per cent., the calculation of present value was not seriously vitiated. But when the estimated beneficial or dividend rate rose to 20 or 25 per cent. for home mines and from 25 to 35 per cent. for foreign mines, the prudent rate to be reckoned for the redemption-fund could not possibly be so high. If it were so taken, the resulting estimate of the necessary annual contribution to this fund would be too small; and the calculations of the present value, thus involving too small an annual charge for redemption, would give too large a result in the present value. Many years ago, discovering this source of error, the writer was led to frame new tables of valuation, which were published in 1877. These tables contemplated two different rates of interest: the variable or beneficial (dividend) rate, and the rate of redemption—the latter being assumed at such a low figure as could be obtained on deposits made for a period of years in permanent financial institutions.

Formulas for Determining Present and Deferred Money-Values.

Every beneficial interest or dividend of constant amount to be paid periodically, at the end of each year, may be considered as yearly income or annuity, either to terminate with the life of an individual or in a number of years given, or to be perpetual. Any such sum of money left unpaid for a number of years is called an annuity in arrears, and when not payable until after a fixed number of years it is said to be a deferred income or annuity. In either case the annuity is transferable, and may be purchased on certain agreed terms; but each class of annuity must receive a particular mode of treatment, according to the special circumstances.

If money could not be put to use and interest obtained for it, the value of an annuity would be equal to the amount for one year multiplied by the whole number of years the annuity had to run; but as simple or compound interest is involved in every case, it is clear that if A desires to sell to B any annuity which has to run a certain number of years, a definite interest or discount upon every yearly payment of the annuity must be allowed to B.*

To determine the total amount of an annual payment of \$1 for n years, with compound interest, the following formulas may be employed.

Let $r =$ the interest (say, at 3 per cent.) on \$1 for one year (or, say, \$0.03); let $R =$ the amount of \$1 with one year's interest, $= 1 + r$; whence, $R - 1 = r$; and let $n =$ any integral number of years.

At the end of the first year, the first payment of \$1 would be due. At the end of the second year, this amount would

---

* *The Engineer's Valuing Assistant, or a Practical Treatise on the Valuation of Collieries and Other Mines.* H. D. Hoskold, London, 1877.

have increased to $R = 1 + r;$ at the end of the third year to $R^2;$ at the end of 4 years to $R^3;$ and at the end of n years to

$$R^{n-1} = (1 + r)^{n-1}. \qquad (1)$$

For instance, if the term be 21 years, and $r = \$0.03$ (3 per cent.), the amount would be $R^{20} = (1.03)^{20} = \$1.806$, and so on, until for $n = 101$ the amount would be $\$19.219$.

The second annual payment of $1 would, at the end of the period of n years, amount in the same way to as much as the first payment in $n - 1$ years; the third, to as much as the first in $n - 2$ years, and so on.

Let $M_n =$ the sum of all these amounts, $i.e.$, the grand total of all annual payments for n years with compound interest. Then,

$$M_n = 1 + R + R^2 + R^3 + \ldots \ldots + R^{n-1},$$

or the sum of a geometrical series of n terms beginning with unity and with the common ratio R. Multiplying both sides of this equation by R, and then subtracting it from the new equation thus formed, we have

$$M_n R - M_n = R^n - 1,$$

and, dividing both sides by $R - 1$, we have

$$M_n = \frac{R^n - 1}{R - 1};$$

or, since $R - 1 = r$,

$$M_n = \frac{R^n - 1}{r}. \qquad (2)$$

Thus, at the end of the first year and at 3 per cent., the amount would be $\dfrac{R^1 - 1}{r} = \dfrac{1.03 - 1}{.03} = \$1$. At the end of 20 years ($n = 20$), it would be $\dfrac{R^{20} - 1}{r} = \dfrac{1.806 - 1}{0.03} = \$26.87$; and for $n = 100$, $M_{100} = \$607.288$.

Let $V_n =$ the present value of $1 due n years hence. In determining this, we simply reverse the above calculation for the amount of the first annual payment with compound interest. At the end of the nth year, that amount was found (1) to be

11

R^{n-1}, and at the end of the $n-1$st year, $\dfrac{R^{n-1}}{R}$. Since, by the terms of the problem, $R^{n-1} = \$1$, the value at the end of $n-1$ years would be $\dfrac{1}{R}$; at the end of $n-2$ years, $\dfrac{1}{R^2}$, and so on to the beginning of the term when it would be $\dfrac{1}{R^n}$. Hence,

$$V_n = \frac{1}{R^n}. \tag{3}$$

The present value at 3 per cent., compound interest, of $\$1$, payable after 20 years, would therefore be

$$V_{20} = \frac{1}{(1.03)^{20}} = \$0.554 \,;$$

and, similarly, for a term of 100 years, the present value of $\$1$ would be $\$0.052$.

We may compute as follows the sum required to redeem the amount of $\$1$ in a given period; that is, the annuity or yearly sinking-fund which, at compound interest, will amount to $\$1$ at the end of n years.

Let S_n be the yearly payment for the sinking- or redemption-fund for a term of n years, to amount to $\$1$ at the end of that term. Formula (2), above, gives $M_n = \dfrac{R^n - 1}{r}$ as the amount of annual payment of $\$1$. To make this amount, whatever·it be, $\$1$, the annual payments of $\$1$ with compound interest must be divided by $\dfrac{R^n - 1}{r}$; hence

$$S_n = \frac{r}{R^n - 1}. \tag{4}$$

If the rate be 3 per cent., and the amortisation is to be effected in 20 years, we have

$$S_{20} = \frac{0.03}{R^{20} - 1} = \$0.0372$$

as the annual payment required.

Similarly, for 30 years at 3 per cent., the annual payment would be $\$0.02101$ for each $\$1$ to be redeemed. At a rate of only 2.5 per cent., and a period of 30 years, it would have to be $0.0228—an illustration of the obvious and important proposition that the lower the rate of interest for redemption, the larger the annual payment required.

Let us now consider the question, how large an income or annuity (A) $1 would buy at different regular rates (r') of annual profit (or beneficial interest, or dividends), allowing at the same time for the eventual redemption of the $1.

We have evidently

$$A = S_n + r'. \tag{5}$$

That is to say, the beneficial rate or dividend, r', added to the necessary annual payment to the sinking-fund, will give the total income to be bought for $1. For example, the annual payment to the sinking-fund, determined by formula (4) for a term of 20 years, at 3 per cent., would be $0.0372 for each $1 to be redeemed. Then, if the rate of annual net profit or dividend be 20 per cent., we have

$$A = \$0.0372 + \$0.20 = \$0.2372,$$

the total annual income to be purchased by the investment of $1.

Clearly, also, for each dollar or unit of total income, or the annual dividend including the annual sinking-fund payment, the corresponding purchasing price, or year's purchase, or present value, P_n, would be the reciprocal of (i.e., unity divided by) the sum of the yearly total income thus defined; for if in formula (5) the second term $S_n + r'$ be made equal to unity, then the first term must be made $\dfrac{A}{S_n + r'}$; and, for $A = \$1$, we have

$$P_n = \frac{1}{S_n + r'}. \tag{6}$$

This is a very simple and useful principle; but, if known, it had not found its way into English technical literature when the writer introduced it in 1877.

What has been said thus far is only a concise statement of the basis upon which the computation of any series of present and deferred values must proceed. Under the old method of valuing mines in England, when the interest required upon the capital amounted to 20 per cent. per annum, it was assumed that the investment could, as previously noted, be eventually redeemed by sinking-fund payments at the same rate. The present cash value of unity, or a year's purchase, so calculated, was, of course, excessive, because the annual payment for the sinking-fund was made too small. It is unnecessary to illustrate this proposition by examples.

But the foregoing discussion contemplates an income, dividend or annuity which can be expected to begin immediately; whereas, the practical problem frequently involves a period of delay before the series of annual benefits begins, thereafter to continue for a certain number of years. In this case, if n be the number of terms in the series, the value of the series at the time of commencement will be, by formula (6),

$$P_n = \frac{1}{S_n + r'}.$$

To find the present value of P_n, recourse is had to formula (3),

$$V_n = \frac{1}{R^n},$$

which gives the present value of \$1 due at the end of n years. For this case, let t be the number of years before the beginning of the annuity, and the whole period to the end of the annuity be $t + n$ years. Hence

$$V_t = \frac{1}{R^t}$$

for the present value of \$1 due after t years, and for the present value of P_n dollars,

$$P_{t+n} = \frac{P_n}{R^t} = P_n\, V_t. \qquad (7)$$

Since the capital is subject, during the period of deferment, to the risks of mining, R in this formula should be taken as $1 + r'$.

Substituting in (7) the value of P_n from (6), namely, $\dfrac{1}{S_n + r'}$,

$$P_{t+n} = \frac{1}{S_n + r'} \cdot V_t, \qquad (8)$$

which is the present value of a dividend of \$1 per annum for n years, beginning after t years; interest being allowed on capital at one rate, r', and for redemption at another rate, r.

Suppose the period of payments, n, to be 20 years; the deferred period, t, 3 years; the rate of dividend, r', 20 per cent.; and the rate of interest for the sinking-fund, r, 3 per cent. Then

$$P_{t+n} = P_{3+20} = \frac{1}{S_{20} + 0.2} \cdot V_3, \qquad (9)$$

in which $V_3 = \frac{1}{R^3} = \frac{1}{(1.20)^3} = \frac{1}{1.728} = 0.5787$; and the cash-value, P_n, or $\frac{1}{S_{20} + 0.20}$, at the beginning of the dividend period, as determined from formula (6), is 4.21557. Then, since $P_{t+n} = P_n V_t$, we have $4.21557 \times 0.5787 = 2.4395$; that is, \$2.4395 is the present cash value of the yearly income of \$1 for 20 years, beginning at the expiration of 3 years.

The old method of treating this problem was as follows:

The present value of 23 years of the estimated or regular annual dividends at 20 per cent., according to the old tables, was \$4.92453, and the present value of the dividends omitted during the first 3 years was \$2.10648. Subtracting the latter from the former, the present net value of the 20 deferred dividends was \$2.81805, an excess over the true value, as above determined, of \$0.37855. This difference would, of course, vary according to the length of the two periods and the rates of interest and benefit assumed.

Evidently, half-yearly or quarterly payments can with some variations be subjected to calculations under the same formulas, which apply to all certain payments at regular intervals.

It is hoped that the foregoing rules and principles, which have been condensed from an ample statement in the writer's book, already cited, and have been proved for many years and acknowledged to be sound in practice, as in theory, may be of service to those who are called upon to make calculations of this nature.

The Need of a Redemption-Fund.

The redemption of the investment is a question affecting particularly the holder of an entailed estate or an estate in trust; and, if he leases, it affects also the lessee. For the mine at length becomes exhausted; and, unless the holder has invested annually a certain portion of the royalty income, the heirs-at-law would be injured by the total loss of the mineral-title or fee, through such exhaustion. The owner or lessor, under such circumstances, has no right to enjoy and consume, in his own life-time, the whole income from the royalty on the mineral extracted from the estate. If suitable provision be made for redemption at the end of the lease, the annual installments will accumulate until the fund covers the original value of the royalty, and the lessor or his heirs will come again into possession of whatever amount was originally invested in the real estate, the value of which would be continued in another form, for the benefit of the legal successors.

The case of the lessee is somewhat different. In addition to his annual rent or royalty, he needs an annual profit. Yet, as the remainder of his unworked mineral is, year by year, further from the station of surface delivery, the expense of working continually augments. Moreover, the lapse of time may bring changes in commercial conditions, so that the income of the mine is more precarious. Still, when practicable, it is a good principle to count, from the start, upon a redemption of the capital. In assessing the probable annual income from a mine it is customary to estimate so much profit per ton as probably to be realized. But, even with constant markets, the working expenses year by year are greater, and it would consequently be well to consider the income as a yearly decreasing amount, and to take the average of a series of years as a sum to be reckoned in determining the total present or deferred value.

It is interesting to note that in a discussion of this subject provoked by the writer, one of the gentlemen* put the case in the following quaint manner:

"He thought that the foundation of the matter was simply this,—was it intended to eat one's cake, or have it? If one determined to eat one's cake, and

---

* Mr. Alexander Smith, North of England Institute of Mining Enginéers.

that was understood definitely from the commencement, then there was no need of depreciation (redemption) fund ; but supposing it was wanted to enjoy the cake and still have it, then it was necessary to provide for depreciation. He could hardly agree with Mr. Simpson, when he, speaking perhaps feelingly as a director of collieries, 'said that it was a difficult thing to set aside any fixed amount of money for a depreciation fund in bad times, when no profit was being made.'

"If they left the concern (supposing it to be a limited company), as had to be done very often, in the hands of accountants or professional valuers, then the latter would take care that the depreciation fund was provided for, and then it was rather hard for the directors to find the little bit of profit appropriated, or a little more loss made for them, in order to provide for the depreciation fund. At the same time, . . . he thought it was very difficult to lay down a fixed principle of depreciation on collieries."

On the same occasion :

"The President said there was one thing that he never could understand, but perhaps some member present could enlighten him upon the question, and that was the question of a depreciation fund. They saw various statements in the accounts of collieries,—not only of private collieries, but in the accounts of those belonging to public companies. There was, perhaps, a certain sum of money written off, as it was called, for depreciation. He could never find out where the sum that was taken off for depreciation went. Another gentleman* said that the proper place for depreciation (redemption) was to have a reserved fund, and the money so reserved should be invested in consols at 3 per cent. He, however, had never heard of a colliery where there was an investment of that kind. Depreciation was generally a sum that was floating about which was at the beck and call of the company when they wanted money. The proper system would be to put by every year the money—redemption annual fund—into some substantial undertaking where it would be quite certain that it could be recovered when it was wanted.

"The President said the depreciation fund was a thing that went off, and nobody saw anything more of it. Certainly, if there was a depreciation fund, then the sum which was taken off the profits should be invested somewhere. As far as his experience went it was simply mentioned in the accounts, and nothing more was heard of it. It was something like the reserved fund which on one occasion he found on examining the accounts of a large gas company. There was a heading not entitled depreciation, but 'reserved fund.' He asked, as was natural for anybody examining the accounts, where the reserved fund was. In his ignorance of the matter he thought it was invested in some bank or some railway company, but the accountant told him 'the reserve fund is what we carry on the concern with.' Since then he had always had a distrust of a reserved fund, and of the wonderful thing called the depreciation fund."

Here we have the opinion of some of the leading and most important people connected with the mining and metallurgical industries of the British Empire; and it is not too much to say

---

* Mr. J. B. Simpson.

that, although it was proved that in usual practice no fund was reserved and accumulated in a bank to redeem capital invested, still the balance of judgment was in favor of such a system.

For reasons broadly stated in the discussion, and other general opinions, members, directors and accountants of public companies would prefer to employ their annual benefit, as is natural, as they pleased, in order to continue a system of re-investing in other shares and in speculation, as has been formerly noted. Besides, there was no law to compel the redemption of capital, and it was "held by the high court of justice in England, in the famous case of Lee *vs.* Neuchatel Company, that it was not obligatory on the directors and shareholders of a company to make any reserve from profits for the purpose of meeting the ultimate loss of the capital which would ensue on the expiring of a lease, or the ultimate exhaustion of minerals."*

There being no law for this object, the questions of prudence and necessity are completely ignored, and consequently there is no attempt to provide for redemption of capital in the form which has been indicated. If, however, such a system were made compulsory, it would tend in a great measure to check immoderate speculation, for it cannot be doubted that, under the system which now rules, many millions of pounds sterling are annually lost.

It is highly probable that in the near future such a law as that indicated may be introduced, and then the question of the practical and proper mode of valuing mines and redeeming capital would have to be considered in the most serious manner at the commencement of every mining undertaking. To-day it is a scientific element entering into every valuation; but then it would become a legally authorized necessity.

---

* Messrs. John H. Armstrong and Thomas Harrison, Discussion, North of England Institute of Engineers.

2

AN ECONOMIC EVALUATION OF AN ORE BODY

EDWARD E. JOHNSON[1]

HAROLD J. BENNETT[2]

The Bureau of Mines is engaged in studies concerning the availability of
and demand for mineral resources that will help government and industry
better understand the total mining problem. The Mineral Resource Evaluation
Division is concerned with the economic availability of and demand for
mineral commodities. In determining the availability of a mineral commodity, it is necessary to make an economic evaluation of the resource
whether it be developed or undeveloped. In the case of the undeveloped
resource it is necessary to determine the "in-ground" value of the mineral
deposit, in which instance it would be necessary to estimate the cost of
producing a marketable product.

The complexity and interrelationship of the various phases of a mining
operation present problems that can be best analyzed by the computer. Applications of the computer for most mine-related problems are in the following
areas: (1) Ore reserve and inventory computation, (2) cost and equipment
analyses, (3) production scheduling, (4) development and mine planning,
(5) profitability analyses, (6) resource evaluation, and (7) economic
feasibility.

The government is most concerned about cost and equipment analyses,
profitability, and measuring the economic value of a resource as it exists in
the ground. In this paper we shall concentrate upon computer applications
in these areas. However, computer applications in the other areas will not be
completely ignored because in many instances the information derived from
these areas of application may be used as inputs to the applications that we
are principally concerned about.

The cost of developing and implementing new methods can be compared
to the reduction in cost that is produced, which provides a meaningful
measure of the benefits that can be derived from research. The rate of return
is a measure of the profitability of an operation; the impact of alterations in
the operations can be evaluated by its effect upon the rate of return. An

[1]*Economist, Bureau of Land Management, Washington, D.C.*
[2]*Mining Engineer, Bureau of Mines, Mineral Resource Office, Denver, Colorado*

evaluation of mineral resources especially as they exist in the ground is needed for many purposes; for example, when deposits are sold or when bids are made for exploitation rights.

A measure of the economic feasibility of exploiting a mineral resource is the desired result of a mineral deposit evaluation. This measure should provide a common basis to compare the results of several evaluations and to indicate the degree of desirability between the various alternatives. The results of an economic evaluation of a mining operation can be measured by a discounted cash flow rate of return. Although this is not the only method that can be used to evaluate a mining operation, it is the one that seems to be most useful for this type of analysis.

Some of the common methods used to evaluate a mineral resource are the average annual return, payback period, present value, and discounted cash flow. Each method provides information that an investor may desire. The discounted cash flow rate of return provides a more meaningful tool for analyzing the mining venture than the average annual return, payback period or present value method because it takes into account the time value of money, accommodates actual as opposed to average annual earnings, produces a common unit of measurement, and evaluates the entire expected stream of income.

The widely used payback method does not consider the entire stream of earnings expected over the life of the property and is only concerned with the time required to recoup the original capital investment. Once the original capital investment has been recovered, the additional income derived from the continuation of the operation is not relevant to the payback analysis. To make a valid evaluation of a mining operation, the additional stream of earnings obtained after the capital investment has been recovered must be considered. The most serious disadvantages of the payback period as a measure of investment desirability are that it ignores the remaining economic life of the operation, provides no measure of anticipated earnings, and ignores the "time value" of money. Time value of money considers the fact that dollars in hand today are worth more than the promise of dollars in the future. If all other factors are equal, projects that return money sooner are more desirable. Early return of profits paves the way for faster reinvestment and higher profit return.

The payback period can be used as a screening device to eliminate projects that do not meet companies requirements as to time to recover the investment. It can also be used to measure the time during which the original investment is at a risk.

The computer model which was developed to perform an economic evaluation of an ore body uses estimates of production costs as input data to arrive at the cash flows of an operation. A graphic illustration (fig. 1) shows the initial investment and the cash flow or expected future earnings resulting from the investment. In the computer model the discounted cash flow rate of return method computes the present value of the cash flows which occur at various time periods. These present values are compared with the capital investment which is present value, and a rate of return is computed. This rate is the return an investor can expect on his investment. The following continuous discounting method is used:

$$P = F \times \frac{e^r\text{-}1}{re^{rt}}$$

where P = present value
F = future sum
e = 2.71828 (approximate)
r = annual rate of interest as a decimal
t = number of years from zero date

The rate of return can also be used to indicate the effect that cost reductions as a result of research or changes in the method of operation have on the profitability of the operation. The model, by use of a computer, is able to produce results easily and quickly when cost factors are varied. Cost factors are varied as a result of changes in improved technology or of changes in different approaches or methods to accomplish a particular segment of the mining and processing operation.

An example indicating how the rate of return can be used to evaluate these changes before actually implementing the changes follows. Assumptions might be made as to alternative methods of processing an ore. The alternatives would result in decreased capital investment and operating costs and a decrease in the recovery of values from the ore. These alternatives would affect the mining operation and the return on investment as illustrated in figure 2. A 16-percent rate of return could have been achieved by mining and processing ore averaging 0.065 ounce of gold per ton using assumption 3 versus 0.115 ounce of gold per ton using the initial operational method. Under one circumstance the lower grade ore could not have been mined and processed and a 16-percent rate of return maintained.

The model consists essentially of four major segments: (1) input data (initialization), (2) accounting procedure (depreciation through production levy), (3) rate of return procedure and, (4) output or operations summary (fig. 3).

The inputs (fig. 4) are the result of estimates based on experience and data concerning the mineral deposit. The percentage error in these estimates

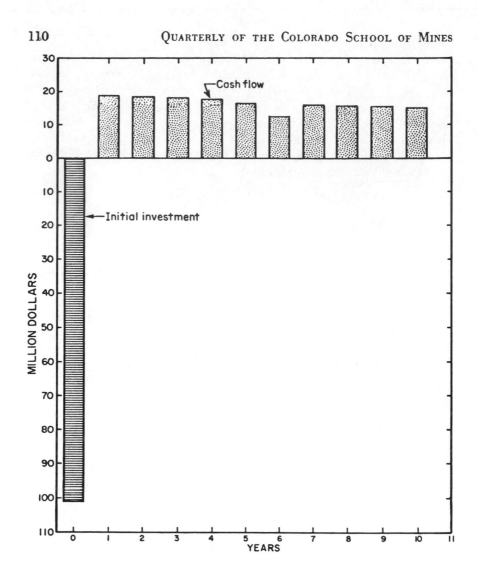

FIGURE 1.-Time Profile Cash Flow of an Operation.

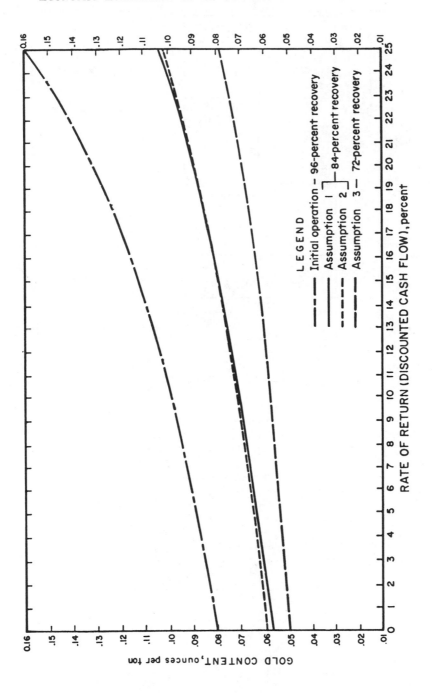

FIGURE 2.—Comparison of Rate of Return of Initial Operation and of Operation Under Each of Three Assumptions.

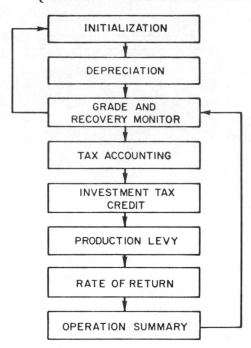

FIGURE 3.-GENERALIZED FLOW DIAGRAM OF
THE COMPUTER PROGRAM

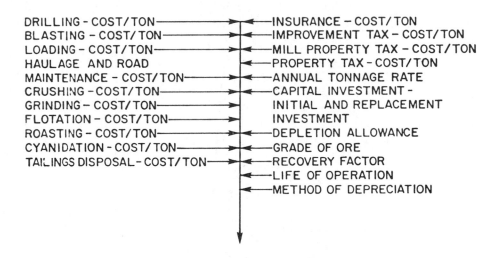

FIGURE 4.- INPUTS TO THE COMPUTER PROGRAM

can be reduced by a careful appraisal of each segment of the operation. Subroutines could be programed to facilitate the calculation of these estimates.

The inputs are certain control factors for decision making as well as investment and operating costs. A brief review of factors affecting the selection of the mining equipment and the interrelationship of the segments of the mining phase follows:

DRILLING AND BLASTING

The selection of drilling equipment for the operation is determined by the rock characteristics of the ore body. The term "drillability" is often mentioned when determining the type and performance of a drill. Factors affecting drillability are the abrasiveness, hardness, specific gravity, and friability of the specimen. Most representatives of the drilling equipment manufacturers will determine the drillability of the rock prior to recommending a particular type of drill and stating its expected performance. The drillability is currently determined using testing procedures in a laboratory. These factors can vary from one deposit to another and therefore cause a difference in the productive capacity of a particular piece of equipment, since each deposit requires separate treatment.

The desired degree of fragmentation is the criteria which influences the blasting as well as the drilling, loading, haulage, and crushing phases of the operation.[1] Indicators that could be used to indicate the degree of fragmentation desirable are shovel loading speed, quantity of secondary breaking required, hauling costs, and bridging delays at the crusher. Drilling costs may or may not be affected by an increase or decrease in the fragmentation desired since the type of explosive, the hole spacing, and depth can all be changed. Loading costs may be affected since an increase in fragmentation may increase the productivity and decrease the wear and tear. Hauling costs, under similar conditions of haul, lift, size, and type of truck and haul road conditions may decrease due to increased fragmentation because of faster loading rates at the shovel and consequent decrease in cycle time. Crushing costs will decrease due to increased fragmentation. More material may pass through as undersize; liner costs, repair and maintenance, and bridging time will decrease and the crushing rate per hour will increase. The decrease in bridging time is twofold in that it does cut down on truck delay time at the crusher which in turn gives higher truck and shovel productivity. Optimizing the blast could be ascertained by evaluating the cost of each operation against the degree of fragmentation under which it occurred.

The computer model can reflect these changes if there is a change in the inputs (operating cost of the drilling, blasting, loading, haulage, and crushing segments) and indicate an increase or decrease in profitability by a change in the rate of return. A more efficient or an optimum mining system could be approached in this manner.

HAULAGE

The primary function of a haulage system is to provide a smooth, regular, and reliable flow of material at a predetermined rate from the pit to the concentrating plant. Several haulage methods—trains, truck, skip, and conveyor belt—or a combination of haulage methods are available, and the selection of a particular method should be based upon an economic evaluation. The following factors relative to the ore body must be considered in these evalutions: (1) size, depth, and shape of the ore body, (2) uniformity of the material, (3) annual production and life of the mine, (4) ratio of waste to ore, (5) haulage distances for waste material and ore, (6) temperature and seasonal problems, and (7) flexibility of the system.

The truck haulage system lends itself to computer programing to develop a subroutine to arrive at a truck selection based on its performance.[2] In such an analysis, all of the performance capabilities of the various machines are programed into a computer system. Data describing one or a multitude of specific haul profiles of the operation is then fed into the computer which will print out the productive capability of each machine studied for each haul profile. Input data as to truck performance is usually available from various truck manufacturers.

LOADING

Selection of the loading equipment depends upon the size of the haulage equipment which in turn is a function of the capacity of the operation. As a rule of thumb in pit operations, the haulage unit should be loaded with four to six buckets of the loading equipment; the haulage and loading units should be balanced so that there is very little slack time in the load-haul cycle; and loading equipment should be fully utilized.

The computer can be used in selecting the most efficient and economic truck-shovel combination.[2] The operating capabilities and capital and operating costs of the truck and shovel are programed into the computer. An analysis is performed on the data varying the truck and shovel combination. Output data would indicate the costs of purchasing and operating the combination from which one would select the truck-shovel combination that would be most economic or suitable for a particular operation.

FINANCIAL MANAGEMENT

When evaluating an operation, most of the emphasis has been placed upon methods of reducing the cost of operating the physical plant and equipment. The use of financial management techniques to reduce costs has often been neglected. For example, certain cost reductions may be achieved through a detailed study of the total tax structure. Lower state and Federal taxes may be obtained using the allowances for investment credit, depletion, and depreciation. The impact of various financial management techniques may result in altering the rates of return significantly.

In the model the ore grade and/or recovery are varied to determine the revenue and note the effect that mining material at various grades and recovery has upon the profitability of an operation. This would be significant when determining cutoff grade at various recovery rates.

The accounting procedure (fig. 3) deals with the computation of depreciation, depletion, and taxes.

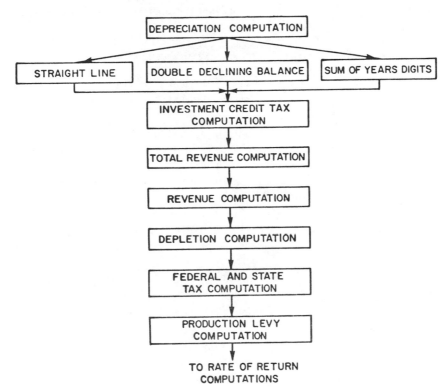

FIGURE 5.- ACCOUNTING PROCEDURE USED IN COMPUTER PROGRAM

DEPRECIATION

The mine owner may deduct from revenue each year, as depreciation, an amount which represents a reasonable allowance for the exhaustion, wear and tear, and obsolescence of depreciable property used in the operation. This allows the mine owner to recover the cost of depreciable property during its estimated useful life.

The allowance for depreciation may be calculated using a number of different methods, of which the three most commonly used are (1) the straight line, (2) the double declining balance, and (3) the sum of years digits.

The amount of a tax reduction resulting from depreciation will vary with the method elected to calculate depreciation. The selection of a depreciation schedule is influenced by the time occurrence of anticipated earnings. The concept of the time value of money implies that the earlier earnings are obtained the greater will be the rate of return. But for many new operations the income for the early years is low. The maximum benefit from depreciation will be obtained as long as net income is positive. When the cost of operation and the amount of depreciation is such that the operation sustains a loss, the full benefit of the depreciation is not being utilized.

DEPLETION ALLOWANCE

The depletion allowance is a tax credit which is granted to owners of mineral properties, and includes two methods of computing depletion: cost depletion and percentage depletion.[3]

In general, cost depletion is computed by dividing the cost of the mineral property by the number of recoverable units in the mineral deposit and multiplying by the number of units for which payment is received during the tax year. The adjusted cost is the original cost of the mineral property less all the depletion allowed or allowable on the property.

Percentage depletion is a certain percentage of gross income from the property during the tax year. The deduction for depletion under this method must not exceed 50 percent of the taxable income from the property computed without the deduction for depletion.

The taxpayer should select as his allowable depletion for any given year the larger of percentage or cost depletion.

For accounting purposes, the depletion allowance is treated as a cost of operation and is subtracted from total revenue, which results in reducing the amount of taxable income. The depletion allowance for domestic mineral production varies from 5 to 27½ percent, not to exceed 50 percent of the net income. The taxpayer does not benefit by the full amount of the depletion allowance as he does from the investment credit.

STATE AND FEDERAL TAXATION

The state taxes imposed on mining operations, in Colorado for example, are the corporation income tax, production levy, improvement tax, and land tax. The state corporation tax is assessed at 5 percent of net income. These taxes will vary from state to state.

The Federal income tax for all corporations consists of a normal tax of 22 percent on the total amount of taxable income and a surtax of 26 percent on taxable income in excess of $25,000. The allowances that may be used to reduce taxes are depreciation, depletion, and the investment credit. Depreciation is used to reduce both Federal and state income taxes. Although depletion is primarily a Federal credit, many states, including Colorado, allow a depletion allowance when calculating state taxes.

The investment credit is a tax relief designed to stimulate new investment in plant and equipment. As much as 7 percent of the cost of new capital investment may be deducted from the Federal income tax burden. If the tax burden is not large enough to absorb the entire credit, the credit may be carried back 3 years. If the credit is not absorbed in the carryback period, the remainder of the credit may be carried forward 7 years.

The printout of the computer is the annual values for depreciation, depletion, taxes, investment tax credit used, net income, and cash flow (table 1). The rate of return is calculated by discounting the annual cash flow by the appropriate discount factor.

A flow diagram of the computer model is contained in appendix A; appendix B contains a computer printout of the compiler sheet.

CONCLUSION

An analysis of an ore body involves both an engineering and economic study. The computer is especially useful because of the ease of handling the multiple interrelation between the mining techniques and associated economic considerations. The computer model, as developed can be used to evaluate the feasibility of economically exploiting an ore body. It can also be used to indicate the effect cost reductions, as a result of research or changes in the method of operation, have on the profitability of the operation. A more efficient or an optimum mining system could be approached using the rate of return to indicate the result that changes in various phases of the mining or processing system have upon the profitability of an operation.

The sequence of statements in the program may vary due to the state tax laws which change from state to state. Taxation laws are sometimes difficult to interpret and one may need the advice of Federal and state tax authorities.

TABLE 1.—Computer print out

Period year	Volume tons	Investment dollars	Depreciation dollars	Depletion dollars	Production levy, dollars	State tax dollars	Federal tax dollars	Credit dollars	Net income dollars	Cash flow dollars	Rate of return
0		101,342,700									
1	11,680,000		10,322,705	4,131,569	516,446	211,875		1,976,653	4,131,569	18,585,843	0.000
2	11,680,000		10,322,705	4,069,570	508,696	208,696		1,946,893	4,069,569	18,461,844	.000
3	11,680,000		10,322,705	4,000,801	500,100	205,169		1,913,884	4,000,800	18,324,306	.000
4	11,680,000		10,322,705	3,961,737	415,395	203,166	638,576	1,256,558	3,323,161	17,607,804	.000
5	11,680,000		10,322,705	3,950,682	257,607	202,599	1,889,828		2,060,855	16,334,242	.000
6	11,680,000	2,184,479[1]	10,382,731	2,781,901	214,040	142,662	1,175,899	152,914	1,606,002	12,586,154	.001
7	11,680,000		10,382,731	3,728,721	243,179	191,216	1,783,28€		1,945,435	16,056,887	.041
8	11,680,000		10,382,731	3,619,935	236,108	185,638	1,731,06€		1,888,866	15,891,532	.067
9	11,680,000		10,382,731	3,502,836	228,497	179,633	1,674,861		1,827,975	15,713,542	.086
10	11,680,000		10,382,731	3,377,176	220,329	173,189	1,614,544		1,762,631	15,522,538	.100

[1]Investment in sixth year is the amount required to replace trucks .

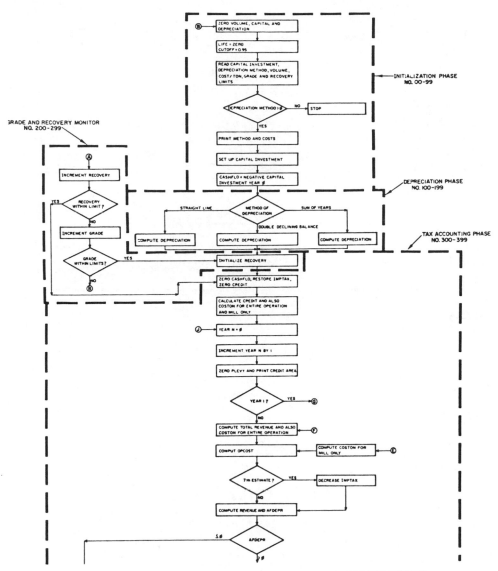

GRADE AND RECOVERY MONITOR
NO. 200-299

INITIALIZATION PHASE
NO. 00-99

DEPRECIATION PHASE
NO. 100-199

TAX ACCOUNTING PHASE
NO. 300-399

APPENDIX A-FLOW DIAGRAM FOR COMPUTER MODEL USED TO CALCULATE DISCOUNTED RATE OF RETURN.

31

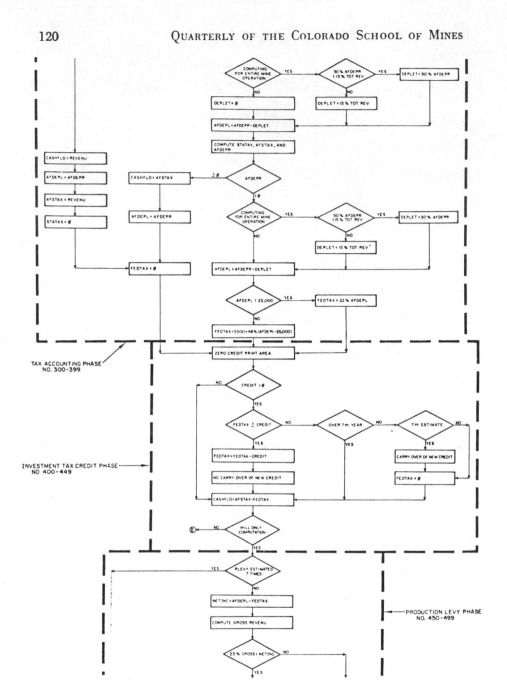

APPENDIX A — FLOW DIAGRAM FOR COMPUTER MODEL USED TO CALCULATE DISCOUNTED RATE OF RETURN (CONT.)

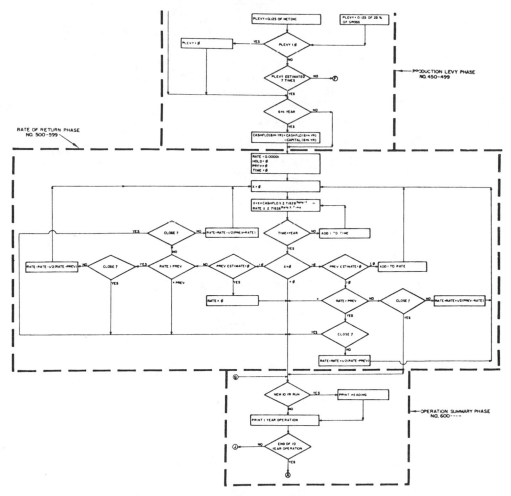

NO. 00 —----- REFER TO STATEMENT NUMBERS ON COMPUTER PRINT OUT OF COMPILER SHEET

APPENDIX A – FLOW DIAGRAM FOR COMPUTER MODEL USED TO CALCULATE DISCOUNTED RATE OF RETURN (CONT.)

```
         C      PROGRAM MINSIM
         C
         C      GENERALIZED MINE SIMULATOR - J G THOMPSON
         C              COLORADO OPERATION
         C
0001     DIMENSION CAPTAL(2,11),DEPREC(2,11),CSHFLO(2,11),CREDIT(2,11),
         1          VOLUME(11),COSTON(2),OPCOST(2),REVENU(2),AFDEPP(2),
         2          DEPLET(2),AFDEPL(2),STATAX(2),AFSTAX(2),FEDTAX(2),
         3          INSUR(2),IMPTAX(2),IMPROV(2),CR(2)
0002     INTEGER R,W
0003     REAL INVSTA,INVSTB,INVSTM,LOAD,INSUR,IMPTAX,IMPROV,NETINC
0004     CALL XCLOCK (1,KSEC,KOLD,3)
0005     R=2
0006     W=3
         C
         C              INITIALIZATION
         C
0007   1 DO 10 N=1,11
0010       VOLUME(N)=0.
0011       DO 10 M=1,2
0012           CAPTAL(M,N)=0.
0013           DEPREC(M,N)=0.
0014  10   CONTINUE
0015     LIFE=10
0016     LIFE=LIFE+1
0017     CUTOFF=.95
0020     READ (R,20) INVSTA,INVSTB,INVSTM,METHOD
0021  20   FORMAT (3F10.0,I1)
         C
         C              FIRST PARAMETER CARD
         C                  COL  1-10 CAPITAL INVESTMENT ( 5 YEAR)
         C                  COL 11-20 CAPITAL INVESTMENT (10 YEAR)
         C                  COL 21-30 CAPITAL INVESTMENT (MILL ONLY)
         C                  COL 31    METHOD OF DEPRECIATION
         C                      1 = STRAIGHT-LINE
         C                      2 = DOUBLE DECLINING BALANCE
         C                      3 = SUM OF YEARS-DIGITS
         C
0022     IF (METHOD) 30,30,40
0023  30   CONTINUE
0024     CALL XCLOCK (3,KSEC,KOLD,3)
0025     STOP FINIS
0026  40 READ (R,50) (VOLUME(N),N=2,11)
0027  50   FORMAT (10F8.0)
         C
         C              SECOND PARAMETER CARD
         C                  CONTAINS TEN 8-COLUMN VOLUMES
         C
0030     READ (R,60) DRILL,BLAST,LOAD,HAUL,CRUSH,GRIND,FLOT,ROAST,
         1    CYANID,TAIL,(INSUR(M),M=1,2),PROPTX,(IMPTAX(N),N=1,2)
0031  60   FORMAT (15F5.5)
         C
         C              THIRD PARAMETER CARD
         C                  CONTAINS COST PER TON IN .XXXXX FORMAT
         C                      COL  1- 5 DRILLING
         C                      COL  6-10 BLASTING
         C
         C                      COL 11-15 LOADING
         C                      COL 16-20 HAULAGE + ROAD MAINTENANCE
         C                      COL 21-25 CRUSHING
         C                      COL 26-30 GRINDING
         C                      COL 31-35 FLOTATION
         C                      COL 36-40 ROASTING
```

34

```
      C                        CCL 41-45 CYANIDATION
      C                        CCL 46-50 TAILINGS DISPOSAL
      C                        CCL 51-55 INSURANCE (TOTAL)
      C                        CCL 56-60 INSURANCE (MILL ONLY)
      C                        CCL 61-65 PROPERTY TAX
      C                        CCL 66-70 IMPROVEMENT TAX (TOTAL)
      C                        CCL 71-75 IMPROVEMENT TAX (MILL ONLY)
      C
0032              READ (R,70) GRADE,GEND,GINC,RBEGIN,REND,RINC
0033        70        FORMAT (6F4.2)
      C
      C                 FOURTH PARAMETER CARD
      C                    GRADE + RECOVERY (BEGINNING, ENDING + INCREMENT)
      C                    EACH FIELD IS IN X.XX FORMAT
      C
0034              WRITE (W,90) METHOD,DRILL,BLAST,LOAD,HAUL,CRUSH,GRIND,FLOT,
           1          ROAST,CYANID,TAIL,(INSUR(M),M=1,2),PROPTX,(IMPTAX(N),N=1,2)
0035        90        FORMAT (1H1,
           1          21H DEPRECIATION METHOD ,I1////16H COST BREAKDOWN //
           2          16H     DRILLING    ,F8.5//16H     BLASTING     ,F8.5//
           3          16H     LOADING     ,F8.5//16H     HAULING      ,F8.5//
           4          16H     CRUSHING    ,F8.5//16H     GRINDING     ,F8.5//
           5          16H     FLOTATION   ,F8.5//16H     ROASTING     ,F8.5//
           6          16H     CYANIDATION ,F8.5//16H     TAILINGS DISP.F8.5//
           7          16H     INSURANCE (T),F8.5//16H    INSURANCE (M),F8.5//
           8          16H     PROPERTY TAX ,F8.5//16H    IMPROV TAX(T),F8.5//
           9          16H     IMPROV TAX(M),F8.5)
0036          CAPTAL(1,1)=INVSTA+INVSTB
0037          CAPTAL(2,1)=INVSTM
0040          CAPTAL(1,7)=INVSTA*(1.03**5)
0041          CSHFLO(1,1)=-CAPTAL(1,1)
      C
      C            DEPRECIATION SCHEDULE
      C
0042          IF (METHOD-2) 100,120,140
      C
      C                 STRAIGHT-LINE METHOD
      C
0043       100    DO 115 N=2,11
0044              IF (N-7) 110,105,110
0045       105        INVSTA=INVSTA*(1.03**5)
0046       110    DEPREC(1,N)=INVSTA/5.+INVSTB/10.
0047              DEPREC(2,N)=INVSTM/10.
0050       115 CONTINUE
0051          GO TO 200
      C
      C                 DOUBLE DECLINING BALANCE METHOD
      C
0052       120 HOLD=INVSTA
0053          DO 135 N=2,11
0054              IF (N-7) 130,125,130
0055       125        INVSTA=HOLD*(1.03**5)
0056       130    DEPREC(1,N)=.4*INVSTA+.2*INVSTB
0057              DEPREC(2,N)=.2*INVSTM
0060          INVSTA=INVSTA-.4*INVSTA
0061          INVSTB=INVSTB-.2*INVSTB
0062          INVSTM=INVSTM-.2*INVSTM
0063       135 CONTINUE
0064          GO TO 200
      C
      C                 SUM OF YEARS-DIGITS METHOD
      C
0065       140 DO 155 N=2,11
0066              HOLD=N
0067              IF (N-7) 141,142,143
```

```
0070        141          HOL=N
0071                     GO TO 150
0072        142          INVSTA=INVSTA*(1.03**5)
0073        143          HOL=N-5
0074        150          DEPREC(1,N)=(7.-HOL )*INVSTA/15.+(12.-HOLD)*INVSTB/55.
0075                     DEPREC(2,N)=(12.-HOLD)*INVSTM/55.
0076        155  CONTINUE
            C
            C
            C            ORE GRADE & RECOVERY MONITOR
            C
0077        200  RECOV=RBEGIN
0100                     GO TO 300
0101        210  GRADE=GRADE+GINC
0102                     IF (GRADE-GEND) 200,200,1
0103        220  RECOV=RECOV+RINC
0104                     IF (RECOV-REND) 300,300,210
            C
            C
            C            TAX ACCOUNTING SECTION
            C
0105        300  DO 305 M=1,2
0106                     IMPROV(M)=IMPTAX(M)
0107                 DO 305 N=2,11
0110                     CSHFLO(M,N)=0.
0111                     CREDIT(M,N)=0.
0112        305  CONTINUE
0113             CREDIT(1,2)=.07*(.33*INVSTA+INVSTB)
0114             CREDIT(1,7)=.07*(.33*INVSTA)
0115             CREDIT(2,2)=.07*CAPTAL(2,1)
0116             COSTON(1)=DRILL+BLAST+LOAD+HAUL+CRUSH+GRIND+FLOT+ROAST+
                1      CYANID+TAIL+INSUR(1)+PROPTX
0117             COSTON(2)=CRUSH+GRIND+FLOT+ROAST+CYANID+TAIL+INSUR(2)+PROPTX*
                1      (INVSTM/(INVSTA+INVSTB))
0120             DO 700 N=1,LIFE
0121                     PLEVY=0.
0122                     CR(1)=0.
0123                     IF (N-1) 600,600,310
0124        310  DO 480 K=1,7
0125                     TOTREV=GRADE*VOLUME(N)*RECOV*35.
0126                 DO 440 M=1,2
0127                     OPCOST(M)=VOLUME(N)*(COSTON(M)*1.03**(N-2)+IMPROV(M))+PLEVY
0130                     IF (K-7) 320,315,315
0131        315              IMPROV(M)=IMPROV(M)-IMPTAX(M)/10.
0132        320          REVENU(M)=TOTREV-OPCOST(M)
0133                     AFDEPR(M)=REVENU(M)-DEPREC(M,N)
0134                     IF (AFDEPR(M)) 390,390,325
0135        325          IF (M-1) 330,330,345
0136        330          IF (.5*AFDEPR(M)-.15*TOTREV*CUTOFF) 335,340,340
0137        335              DEPLET(M)=.5*AFDEPR(M)
0140                     GO TO 350
0141        340              DEPLET(M)=.15*TOTREV*CUTOFF
0142                     GO TO 350
0143        345              DEPLET(M)=0.
0144        350          AFDEPL(M)=AFDEPR(M)-DEPLET(M)
0145                     STATAX(M)=.05*AFDEPL(M)
0146                     AFSTAX(M)=REVENU(M)-STATAX(M)
0147                     AFDEPR(M)=AFSTAX(M)-DEPREC(M,N)
0150                     IF (AFDEPR(M)) 395,395,355
0151        355          IF (M-1) 360,360,375
0152        360          IF (.5*AFDEPR(M)-.15*TOTREV*CUTOFF) 365,370,370
0153        365              DEPLET(M)=.5*AFDEPR(M)
0154                     GO TO 375
0155        370              DEPLET(M)=.15*TOTREV*CUTOFF
0156        375          AFDEPL(M)=AFDEPR(M)-DEPLET(M)
0157                     IF (AFDEPL(M)-25000.) 380,380,385
0160        380              FEDTAX(M)=.22*AFDEPL(M)
```

```
0161                      GO TO 400
0162          385         FEDTAX(M)=5500.+.48*(AFDEPL(M)-25000.)
0163                      GO TO 400
0164          390         AFDEPL(M)=AFDEPR(M)
0165                      AFSTAX(M)=REVENU(M)
0166                      DEPLET(M)=0.
0167                      STATAX(M)=0.
0170                      FEDTAX(M)=0.
0171                      GO TO 400
0172          395         AFDEPL(M)=AFDEPR(M)
0173                      DEPLET(M)=0.
0174                      FEDTAX(M)=0.
              C
              C                 INVESTMENT TAX CREDIT
              C
0175          400         CR(M)=0.
0176                      IF (CREDIT(M,N)) 435,435,405
0177          405         IF (FEDTAX(M)-CREDIT(M,N)) 410,430,430
0200          410         IF (N-9) 415,415,435
0201          415         IF (K-7) 425,420,420
0202          420         CREDIT(M,N+1)=CREDIT(M,N+1)+CREDIT(M,N)-FEDTAX(M)
0203                      CR(M)=FEDTAX(M)
0204          425         FEDTAX(M)=0.
0205                      GO TO 435
0206          430         FEDTAX(M)=FEDTAX(M)-CREDIT(M,N)
0207                      CR(M)=CREDIT(M,N)
0210          435         CSHFLO(M,N)=AFSTAX(M)-FEDTAX(M)-CAPTAL(M,N)
0211          440 CONTINUE
              C
              C                 PRODUCTION LEVY COMPUTATION
              C
0212          450         IF (K-7) 455,480,480

0213          455         NETINC=AFDEPL(1)-FEDTAX(1)
0214                      GROSS=TOTREV-DEPPEC(2,N)-STATAX(2)-FEDTAX(2)-VOLUME(N)*
              1               (IMPROV(2)+COSTON(2)*1.03**(N-2))
0215                      IF (.25*GROSS-NETINC) 465,460,460
0216          460         PLEVY=.125*.25*GROSS
0217                      GO TO 470
0220          465         PLEVY=.125*NETINC
0221          470         IF (PLEVY) 475,480,480
0222          475         PLEVY=0.
0223                      GO TO 500
0224          480 CONTINUE
              C
              C                 RATE OF RETURN - JGT
              C
0225          500         RATE=.00001
0226                      HOLD=0.
0227                      PREV=0.
0230          505         X=CSHFLO(1,1)
0231                      DO 510 K=2,N
0232                      TIME=K-1
0233                      X=X+CSHFLO(1,K)*(2.71828**RATE-1.)/
              1               (RATE*2.71828**(RATE*TIME))
0234          510 CONTINUE
0235                      IF (X) 515,600,550
              C
              C                 RATE ESTIMATE HIGH
              C
0236          515         IF (HOLD) 520,520,525
0237          520         RATE=0.
0240                      GO TO 600
0241          525         PREV=HOLD
0242                      HOLD=RATE
0243                      IF (RATE-PREV) 530,600,540
0244          530         IF (PREV-RATE-.0005) 600,535,535
```

```
0245        535        RATE=RATE-(PREV-RATE)/2.
0246                   GO TO 505
0247        540        IF (RATE-PREV-.0005) 600,545,545
0250        545        RATE=RATE-(RATE-PREV)/2.
0251                   GO TO 505
            C
            C
            C          RATE ESTIMATE LOW
            C
0252        550        IF (PREV) 555,555,560
0253        555        HOLD=RATE
0254                   RATE=RATE+1.
0255                   GO TO 505
0256        560        PREV=HOLD
0257                   HOLD=RATE
0260                   IF (RATE-PREV) 565,600,575
0261        565        IF (PREV-RATE-.0005) 600,570,570
0262        570        RATE=RATE+(PREV-RATE)/2.
0263                   GO TO 505
0264        575        IF (RATE-PREV-.0005) 600,580,580
0265        580        RATE=RATE+(RATE-PREV)/2.
0266                   GO TO 505
            C


            C          OPERATIONS SUMMARY
            C
0267        600        K=N-1
0270                   IF (K) 610,610,630
0271        610        WRITE (W,620) GRADE,RECOV,K,CAPTAL(1,N)
0272        620        FORMAT (1H1,8H GRADE =,F5.2,10X,8H RECOV =,F5.2/////1X,
            1               40H   VOLUME   INVESTMENT   DEPREC     DEPLET,
            2               40H   P-LEVY    STATAX    FEDTAX    CREDIT,
            3               28H  NETINC   CSHFLO  RETURN///
            4               1X,I2,9X,F11.0/)
0273                   GO TO 700
0274        630        WRITE (W,640) K,VOLUME(N),CAPTAL(1,N),DEPREC(1,N),
            1               DEPLET(1),PLEVY,STATAX(1),FEDTAX(1),CR(1),NETINC,
            2               CSHFLO(1,N),RATE
0275        640        FORMAT (1X,I2,10F10.0,F6.3/)
            C
            C          END OF OPERATION
            C
0276        700    CONTINUE
0277               GO TO 220
0300               END
```

APPENDIX B

TABLE B-1. — *Symbols used in computer program*

Variable name	Description
CAPITAL	Capital investment
DEPREC	Depreciation
CASHFLO	Cash flow
CREDIT	Investment credit (total)
VOLUME	Annual tonnage
COSTON	Cost per ton
OPCOST	Operating cost
REVENU	Revenue
AFDEPR	Net after depreciation
DEPLET	Depletion
AFDEPL	Net after depletion
STATAX	State corporation income tax
AFSTAX	After state corporation income tax
FEDTAX	Federal income tax
INSUR	Insurance cost
IMPTAX	Improvement tax (constant)
IMPROV	Improvement tax (destructive area)
CR	Investment credit (annual) (print area)
INVSTA	Capital investment (5 year)
INVSTB	Capital investment (10 year)
INVSTM	Capital investment (mill)
NETINC	Net income
LIFE	Life of operation
DRILL	Drilling
BLAST	Blasting
LOAD	Loading
HAUL	Hauling and road maintenance
CRUSH	Crushing

APPENDIX B

TABLE B-1. — *Symbols used in computer program* (Cont.)

Variable name	Description
GRIND	Grinding
FLOT	Flotation
ROAST	Roasting
CYANID	Cyanidation
TAIL	Tailings disposal
PROPTX	Property tax
GRADE	Grade of ore
GEND	Grade (ending)
GINC	Grade (increment)
RECOV	Recovery
RBEGIN	Recovery (beginning)
REND	Recovery (ending)
RINC	Recovery (increment)
PLEVY	State production levy
TOTREV	Total revenue
CUTOFF	Percentage of total revenue eligible for depletion allowance
RATE	Rate of return
TIME	Time (years)
X	Present value
PREV	Previous rate

REFERENCES

1. Engineering and Mining Journal, 1966, Factors to consider for the drill system and proper blast design can yield economics in the pit: p. 122-126, Sept.
2. Gibbs, L. W., Gross, J. R., and Pfleider, E. P., 1967, Systems analysis for truck and shovel selections: Am. Inst. Mining Metall. Petroleum Engineers Trans., v. 238, p. 354-359, Dec.
3. U.S. Treasury Department, Internal Revenue Service, 1966, Depreciation, investment credit, amortization, depletion: Document no. 5050 (11-66), 20 p.

3

Reprinted from *Compass* **49**:99-105 (1972), by permission of Sigma Gamma Epsilon

PROFITABILITY INDEX (I) AND OTHER METHODS OF MINERAL PROPERTY EVALUATION

John M. Olle[1] and William A. Kneller
DEPARTMENT OF GEOLOGY, UNIVERSITY OF TOLEDO, TOLEDO, OHIO 43606

Abstract

As more and more mineral deposits are zoned out of existence, the geologist must familiarize himself with the methods of mineral valuation that mineral producers use to determine the dollar value of mineral reserves. This demand in part is due to increased population growth and pollution controls. The estimation of the in-ground value of a mineral deposit, especially one of large bulk volume and low unit cost, is essential for the proper development and economic exploitation of any mineral resource. This paper reviews some valuation methods in current use, recommends a modification of the transportation advantage method suggested by P. P. Hudec, *et al.*, (1970), and proposes that individual mineral development projects be compared by the profitability index:

$$I = \frac{Present\ Value}{Total\ Cost.}$$

A high profitability index indicates the project is probably a good investment, whereas one with an index of less than 1 is a very risky or poor investment.

Introduction

The metallic and the non-metallic industries are finding it increasingly difficult to locate new sources of premium grade raw materials near major market centers due to a rapid growth in population as well as an increase in the severity of pollution controls. Many prime deposits are being zoned out of existence, making it necessary for the mineral industries to turn to less desirable deposits farther from the central marketing

[1]Now at Edward C. Levy Co., Detroit, Michigan 48209

area. This deterioration of natural resources by zoning laws, the expanding megapolis and pollution controls, demands that sound economic business principles be employed to select, develop, and rehabilitate any mineral property. This paper reviews some methods that are now being employed to determine the in-ground value of mineral deposits.

Estimating the in-ground value of mineral resources is difficult to ascertain. A major problem in evaluating a mineral deposit is the determination of the true value. The true value is based on the sale of the rehabilitated land and the quality and quantity of usable material and therefore cannot be determined accurately until the deposit is depleted. Land rehabilitation increases the value of any mineral deposit. In cases where a deposit is marginal, the monies obtained from the resale of the reclaimed land can amount to more than those realized from the extraction of the mineral product. Klosterman (1961, p. 235) recommends that mining and reclamation be executed simultaneously because they are interdependent. First, the geometry and character of a deposit must be determined by drilling. This permits planning to take into account the many factors that are involved in mining operations. Such planning often results in the determination of a program that permits mining operations and land reclamation to be executed in the most economical way.

Several methods have been developed in order to meet the need for evaluating a mineral resource. Some of these methods permit the use of the socio-economic factor of land reclamation. Some of the more common methods of mineral resource evaluation are: (1) Comparable Sales, (2) Discounted Cash Flow, (3) Royalty Income, and (4) the Transportation Advantage suggested by Hudec, *et al.*, (1970, p. 89).

COMPARABLE SALES

The comparable sales method evaluates a resource by comparing its value to a value placed on a similar resource in the immediate past. Many factors must be used for evaluating a low value mineral resource. Hudec, *et al.*, (1970, p. 87) list some of these as: (1) the quality of the deposit; (2) the size of the reserves; (3) the demand; (4) the time at which mining begins; (5) the rate of mining; and (6) the transportation costs. Any two resources are rarely identical; therefore, no direct comparison of "similar" deposits can be made by this method.

DISCOUNT CASH FLOW

The process of dividing future income or cash flow by an appropriate interest rate is termed discounting. The capital that remains after expenditures for operating cost and income taxes is the cash flow. The discounted cash flow is an appraisal method used to evaluate an income property that takes the cash flow in each year and discounts it back to the present. This process gives a present value figure for each year. The sum of the individual present value figures for the life of the deposit yields the total value of the deposit.

The discount cash flow method is considered too speculative by the courts when applied to the evaluation of an aggregate resource (Hudec, *et al.*, 1970, p. 89). The courts are skeptical of the use of this method because the following values (Colby and Brooks, 1969, p. 17) are difficult to determine.

"1. Cost of capital improvements
2. Annual operating costs
3. Life of capital improvements
4. Life of resource at rate of production provided by the capital improvements
5. Annual gross revenue at that same rate of production
6. Applicable tax rates."

ROYALTY INCOME

Royalty valuation of a property is not the best means of determining present value for two reasons. First, a royalty in a given area is determined by the highest prevailing royalty or it may vary greatly in widely scattered areas. The fallacy of this method is that the prevailing royalty may not be applied fairly to an individual deposit. Secondly, the owner of the mineral rights is often in a poor position to bargain and thus may not receive an equitable royalty on the resource (Colby and Brooks, 1969, p. 30).

Hudec, *et al.* (1970, p. 89) suggest that if a royalty valuation must be used, it should be a graduated percentage based on the size of the reserves, time and rate of mining, and the distance to the market.

TRANSPORTATION ADVANTAGE

Cost of transportation is important in any valuation method. The transportation expense becomes the primary cost item of the aggregate to the consumer as the distance between the aggregate producer and the consumer increases. The expense of moving a ton of material one mile can amount to five percent of the sale price (Hudec, *et al.*, 1970, p. 89). Therefore, when a producer is forced to move away from the central market, the increased transportation costs may exceed the economic feasibility of the new operation.

The present value must be determined to evaluate the transportation advantage between two similar resources. Dunn, *et al.* (1970, p. 86) determined the present value of the transportation advantage by the formula:

$$V_p = (A\text{-}T\text{-}I)a_n + P + (H)V_n$$

where:

V_p = The present value of the land and mineral resource

A = Annuity payments (transportation advantage, cash flow to producers, government or royalty owner)

P = Principal or present price of the land (purchase price or estimated value of comparable land not considering the minerals)

n = Life of deposit or operation
r = Safe interest rate
r' = Speculative or risk interest rate
T = Property taxes

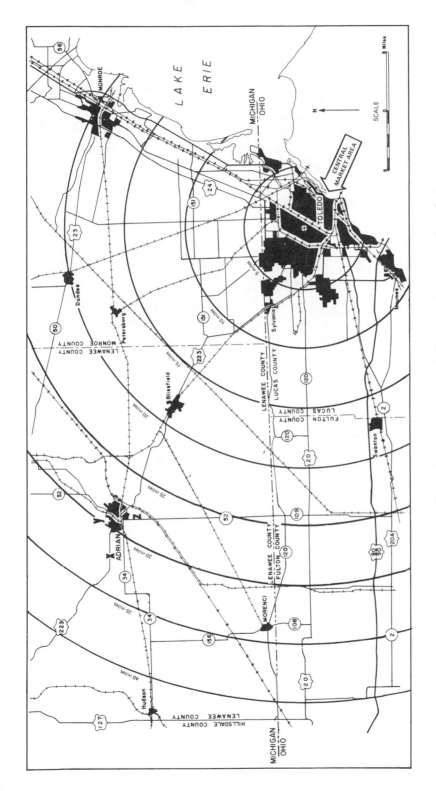

Figure 1.—Map of principal transportation routes and distances from central marketing area, illustrating three sand producing sites (X, Y, Z) for comparison of their respective present values.

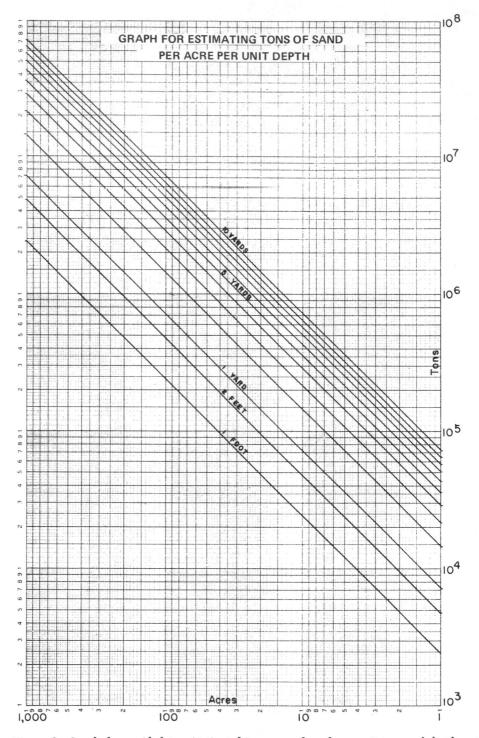

Figure 2.—Graph for rapid determination of tonnages of sand per unit area and depth.

a_n = Inwood ordinary annuity coefficient (speculative rate r' in this case)

I_r = The interest on the principal (P) at the safe rate r

H = Value change of the land as the result of the operation

V^n = Reversion factor. The present value of one dollar to be collected at a given future time when discounted at the effective interest rate for the number of periods (n) from now to the date of collection (at the speculative rate r' in this case)

The authors and John J. Dran, Jr., Assistant Professor of Finance (personal communication) at the University of Toledo, believe the formula should be modified to read $Vp = (A-T-I)a_n + (P + H) V^n$. The modification of P + (H)V^n to (P + H)V^n is necessary because P is the present value of the property to the person from whom the land is purchased. The value of the land to the producer is the future value of the land (P + H). This value is the net cash flow to be realized by the developer from the future sale of the rehabilitated property. The present value of the property is the net cash flow from the property discounted at the speculative interest rate for the number of periods (n) from now to the date of sale of the land.

The following examples will illustrate the application of the modified transportation advantage formula. It is assumed that three similar deposits exist. These are located at "X", "Y", and "Z", which are 32, 30, and 28 miles from the center of the market, respectively (Figure 1). Site "Z" has a $0.20 per ton transportation advantage over site "X" and $0.10 per ton over site "Y", assuming a shipping cost of $0.05 per ton per mile. Site "Y" has a $0.10 per ton advantage over the location at "X". The following values are assumed, using the modified equation:

Vp = To be calculated

A = Respective transportation advantage ($0.20 for site "Z" over site "X", $0.10 for site "X" over site "Y" and for site "Y" over site "Z") for annual production of 700,000 tons

P = $80,000 purchase price of land

r = Safe interest rate of 8%

r' = Speculative interest rate of 12%

T = Property tax, $10,000 per year

a_n = Inwood ordinary annuity coefficient at 12% for 4 years = 3.037 from tables

I_r = Interest on principal ($80,000 at 8%) $6,400 annually.

The principal P is not repaid until the end of the operation.

H = $150,000 increase in land value due to rehabilitation

V_n = Reversion factor = 0.636 (from tables)

Thus, the transportation advantage of "Z" over "X" is:

VP = ($0.20 × 700,000 — 10,000 — 6,400)3.037 + (80,000 + 150,000) 0.636 =

Vp = (162,783) + (146,280) = $309,063

This formula can also be used to evaluate a mineral resource for a producer by making "A" equal to the cash flow. The cash flow is determined from basic accounting as follows:

Gross revenue — operating cost — depreciation — depletion = taxable income

Taxable income — income tax = net income

Then Net income + depreciation + depletion = cash flow

Before any example is presented six assumptions must be made in order to obtain the necessary values for the factors involved in the formula. They are:

(1) Three 100-acre plots are located at sites "X", "Y", and "Z", with material present to depths of 30 feet, 15 feet, and 12 feet, respectively. The tonnage for each site was obtained from Figure 2.

(2) The rate of production must be considered as constant. Since the rate of production is not constant, an average annual production for each site based on the demand of the central market must be estimated.

The demand for the producer is based on the consumption figures of aggregate in the central market. Anticipated growth of the central market also must be considered. The projected tonnage demands were calculated for a 15-year period based on an initial year consumption of 625,000 tons for the central market. These figures were obtained by using simple-compound rate of increase of $2\frac{1}{2}\%$ (Table 1). A constant rate of production was obtained by comparing the total tonnage of a deposit (Figure 2) to the proposed simple-compound increase in production (see Table 1). This yields the total

TABLE 1

PROJECTED TONNAGE INCREASE AT $2\frac{1}{2}\%$ ANNUAL SIMPLE-COMPOUND INCREASE

YEARS	TOTAL TONNAGE
i	625,000
1	640,625
2	656,625
3	673,062
4	689,875
5	707,125
Subtotal	3,992,312
6	724,812
7	742,937
8	761,500
9	780,550
10	800,062
Subtotal	7,802,173
11	820,062
12	840,550
13	861,562
14	883,125
15	905,187
Subtotal	12,112,659

number of years the deposit is expected to produce at the $2\frac{1}{2}\%$ annual simple-compound increase in production. The total tonnage was then divided by the expected life of the deposit in order to obtain the annual average production.

(3) An average purchase price for the land is assumed to be $800 per acre.

(4) The safe rate of interest is taken as 8% and the risk rate as 12% from Hudec, *et al.* (1970, p. 87).

(5) The cash flow is assumed to be $0.40 per ton.

(6) The value change of the land (H) is estimated by obtaining values for lake front lots from realtors. These values vary widely depending on the improvements accompanying the lots. A minimum approximate value appears to be $3,000 per acre. It is assumed that 50 acres are consumed by development of a lake, leaving 50 one-acre lots with a total value of $150,000. This assumes that all expenditures for rehabilitation have been subtracted before the final value of $150,000 is obtained.

The following is a sample calculation illustrating the use of the modified transportation advantage formula to find the present value for the expected market growth of $2\frac{1}{2}\%$:

$A = 699,000 \times \$0.40$ (Average annual tonnage $\times$ cash flow)

$P = \$80,000$

$n = 10.03$ years (Life of deposit at $2\frac{1}{2}\%$ annual simple-compound production increase)

$r = 8\%$ Safe interest Rate

$r' = 12\%$ Speculative interest rate

$T =$ Included in cash flow

$I^r =$ Included in cash flow

$a^n = 5.658$ Inwood coefficient from tables. Figure based on life of deposit at $2\frac{1}{2}\%$ annual simple-compound production increase

$H = \$150,000$ (see assumption 6 above)

$V^n = 0.321$ Reversion factor from tables based on life of deposit at $2\frac{1}{2}\%$ annual simple-compound production increase

$V_p = (699,000 \times \$0.40)5.658 + (80,000 + 150,000)0.321$

$V_p = 790,988 + 73,830 = \$1,655,806$

Table 2 summarizes the present value calculations for 100-acre plots at sites "X", "Y", and "Z".

The present value is a term used to designate the amount of capital that would have to be invested immediately at the speculative interest rate to produce the future cash flow to be received as a result of the investment. However, the present value figure is not the amount of capital that must be invested in the operation, but is the maxi-

TABLE 2

SUMMARY OF PRESENT VALUE CALCULATIONS
FOR THREE SELECTED 100-ACRE PLOTS

Site	Depth of Deposit (feet)	A (Tonnage x Cash Flow)	Life of Deposit (n years)	r	r'	a_n	P	H	Vn	Vp
"X"	30	699,000 x $0.40	10.03	8%	12%	5.658	$80,000	$150,000	0.321	$1,655,806
"Y"	15	676,691 x $0.40	5.32	8%	12%	3.767	$80,000	$150,000	0.547	$1,565,808
"Z"	12	669,014 x $0.40	4.26	8%	12%	3.184	$80,000	$150,000	0.618	$ 994,194

mum value that could be invested and still realize the speculative rate of return.

Johnson (1971, p. 160) discusses a profitability index that permits the comparison of individual projects. The profitability index (I) is obtained by dividing the present value of the project by the total cost of the project. This can be expressed as follows:

$$I = \frac{\text{Present Value}}{\text{Total Cost}}$$

A high profitability index indicates the project is probably a good investment. Conversely, as the profitability index approaches one, the investment becomes less desirable. If "I" is below 1.0, the project should not be considered.

The profitability index is not incorporated into Table 2 because accurate values for the total cost of the individual projects are not available. It is recommended that this method be employed when comparing individual projects where reliable values for the present value and total cost of the individual projects can be determined.

SUMMARY

Evaluation of any mineral resource is a complex and involved process. Several valuation methods are reviewed and a modifica-

tion of the transportation advantage method is proposed. Comparison of individual projects is essential to obtain a maximum return on capital invested. Determination of the present value of several individual deposits and the total costs of their exploitation permits the calculation of the profitability index. This index enables a geologist to decide which project possesses the greatest economic potential.

REFERENCES

COLBY, D. S., and D. B. BROOKS, 1969, Mineral resource valuation for public policy: U. S. Bureau of Mines, Information Circular 8422, 34 p.

DUNN, J. R., P. P. HUDEC, and S. P. BROWN, 1970, How valuable are your mineral reserves: Rock Products, v. 73, no. 9, p. 83-87, 132-133.

HUDEC, P. P., J. R. DUNN, and S. P. BROWN, 1970, Transportation advantage — a unifying factor in mineral aggregate valuation; *in* Proceedings Sixth Forum on Geology of Industrial Minerals, Kneller, W. A., editor: Michigan Geological Survey, Misc. Pub. No. 1, p. 87-94.

JOHNSON, R. W., 1971, Financial Management: Boston, Allyn and Bacon, Inc., 583 p.

KLOSTERMAN, G. E., 1969, Mineral aggregates, exploration, mining, and reclamation; *in* Proceedings Fifth Forum of Geology of Industrial Minerals, Hoover, K. E., editor: Penn. Geol. Bureau of Topographic and Geologic Survey, Misc. Resources Rept. M64, p. 233-241.

4

EVALUATION OF IRON ORE DEPOSITS

E. L. Ohle

Abstract

The evaluation of iron ore deposits involves several types of geological data additional to those normally collected in appraising other mineral deposits. In order to assist geologists who may be faced with the task, all of the various factors, both geological and non-geological, which can influence the viability of an iron ore resource are discussed. Essentially the procedure involves reducing of each of the items to a monetary value so that the pluses and minuses can be totalled. Thus, the relative economic potential of different deposits can be compared.

THE geology of iron ore deposits is so diverse that all sorts of geologic and geophysical techniques are used in their exploration and evaluation. Also, their geographic distribution is so wide that every continent has important productive areas. As a result, the search must be carried on under a wide variety of conditions, from the artic to the tropics, and from well-developed and highly civilized areas where manpower and supplies are readily available, to remote, primitive locations where advance planning is of utmost importance.

The past two decades have seen a revolutionary change in the quality of material demanded by blast furnace operators. High-grade iron concentrates and pellets made from low-grade iron formations are now the object of most exploration work in North America, although in other areas large deposits of direct shipping ore are still being developed. The new importance of what was formerly regarded as unenriched and therefore unattractive iron formation has caused a reappraisal of all the older districts, and major changes in property evaluations have resulted.

Methods used in the search, discovery, and evaluation of iron ore deposits are broadly similar to those applied to base metal and precious metal exploration but with particular emphasis on certain tools which have proved to be especially useful. Programs involving regional reconnaissance geology and geophysics followed by detailed studies of selected areas obviously have been very successful, as is evidenced by the present widespread abundance of newly developed reserves in all parts of the world.

The new emphasis on ore amenable to concentration has made unusually detailed geologic study of the deposits increasingly important, for only these detailed studies can provide the information which is of vital importance to proper evaluation and profitable exploitation. Characteristics such as mineralogy, texture, concentratability, grindability, manner of distribution of ore types, etc. now are routinely investigated, as well as other more traditional factors such as transportation cost, fuel cost, taxes, and power rates. In the evaluation procedure each of these is translated into dollars and cents so that the relative importance of the positive and negative aspects of the various deposits can be compared. Only by complete monetary appraisal of each item can the relative overall profit potential of different deposits be determined.

The profit margin estimate is most often the product of a team effort. The geologist has gathered the basic data, and indeed may be the only one who has actually seen the deposit, but important facts have been developed by engineers and metallurgists working on data and samples the geologists have gathered. Some of the types of knowledge that are applied may seem far removed from the normal expertise of a geologist; and in the more sophisticated details they are, but at least a general comprehension of all of them is essential for the practicing professional economic geologist. The geologist is customarily the member of the team with the best knowledge of the basic environment and physical make-up of the ore in the ground. As such he is in a position to see that the talents of his colleagues are used to the best advantage and often he serves as the general coordinator of the project up to the engineering and construction stage.

Evaluation Procedure

Table 1 lists the more important characteristics of an iron deposit which enter into the evaluation. Many of the factors would also apply to direct shipping ore deposits. The list is divided into those factors which are strictly geological and those which are non-geological. A successful score on the first list is absolutely essential if an iron deposit is to be an eonomic ore body. Without the technical requirements of tonnage, grade, treatability, etc., the availability of a market, cost of power, or the amount of tax burden are only incidental. However, once a successful combination of geologic factors has been established, these other items become of prime importance and most geologists can point to fine mineral deposits which unfortunately do not pass the tests in the second list and consequently remain undeveloped.

For a successful score on the first list, it is not necessary for the deposit to be perfect in every regard. None are. Nearly all deposits have strong plus factors but all of them also have weak points. The crude grade may be good but the grindability difficult; the tonnage may be enormous, but the shape and attitude of the bodies such that more expensive underground mining is required, etc.

Thus the evaluation proceeds with the pluses having to outweigh the minuses to make development attractive. There are so many variables in both the A and B lists that it is necessary to make a complete profit margin estimate, putting actual dollar amounts beside each item to ascertain just what the ultimate value per ton may be. Only after all the various factors have been reduced to common terms—dollars and cents—can a determination of the true profitability of a deposit be made.

No economic estimate is likely to be valid for a long period of time. Technological changes, changes in transportation accessibility or rates, and changes in political factors, to name just a few, cause a constant juggling in the relative economic standings. The great development in autogenous grinding, for example, has caused a reappraisal of grinding costs. Some deposits do not respond well to this method, but others do and a substantial cost saving results.

Likewise, the construction of a major, new rail line leading to one deposit in an area may result in the development of other mines because a major hurdle has been cleared. The minimum tonnage necessary for the second mine is less than for the first.

The development of the St. Lawrence Seaway made it feasible to ship Labrador ore to central U. S. steel plants and thus affected the competitive position of every other deposit supplying these plants. Construction of a larger canal at Panama, enabling larger ships to carry ore from Peru, Chile, or Australia into the Atlantic would have a similar effect, though the recent introduction of huge bulk carriers has tended to equalize ocean shipping rates to points throughout the world.

In recent years, construction of these huge ocean-going bulk carriers has had a great effect on world iron ore traffic. It is now possible for ore to be hauled halfway around the world in 100,000 to 200,000 ton capacity ships and delivered at competitive rates into ports in Europe and North America. Thus in a greater sense than ever before, every iron ore deposit is competing with every other source of iron units and there are few "geographically captive" markets.

The trend toward larger and larger ocean-going ships, of course, requires enlarged port facilities and deeper drafts. This again affects the relative economic standing of nearby ore deposits because some harbors cannot be deepened. A previously less profitable mine may be transformed into a more profitable one than its competitors because the harbor it uses has deep water close to shore and arrangements to accommodate the largest vessels are relatively easy to make.

From the political viewpoint, the main consideration is taxes, although the problems of foreign exchange and repatriation of capital also are important. Farsighted moves by some political entities have tended to attract risk capital. The Canadian tax deferment and liberal depreciation allowances for new mines are examples. Since iron ore has a widespread geographic distribution, mine developers have an unusually wide choice of location and, as a result, tax rates and the other political factors may be expected to exert a great influence on the deposits selected for development in the more immediate future.

Thus it is obvious that profitability estimates are in a constant state of review and it is necessary for every aggressive exploration department to maintain an inventory of world-wide iron resources which is constantly reviewed in the light of changes in the factors that influence profit margins. Such estimates usually pinpoint the negative factors which must be minimized or overcome if the deposits are to succeed. Classic examples of such resources whose development has awaited technological advance are the titaniferous iron ores—both hard rock and beach sands. Enormous tonnages of material containing too much titanium for iron ore and too little titanium for titanium ore exist throughout the world in most advantageous locations from a transportation point of view, but the intimate mixtures of magnetite, hematite, and ilmenite have made separation by physical means uneconomic. A satisfactory solution to this problem could have a considerable impact on the iron ore trade. It is interesting to note that a recent re-evaluation of the effect of titanium on the chemistry of steel-making has concluded that small percentages of titanium may be beneficial rather than detrimental; as a result some of these ores that previously were avoided are being mined and processed.

Some of the following discussion is pertinent only in the evaluation of taconite but most of the factors are as important to direct shipping ore deposits as to concentrating ores.

Ore type and grade

Ore type and grade are interrelated but are not the same thing because the former also involves the amenability of the ore to various kinds of beneficiation. In the case of direct shipping ore, there is, of course, no need for any processing. In the iron ore trade, especially on the Mesabi Range, various ore deposits are classified according to the process used to upgrade the crude to marketable quality; thus the terms Wash Ore, Heavy Media Ore, etc.

Grade of ore, of course, is all-important, as high iron content can compensate for shortcomings in other factors. However, there are wide variations in what can rightfully be called ore, depending on the economics of the other factors in the evaluation. Ores which are easily treated or which make good concentrates may show some profit with a magnetic iron content of only 12 to 15 percent even though the average magnetic iron grade for operating plants in 1971 was in most cases above 20 percent. If ore is extremely difficult and costly to grind, it may take 25 percent or more to break even. Indeed, there are some iron deposits with relatively high assay (40% Fe plus) which are not now ore at all because they are poorly located or will not yield a suitable product with available methods of beneficiation.

A factor in the iron ore trade which is nearly as important as grade is structure. The term "structure" refers to the particle size or screen analysis of the ore and it has become increasingly important as steel mills have modified their furnace practices in their efforts to achieve maximum daily productivity. A friable ore is at a distinct disadvantage in today's market and a factor identified as the "Q" index is closely watched by both buyers and sellers. This index is the product of the percent of $+\frac{1}{4}''$ material before tumbling in a laboratory mill times the percent $+\frac{1}{4}''$ after tumbling. It is related to the basic geologic properties of hardness, toughness, and grain size.

An iron assay report is not complete unless it states whether it represents the total iron content, the acid soluble iron, or the amount of magnetic iron and whether it is wet (natural) or dry. There is a great difference. Normal Mesabi Range taconite runs over 30 percent in total iron but usually is less than 25 percent magnetic iron. Thus a straight magnetic plant usually cannot be expected to recover over 75 percent of the total iron, although it may be recovering 98 percent of the iron present as magnetite. A magnetic roasting plant, on the other hand, will recover a much higher percentage of the total iron present—at least 90 percent if iron silicates are not too abundant. This results from the fact that goethite, hematite, and iron carbonates are all converted to artificial magnetite.

TABLE 1. Evaluation of Iron Ore Deposits

A. *Geological Factors*

1. Type and Grade: Direct shipping, wash, heavy media, taconite, etc.; Rice ratio, impurities, and associated elements.
2. Tonnage: Crude and product, effect on capital cost and recuperation schedules; weight recovery.
3. Grain-Size: Grind size, texture, liberation of ore minerals, elimination of impurities.
4. Grindability: KWH per ton to reduce the ore to concentrating and agglomerating sizes.
5. Mineralogy: Magnetite, hematite, goethite, silicate or carbonate. Impurity mineralogy—effect on the ability to separate the impurities in processing.
6. Distribution of Ore Types: Grades, textures, mineralogies—can selective mining be done?
7. Depth and Nature of Overburden: Sand and gravel, or rock. Open pit versus underground mining.
8. Shape and Attitude of the Ore Body: Tons per vertical foot. Effect on stripping ratio.
9. Location: Topographic effects, climate.

B. *Non-Geologic Factors*

1. Market and Price: Individual company variations in requirements and locations of furnaces.
2. Politics and general business climate.
3. Transportation cost.
4. Labor and Housing: Availability and Cost.
5. Construction cost.
6. Power: Availability and Cost: Fuel
7. Water Supply.
8. Taxes.
9. Royalty Rate.
10. Inflation Factors.
11. Tailings disposal.
12. Environmental Factors: Ecology.
13. Financing.

The question of impurities and associated elements is also involved in the discussion of grade. Relatively minor amounts of certain elements are so noxious in blast furnace operation that the material may be unmarketable even though the iron content is high. Table 2 lists the percentages of different elements which are found in iron ores from various sources available to the American iron and steel industry. It will be noted that there are wide variations from element to element and from deposit to deposit. Some of these assays have a cumulative effect and in general the total amount of minor elements (other than Mn, SiO_2, Al_2O_3, CaO and MgO) in the furnace burden should not exceed 1.0%. The exact figures vary from furnace to furnace and company to company depending on what other ores are available for blending. The limestone and coke, of course, also contribute some trace elements. In iron ore, the presence of an unusually high amount of some minor element usually reduces the marketability of the ore.

Actually, when we speak of the grade of an iron deposit in an economic sense the important consideration really is the weight percent recovery that can be made of a concentrate of a given assay. This defines the amount of saleable product that will result

TABLE 2. Summary of Elements Found in Iron Ores Used by the American Iron and Steel Industry*

Analyses, weight percent

Source	Fe	P	SiO₂	Al₂O₃	Mn	S	CaO	MgO	Ti	V	Cr	Zn	Mo	As	Pb	Sn	Co	Ni	Cu
Cuyuna	55.83	0.367	10.61	4.60	12.92	0.093	0.84	0.35	0.090	0.009	0.019	0.006	0.005	0.006	—	—	—	0.002	0.011
	32.75	0.104	24.82	0.94	0.47	0.008	0.15	0.07	0.005	*	*	*	*	*	—	—	—	*	*
Marquette	61.57	0.180	5.93	4.66	2.46	0.210	0.93	1.53	0.325	0.020	0.010	0.005	0.005	0.006	*	*	*	0.006	0.012
	37.13	0.014	44.98	0.59	0.06	0.008	0.28	0.16	0.03	0.004	0.001	0.002	0.0018	0.002	*	*	*	0.002	0.006
Menominee	59.60	0.465	4.33	3.97	5.69	0.259	1.49	4.20	0.235	0.035	0.015	0.041	0.004	0.014	*	*	*	0.020	0.025
	37.11	0.015	44.57	0.89	0.05	0.006	0.24	0.18	0.065	0.016	0.002	0.012	0.0012	0.005	*	*	*	0.009	0.007
Mesabi	63.20	0.118	3.18	11.48	5.48	0.073	0.72	0.21	0.21	0.010	0.018	0.007	0.005	0.008	*	—	*	0.005	0.019
	49.07	0.028	21.68	0.45	0.07	0.005	0.09	0.07	0.008	*	*	*	*	*	—	—	—	*	—
Steep Rock	60.69	0.033	2.97	2.36	0.28	3.60	0.23	0.22	0.30	0.012	0.072	0.007	0.004	0.03	—	—	0.10	0.02	0.10
	50.80	0.019	10.35	0.98	0.10	0.016	0.08	0.04	0.01	0.006	0.013	*	*	*	—	—	—	*	*
Michipicoten	51.54	0.027	9.81	2.91	2.90	0.098	3.02	8.40	0.075	0.009	0.004	0.010	0.003	0.024	*	0.075	*	0.004	0.025
	49.31	0.016	11.79	1.59	2.79		2.98	7.46	0.056	0.005	0.001	0.005	*	0.020	*	*	*	0.002	0.005
Eastern Penna. (Cornwall)	62.83	0.005	5.10	1.52	0.072	0.031	1.40	1.60	0.08	0.011	0.013							0.013	0.021
Texas (Lone Star)	47.52	0.18	15.34	6.90	0.32	0.10	0.15	0.030	0.026	0.032	0.020	—	0.004	0.013	—	—	—	0.008	0.004
California (Eagle Mt.)	55.00	0.080	11.00	1.60	0.08	0.40	1.80	3.50	0.13	0.003	0.005	—	0.001	0.005	—	0.003	0.020	0.005	0.05
Wyoming (Sunrise)	52.40	0.090	12.68	0.751	0.09	0.036	1.25	0.36	0.07	0.020	0.013	0.058	0.004	0.011	—	—	0.002	0.046	0.005
Utah	57.81	0.310	4.29	1.17	0.06	0.070					0.001			0.019				0.005	0.014
Tenn. C & I (Red Ore)	37.70	0.32	15.50	3.00	0.15	0.30	13.35	0.60	0.080	0.018	0.002	0.001	0.001		0.006	0.001	0.001	0.003	0.004
Tenn. Copper Co. (Sinter)	69.00	0.005	1.50	0.45	0.10	0.07	0.10	0.60		0.008		0.20	0.01				0.05	0.01	0.12
Venezuelan	64.85	0.13	0.98	2.45	0.10	0.043	0.017	0.084	.20	0.006	0.009	—	*	*	—	—	*	—	0.003
	63.00	0.05	0.80	1.80	0.08	0.020	*	*	0.09	—	—	—	—	—	—	—	—	—	*
Chile (Tofo)	60.30	0.042	7.49	1.76	0.09	0.021	1.41	2.00	0.59	0.28	0.008	—	—	—	—	—	—	0.15	0.010
Labrador	58.00	0.058	8.40	0.84	1.30	0.012	0.17		0.005	—	0.001	0.008	—	0.001	—	—	0.001	—	0.005
Brazilian (Itabira)	68.50	0.035	0.39	0.76	0.05	0.008	0.01	0.03	0.014	—	0.008	0.005	0.004	—	—	—	—	0.006	0.018
Swedish (Kiruna)	66.35	0.485	1.97	0.50	0.11	0.006	2.24	0.94	0.040	0.074	—	—	0.010	—	—	—	0.022	0.009	0.004
Liberian (Bomi Hills)	69.50	0.055	0.11	0.19	0.16	0.011	0.03	0.15	0.04	—	0.012	0.002	0.003	—	—	—	—	0.002	0.004

— None
* Trace

* Modified from Jacobs et al. (1954), Table 1.

51

and the value per ton of that product. The raw crude assay is an indication of the ultimate amount of iron that theoretically is available, just as the iron unit recovery is a measure of plant effectiveness, but these are of more use in metallurgical considerations than as economic yardsticks. The weight percent recovery of a certain quality product is the most useful single factor employed in the evaluation process.

Tonnage

Tonnage is important as it governs the practicable size of the mining operation. Generally, amortizations are set up on a 20- to 25-year basis and any shorter term raises the capital charge per ton sharply. An ore reserve larger than that needed for a 25-year life does not affect the economics of depreciation but obviously influences plant expansion consideration. Several taconite plants now operating have an ore reserve that will last over 50 years. In general, in a remote location where costly transportation facilities and a townsite must be constructed, the plant size is made as large as possible so that the construction cost per unit of capacity is as low as possible. However, if the ore reserve is essentially infinite, the initial plant capacity is likely to be set by the size of the available market for the product.

Grainsize and grindability

These two factors are most important in the evaluation of a taconite deposit; usually they do not become involved in evaluating other ore types. Taconite was originally a Minnesota term applied to hard, compact iron formation which required grinding to finer than 20 mesh for concentration but the term has since been applied to many other iron deposits to which fine grinding has to be applied to make a saleable product.

The size of the mineral grains determines the amount of comminution necessary to liberate the ore minerals from the gangue so as to make an acceptable concentrate. The hardness and toughness of the ore determines how much work is necessary to achieve this liberation. Since grinding constitutes the largest single concentrating cost item, any geological characteristic or operating procedure which reduces the grinding effort is most welcome.

For most taconite iron ores there is a special level of grinding size at which there is relatively sharp improvement in the quality of the concentrate. This is related to the average grainsize. Since it is necessary to produce a concentrate with less than 7 percent silica to have good marketability, deposits tend to be rated as to what product screen size it takes to achieve it. The grindability index is an empirical measurement of the amount of horsepower necessary

to accomplish the liberation. In certain Lake Superior area taconites, a grind of 85 to 90 percent minus 325 mesh is normal; in other areas of coarser texture, grinding to 14 mesh may do the job. Not uncommonly, the iron formation texture closely reflects the degree of regional metamorphism as is well illustrated in Figure 1 (after James), which shows two prominent areas of relatively high intensity metamorphism; both areas are now the sites of operating iron ore plants.

Another consideration entering into the estimate of the grinding cost is the fact that, in order to pelletize, the coarse concentrate must be reground to approximately 325 mesh. Thus, the reduced cost of liberation at a coarse size in the mill is not all an economic gain because the 30 to 40 percent of the crude which is recovered as concentrate still has to be ground to agglomerating size. Nonetheless, a coarse-textured ore is regarded as having a distinct advantage. Obviously, it would be particularly attractive to a consuming company able to use unagglomerated concentrate because it owns large sintering capacity.

Another item for consideration in evaluating the grindability of an iron ore is the recent development of autogenous mills which may have definite economic advantages. Not all ores respond well to autogenous grinding but those that do are likely to have lower grinding costs using this method. One of the reasons for this is that autogenous mills tend to liberate the minerals at a coarser average size without overgrinding. Thus, one deposit which was tested both ways was capable of making a 65.3% Fe product at 1707 Blaine Index [1] using autogenous equipment whereas, using conventional methods, it yielded 66.3% Fe concentrate at 2125 Blaine. The relative power consumptions for grinding and concentrating by the two methods were 16.49 kwh/ton and 21.92 kwh/ton, respectively. The reduced amount of grinding in an autogenous mill usually results in reduced consumption of mill liners, as well as power. In addition, the coarser product presents less of a filtering problem with resulting benefits in balling and pelletizing.

The liberating characteristics of an iron deposit are much influenced by the fabric or texture of the ore. Most Precambrian iron formations are banded, with laminae and layers that are iron-rich alternating with those that are silica-rich and iron-poor. It is important whether the iron-rich bands are solid iron oxide or simply concentrated disseminations and whether the lean layers are coarse or fine disseminations of iron oxide in silica as shown in Figure 2

[1] A measure of surface area in square centimeters per gram.

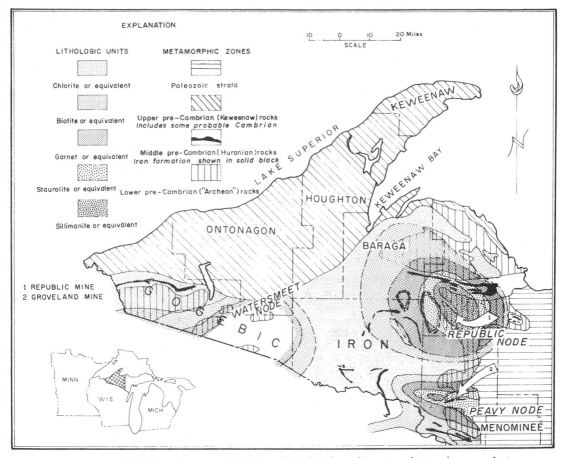

FIG. 1. The distribution of metamorphic facies and the locations of two taconite-type iron ore plants in the Upper Penninsula of Michigan. (After James, 1955.)

(after Gunderson and Schwartz, 1962). These textural qualities obviously affect the liberation that is achieved at various grind sizes. A well-banded ore has a definite advantage over a uniformly and finely disseminated ore with the same iron content.

Part of the attractiveness of banded magnetic ore results from its amenability to magnetic cobbing at various sizes. Obviously, if 15 to 25 percent of the crude can be rejected after simple crushing or crushing plus coarse grinding without appreciable loss of iron units, there are worthwhile savings to be had in fine grinding costs. Some deposits have been tested where it was found that the coarse cobbing reject luckily proved to contain a disproportionately large amount of the hard grinding grains. Thus, cobbing not only eliminated tonnage from the ball mill feed but also got rid of those particles which required excessive grinding power and more frequent mill lining replacement.

Coarse cobbing at quite extreme sizes (up to 12″) has proven possible in at least one area where the iron formation was interlaced with barren igneous dikes. This eliminated the need for close selective mining which would have precluded the use of the largest mining equipment; also it recovered blocks of magnetic formation which otherwise would have had to be discarded due to excessive dilution.

Iron silicate minerals are troublesome in beneficiation for many reasons. Not only do they tie up iron units which are lost in the elimination of silica but they also tend to make ore very hard to grind. On the eastern Mesabi Range, for example, the local development of a felty mass of cummingtonite needles gives rise to some tough and resilient pebbles that are almost impossible to break down. Disseminated grains of magnetite often are entrapped within the cummingtonite network and are lost in the trailings. Likewise in a flotation concentrator, certain silicates tend to float along with the iron oxides and thus decrease the quality of the product.

Mineralogy

The mineralogy of an iron deposit has a great influence on the treatment characteristics and economics. The enriched direct shipping ores of the Lake Superior region are predominantly goethite and hematite. Some other large, high-grade deposits, most notably those in Sweden, are mainly magnetite. At the Helen Mine in Ontario, siderite is produced.

Among the taconites there are all combinations of magnetite, hematite, goethite, siderite, and iron silicates mixed with siliceous gangue. At present, iron silicate material is not classed as ore. Mag-

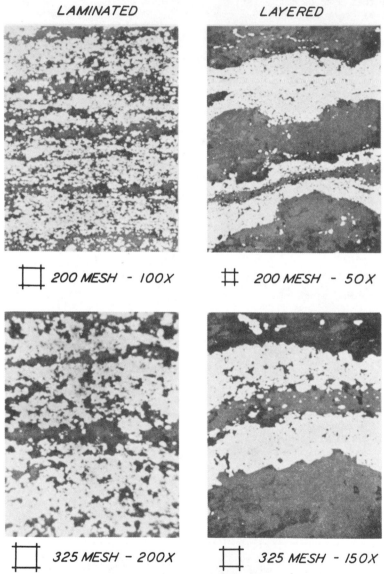

FIG. 2. Microphotographs in reflected light showing typical varieties of iron ore texture and structure. The layered ore shown on the right produces a high-grade concentrate with much less grinding than does the finely laminated and disseminated ore on the left. Dark areas are magnetite; light areas are quartz, carbonate and silicate gangue. After Gunderson and Schwartz, courtesy of the Minnesota Geological Survey.

netite is recoverable by relatively simple and economical magnetic separation; hematite, goethite, and siderite often require more expensive roasting or flotation processes, although when coarse, hematite ores may be treated successfully with relatively low cost spirals.

Magnetite, in addition to the advantage of being more easily concentrated, also has the advantage over hematite of being cheaper to pelletize. During the firing, magnetite in green pellets in oxidized to hematite with the release of a considerable amount of heat. This contributes a substantial part of the total heat necessary in the process and may result in a saving of nearly 400,000 BTU per ton.

The mineralogy of the impurities in iron ore also is important. Phosphorous, as apatite or wavellite, often can be removed in large part by normal concentration practices. Phosphorous bound up in the hematite or magnetite lattice, however, always reports in the concentrate. Titanium commonly is in an intimate grating mixture with hematite and magnetite, and is difficult to remove. Other minor impurities, such as copper, nickel, chromium, vanadium, cobalt, and arsenic are troublesome in some deposits; if they are present as discrete minerals the chance of being able to eliminate them from the concentrate is better than if they are bound up in the iron minerals themselves.

Distribution of ore types in the deposit

It probably is true that none of the taconite deposits presently being worked are homogeneous. All have variations along the bedding and across the bedding in the crude grade, mineralogy, grindability, liberation size, concentrate grade, and other factors. Since concentrating plants operate most efficiently on uniform feed, it becomes extremely important to know the distribution of ore types within the deposit; otherwise costs may be materially affected by unnecessarily wide fluctuations in the properties of the plant feed. For example, at one plant the concentrate output has varied from less than 1,500 tons per day to over 3,000 tons per day, using precisely the same equipment and depending on the part of the pit in which the shovels were located. In this case a variation in crude grade was part of the reason, but ore grindability was an even larger factor. On "bad" days it simply was not possible to crowd more ore through the grinding circuit. The more complicated the flowsheet, the more important it is to have a thorough knowledge of the geological details and the more difficult is the geologist's job of grade control.

During the development and mining stages of an iron ore project, the mine geologist obviously can make a significant contribution to economical and efficient operation. But, even during the early exploration period, detailed mapping and sampling often indicate that various ore types are present and provide a general knowledge of their distribution so that the chances for selective mining or controlled blending can be evaluated. The geologist should be alert to this possibility and aware of its potential importance. Ability to mine a block of higher grade or more readily treatable ore during the early years of an operation may make a significant difference in the present value of the ore body. Plant capital costs can be written off more rapidly and the discounted cash flow rate of return will be higher. Certain of the Mesabi Range deposits are favored by this advantageous situation.

Depth and nature of the overburden

Most deposits require a certain amount of stripping and the cost of development varies widely according to whether the overburden is sand and gravel, or rock. When the amount of stripping exceeds six or eight tons per ton of product shipped, consideration usually is given to underground mining methods, providing of course that other factors would permit it. However, some surprisingly deep rock stripping has been done successfully, as at Marmora where 100 feet of limestone were removed to expose the ore. The mining cost per ton of open pit ore is so much less than the costs of underground mining that it usually pays to make the evaluation both ways if there seems even a remote possibility that an open pit operation might be feasible.

Shape and attitude of the ore body

The shape and attitude of an ore body have an important affect on whether open pit or underground mining is the more suitable. Since open pit mining gives lower operating costs per crude ton, this is an important economic factor. Generally a near-equidimensional ore body will have a lower stripping ratio than a tabular one especially if the latter is dipping steeply. The shape and attitude of the ore also control the tons per vertical foot—a factor which determines the development cost per ton in either open pit or underground mining. A highly irregular underground ore body does not lend itself to low cost, bulk mining methods.

Location

Location affects the evaluation in many ways. Remote, rugged, and inaccessible locations are preventing a number of excellent iron deposits from being developed. Climate is also a factor, with both extremes of hot and cold weather offering problems and affecting the efficiency of operation. Many of these can be and doubtless will be overcome but not until more favorably situated deposits are developed first. In evaluating a deposit, the difficult aspects of an inland location usually are felt first in their affect on transportation costs and the capital outlay necessary to get into production. Transportation is such a large proportion of the delivered cost of most iron ore that a few hundred miles of rail haulage, even in otherwise favorable situations, may be enough to delay development. Thus a deposit which would yield a product worth about $17.00 per ton, may be confronted with a potential transportation cost of $8.00 to $10.00 per ton.

In Ontario, many quoted freight rates equal about $0.01 per ton-mile; in western United States, some operators shipping to Japan pay only $0.007 per mile. Some rates, however, are much higher. Situations which permit installation of a private industrial railroad generally are faced with lower transportation charges per ton but, of course, require large initial capital expenditures to build the road. A relatively recent development which will become increasingly important in the future is the pipeline transport of concentrates; this procedure eliminated the need for a costly railroad in Tasmania and is particularly applicable in such areas of rough topography where weather conditions permit (U. S. Bur. Mines, 1971). A recently developed pipe-line system for off-shore loading of iron concentrates will have a significant effect of the development of deposits near coastlines with poor harbor facilities.

Market and price

Market and price factors in iron ore evaluations are much more complicated now than formerly due to the wide distribution of both sources of ore and potential markets. For many years, the delivered price per long ton for ore delivered at Lake Erie ports was the base from which nearly all North American iron ore prices were calculated. The Lake Erie price is still important but, today, international trade in iron ore is much more common, and increasing amounts of tonnage are sold at prices arrived at after consideration of all of the many cost elements involved (United Nations, 1971). As a consequence, competition price-wise is a very important factor in determining the profit that can be realized on a given deposit. Furthermore, since large deposits of iron ore for the international trade have been proved by exploration in many parts of the world and await only market outlets to justify development, the competition for the available markets has become very keen. None of the large iron ore developments of the past 20 years have been made without prearranged marketing contracts and, in many cases, some of the partners have also been consumers.

Iron ore is a relatively low value mineral commodity. Virtually all shipments are priced at less than $18 per long ton delivered at Lake Erie or other ports, and in many cases they are below $12. Transportation charges represent a sizeable percentage of these prices. Consequently, in order to achieve a profit, very efficient production is imperative and, in the evaluation of new deposits, accurate estimates are imperative.

The acceptability of a particular type of iron ore varies from company to company. Thus a deposit with a high percentage of fines ($-\frac{1}{2}''$) may be very attractive to a steel company with large sinter capacity but much less so to a company without such capacity. Also the content of certain minor elements in a deposit will be less important to a company whose other ore sources are low in these elements so that blending to a satisfactory furnace burden assay is easy. In the future it will be increasingly significant whether a deposit has the capability of being beneficiated to Direct Reduction quality (usually 2–3% SiO_2 or less). The location of a company's blast furnaces obviously also is significant. For such reasons as these there is no clear-cut black or white in the evaluation of many iron ore deposits; much depends on what company is looking at it and where the ore would be used.

Profitability Estimates

To illustrate the method of preparation of profit estimates, four examples of the data sheets and a cost estimate comparison form are shown in Tables 3 and 4. The deposits have not been identified but they are typical of ores of various kinds and they are confronted with various physical, geographic, and political situations. As is usually the case, not all the cost estimates for individual items have the same degree of validity. Some factors are precisely known but others have to be estimated from fragmentary data or by comparison with known costs at other deposits that are similar. The figures will simply represent the best data available at a given time. A fair knowledge of most of the purely geological factors usually is on hand by the time a project has had some widespread reconnaissance drilling and bulk sample testing.

As his contribution to the evaluation team effort, the geologist should provide the best possible factual and interpretive data as to the expected form of the ore body, its composition, and general environment. As engineering and metallurgical data are developed, he should participate in the revisions and modifications of mine layout and flowsheet until the best combination is found. This interplay of disciplines in

TABLE 3. Physical Facts

	Case I	Case II	Case III	Case IV
Type of ore	Magnetic taconite	Magnetic taconite	Oxidized taconite (roast ore)	Direct shipping coarse and fines
Open Pit or underground	Open Pit	Open Pit	Open Pit	Open Pit
Transportation to lake port	400 miles	150 miles	108 miles	—
New railroad required	200 miles	0	0	260 miles
Rail distance to port (other than Great Lakes)				260 miles
Recoverable iron minerals	Magnetite	Magnetite	Hematite Goethite Siderite	Hematite Goethite
Reserves				
Crude, long dry tons $\times 10^6$	450	49	197	240
Total iron (dry)	32.3%	—	32.5%	64.0%
Magnetic iron (dry)	26.9%	22.7%	0	0
Product reserves dry long tons $\times 10^6$	189	15	90	240
Products	Pellets	Pellets	Pellets	75% $+ \frac{3}{8}''$ 25% $- \frac{3}{8}''$
Type of concentrating process	Magnetic	Magnetic	Roast & magnetic separation	None, direct shipping of ore
Grind required (mesh)	200–235	100	—	None
Weight & Recovery	41.9	30.8	46.1	100
Concentration ratio	2.4	3.2	2.2	1.0
Grade of product (natural)				
Fe	63.6	64.0	62.7	64.0
Phos	0.03	0.025		0.05
SiO₂	6.0	6.1	7.3	2.0
Mn	0.10			
CaO + MgO	0.18			
Al₂O₃	0.11	1.2		1.0
S	0.01	0.04		
Rice Ratio*	10.4	8.8		21.1
Stripping ratio Tons/ton of product	1.3	2.7	0.55	0.5
Estimated annual plant capacity, L. T. $\times 10^6$	Min of 3	0.7	2.0	Min. of 6
Estimated capital cost $10^6	$150	$21	$112	$240

*Rice Ratio $= \dfrac{Fe}{SiO_2 + Al_2O_3}$.

a smoothly coordinated organization can be a most satisfying activity for a geologist who sees his field observations and scientific understanding of the orebody translated into profits. A new dimension is added to his thinking. For the experienced exploration geologist it should be second nature to gather the answers to many engineering questions as he goes about his job of assembling the pertinent scientific data.

Case I

This is an example of an excellent magnetic taconite deposit in a poor location. The geological factors are above average, with large tonnage, good crude grade, and a reasonable liberation size. Although only 70 percent of the total iron is magnetically recoverable, the rest being tied up in iron silicate minerals, the 41.9 percent weight recovery of 63.6% Fe pellets is substantially better than most deposits currently being mined. A near vertical dip, favorable topography, and iron formation widths of up to 500 feet result in a reasonable stripping ratio. The total estimated operation cost per ton of pellets would be relatively low.

Unfortunately the factors influenced by location are sufficiently difficult to relegate this deposit to the category of future reserves. Two hundred miles of new railroad, the need for complete new townsite facilities, and the added construction costs far from supply centers result in high capital costs, and the 400-mile rail haul to a lake port erases much of the potential profit.

This type of situation requires constant re-evaluation because it has a good geological potential and it may be favorably affected by entirely independent developments. A new base metal find in the same area, for example, may result in closer rail transportation or new townsites which can be shared.

Capital requirements are thus reduced. The advance of civilization into this currently remote area will make future development of this deposit a certainty.

Case I is an example of a deposit with more ore reserve than it needs—at least so far as the profit per ton calculation is concerned. There is sufficient tonnage available to justify a 10 million-ton-per-year plant and the operating and capital costs per ton for a plant of this size undoubtedly would be lower than for a unit of smaller capacity. The problem then becomes one of marketing such an enormous tonnage, and most producers probably would elect to commence operations at smaller capacity and expand later as markets were developed.

Case II

In contrast to Case I, the deposit in Case II is well located. It also has many other attributes. Unfortunately it is too small to justify the minimum capital expenditures necessary to establish a magnetic taconite operation at the present time. Possibly future technological developments in the grinding, concentrating, and pelletizing processes will reduce the magnitude of these expenditures, or a new, lower cost process will be developed so that the economics of the deposit will become more favorable.

As can be seen in Table 3, a good concentrate is produced at only 100 mesh grind, a relatively coarse size compared to some other ore bodies already developed. No new rail construction is necessary and the total haul to a lake port is not excessive. The royalty is abnormally low ($.05 per ton) but pit development and stripping costs are quite high because the tonnage is distributed among five small pits. Furthermore, there is 15 to 25 percent of pegmatite dike material interspersed within the iron formation and this must be eliminated by selective mining or magnetic cobbing. This resulted at the time the original estimate was made in a development estimate of $1.08 per ton. Since the crude grade is substantially below that in Case I, the concentration ratio is higher, and the beneficiation costs per ton of product was $1.05 cents higher than Case I although the cost per ton of crude is about the same. The coarser liberation size in Case II almost compensates for the inherent higher grinding cost per ton of crude in a small plant but does not equalize the effect of lower weight recovery.

Case III

This is an oxidized iron formation in which, by reduction roasting, the ferric oxide minerals can be transformed into artificial magnetite and then concentrated by normal magnetic means. Most hematitic and geothitic ores can be treated in this manner

TABLE 4. Economic Analysis Form

Cost Item	Case I	Case II	Case III	Case IV
Mining and crushing per ton crude	—	—	—	—
Beneficiation per ton crude	—	—	—	—
Total per ton crude	—	—	—	—
Total per ton product	—	—	—	—
Pelletizing	—	—	—	—
Total per ton pellets or product	—	—	—	—
General expenses	—	—	—	—
Development (stripping)	—	—	—	—
Employee benefits, taxes, and administration	—	—	—	—
Rail freight	—	—	—	—
Terminal costs, lake freight, and unloading	—	—	—	—
Total at Lake Erie per ton product	—	—	—	—
Total at port per ton product				—
Ocean freight				
Market value				
@ Lake Erie price*	—	—	—	—
@ Japan, f.o.b.				—
Depreciation over an appropriate number of years	—	—	—	—
Royalty	—	—	—	—
Income tax (after depletion)	—	—	—	—
After tax profit	—	—	—	—
Cash flow	—	—	—	—
Payback time (years)	—	—	—	—
Return % on investment	—	—	—	—

* In 1972, 28¢ per unit for pellets.

and virtually all of the non-silicate iron is recovered. Usually the weight recovery by this treatment method is higher than it is for straight magnetic separation on the same ore because iron carbonates and ferric oxides are recovered as well as magnetite. Up to this time there are no commerical-sized plants in North America using reduction roasting but most companies have shown in the laboratory that the process is technically feasible. The problem is largely one of determining that the additional weight recovery will pay for the additional processing cost. Enormous tonnages of iron formation amenable to this type of metallurgical treatment are to be found on most of the iron ranges of the world. As in the case of fresh magnetic taconites, there are wide mineralogical and textural variations from deposit to deposit which result in necessary variations in treatment procedure and costs. These variations result both from original sedimentary facies changes and from dissimilar geologic histories, especially as reflected in the oxidation and hydration of the iron minerals. Some iron formations or subunits in the formations, such as the oxidized Upper Cherty on the West Mesabi Range, are very difficult metallurgically because of the intimate mixture of silica and the iron oxides.

Much experimental work has been done which shows that anionic or cationic flotation is a feasible way of beneficiating these ores, as an alternative to roasting. In some cases it will prove to be economically advantageous. However, the example in Case III assumes roasting.

The example in Case III is well located with relatively low rail cost and it has a very low stripping ratio. Treatment costs per ton of crude are relatively high because of a more involved flowsheet but are competitive on a per-ton-of-concentrate basis because of the high weight recovery. The grade of concentrate is, perhaps, below that desired in the next 10 years but it probably can be improved with little iron loss. The geologist can make a distinct contribution in the development of deposits such as this through his correlation of ore types in the field with metallurgical response in the plant. In general, the more sophisticated the treatment procedure the more important it is to have close control of the plant feed. An important consideration in the evaluation of a potential roast ore is the availability of low cost fuel.

Case IV

Case IV is high-grade, direct shipping foreign deposit of large size. This example is typical of deposits under development or in production in several areas of the world, all of them outside of North America. In evaluating all of these foreign deposits, a major consideration must be attached to the political and economic stability of the country in which the deposit is located. The Case IV ore body is in one of the more stable situations.

The tonnage and grade of Case IV are excellent; however, an important consideration is the fact that 25 percent of the ore is minus $\frac{3}{8}$ inch and hence probably would be discounted in price by a few cents per unit. The significance of the property of iron ore called "structure" (particle size distribution) is becoming continually more important and is reflected in the good market reception of pellets with their uniform size and shape.

Case IV can be mined as a broad open pit with low stripping ratio. As no ore treatment is required except screening, operating costs will be extremely low. The location is the principal drawback as it will be necessary to construct 260 miles of railroad and establish a complete town and all related facilities in an isolated area. These constitute the main

development cost items. The estimated rail haulage cost, however, should be low because the railroad will be an industrial road owned by the company and run at cost, rather than a common carrier with rates set by a government body. Once the operation is established it can increase its output at relatively low additional cost so that there is great incentive to generate additional sales.

Summary

The evaluation of an iron ore prospect involves many factors, each of which must be individually assessed in order to arrive at an estimate of the probable profitability of the deposit. Many of these are geological and are inherent in the deposit itself; others are inherent aspects of the environment in which the ore is formed. All are subject to periodic changes. Although the geological character of the ore does not change, the development of new processing techniques may have a great affect on the cost of putting the ore into marketable form. Among the "environmental" factors, the changes over a period of time in the political or taxation climate, the availability of transportation, or the cost of power, for example, can have a marked influence on the competitive standing of an ore body awaiting development. Thus it comes about that iron ore profit margin estimates are in a constant state of revision. By keeping a running tabulation of the influence on total production cost of all of the contributing factors, it is possible to appraise quickly the impact of any change in one of them.

THE HANNA MINING COMPANY,
CLEVELAND, OHIO 44114,
May 2, June 14, 1972

REFERENCES

Gunderson, J. N., and Schwartz, G. M., 1962, The geology of the metamorphosed Biwabik iron-formation, Eastern Mesabi District, Minn.: Minnesota Geol. Survey Bull. 43.
Jacobs, C. B., Elliott, J. E., and Tenenbaum, M., 1954, Significance of minor elements in iron bearing raw materials for integrated steel plants: New York, American Iron and Steel Institute.
James, H. L., 1955, Zones of regional metamorphism in the Pre-cambrian of northern Michigan; Geol. Soc. America Bull., v. 66, p. 1455–1488.
United Nations, 1971, Problems of the world market for iron ore.
U. S. Bur. Mines Inf. Circ. 8512. 1971, Mesabi Range iron ore transportation: feasibility and estimated cost of pipelining.

5

Reprinted from pages 433–460 of *Economics of the Mineral Industries*, 3rd ed., W. A. Vogely, ed., American Institute of Mining, Metallurgical, and Petroleum Engineers, Inc., New York, 1976, 863p.

Valuation of Mineral Property

L. C. Raymond*

Valuations in the mineral industry differ from those of other enterprises because mines and oil wells have a definite life so cannot be considered a perpetuity. This requires that in any mineral-property evaluation the recovery of invested capital is necessary during the life of the property. Capital write-offs for nonextractive industries are assumed to be continuously plowed back to sustain the operation and produce yields on the capital invested. The extractive or mineral industry, however, generally attempts to return to the investor both interest and principal during the life of a property. Normally this is done currently and the dividend includes both items. Where the mineral property possesses large reserves and the life factor approximates a perpetuity, the return on investment is generally comparable to that of industrial stocks.

The valuation of a mineral property, at best, is based on educated estimates and judgment factors. This is particularly true as it relates to estimating selling prices over a long term or assessing the quality of management. Also, it is possible to make various assumptions as to rapidity of working out the deposit, varying time for start-up of operations, varying earning rates, so as to obtain different results with the same basic facts. Thus it is not surprising that two engineers, irrespective of their abilities, may not agree exactly even when given the same facts, but their results should be close enough to be

used in establishing a fair value for trading purposes. The valuation is no better than the source of basic data and the soundness of judgment factors. Fundamentally the basic principles of mineral valuation are similar for base metals, sulfur, oil, or any other mineral. Each mineral, however, requires an engineer to have a knowledge of such factors as its occurrence, distribution, mining and processing methods, and marketing problems in order to gather and analyze all the facts and data that have a bearing on the earning potential of the property.

Present Value of a Mineral Property

In essence the present value of a mineral property is a sum of money in today's dollars that future income from the property are worth. It differs from the total anticipated earnings by the amount of the interest that the unrecouped investment could be expected to earn during the period of recoupment. Thus, if one could accurately estimate what a mineral property will earn during its life, after deducting all operating, administrative, capital, interest, and tax charges, the determination of present value would be a relatively simple mathematical calculation. Normally, where the expected life of the mineral property extends beyond 20 or 25 years, the discounted value of earnings beyond this period is relatively small. The petroleum industry generally discounts the annual cash flow of an operation rather than earnings since, for tax-saving purposes, earnings are kept

* Consulting Engineer, Chappaqua, N.Y.

low by expensing for new exploratory outlays and intangible drilling costs of development wells.

An important aspect of mineral property valuation calculations is to show the investment that has been made in the overall business involving (1) the mineral property, (2) mine equipment and facilities, (3) plant equipment and facilities, and (4) working capital. Also it must be shown that the initial investment and subsequent investments will be fully returned over the life of operations, as well as that the designated percent return on each of the foregoing investments can be made possible out of the net earnings, year by year, over the full period of operations. If the foregoing conditions cannot be met, the valuation is not valid.

Methods of Determining Present Value

Among technical mining men, probably because most mining schools have emphasized the technique, the Hoskold Formula is widely used. Within the industrial and financial sectors of business, however, the straight-discount method is generally used. The Hoskold Formula is more complex mathematically and is not easily applicable to a varying earning rate. More importantly, Hoskold applies a risk interest rate on the total original investment during the life of the property similar to that of a perpetuity. Probably the reason for this concept was Hoskold's thinking that the amount accumulated in a sinking fund at the end of the life of a mine is theoretically used to purchase another mine of the same present value; and the annual earnings from a series of mines, if purchased using money accumulated in a sinking fund, may be considered to comprise a perpetuity. This may be correct theoretically, but well nigh impossible to apply in practice, especially when economic ore bodies are becoming much more difficult and expensive to find. (See page 152 for Hoskold formula.)

The Hoskold concept of maintaining a sinking fund during the life of a property from which an investor will get his capital returned when the property has been depleted, is seldom if ever practiced. The straight-discount concept assumes that the investor is entitled to interest on the unrecouped capital and that the interest so accumulated should not be used to augment a sinking-fund total. This concept is comparable to the practice used in the nonextractive industries.

Straight Discount Method

Numerous formulas for mine valuation have been proposed over the years but probably the simplest and most practical technique is the straight-discount method whereby the future annual earnings are discounted to today's dollars. The technique is applicable to varying income and, more importantly, conforms most closely to normal business practice. The straight-discount method applies only a single interest rate on the outstanding capital of the original investment after deducting that portion of capital returned to the investor. The annual capital write-off increments are such that they accumulate during the life of the mine to the total original investment without the assistance of interest. This conforms to conventional capital write-off procedure.

For an operation of stature and ample reserves, normally the interest rate will not exceed 8 to 12%. Should large risks be involved the present value is simply used as a starting point for negotiations from which additional sums of money are deducted based on the inherent risk factors involved.

Application of Present Value as Applied to Valuing Mining Stocks

Where the mine or mining company's ore reserves are proved to a substantial extent and a good present value of future dividends can be obtained, after adding net current assets and subtracting all prior liabilities ahead of the common shares, such as appraisal becomes a basic tool in valuing such shares. Where the company has a long-established earning record, the present worth of future dividends can be based on the historical record. The actual market price of base metal shares is likely to show a high price-earnings ratio in times of low metal prices and a much lower price-earnings ratio in times of high metal prices. The market price of a particular share may go to extremes of twice or more, or half or less of the appraised value at any given time, but sooner or later it tends to return more closely to a well-reasoned present worth.

Procedure and Objectives of Valuation Report

In the mining industry, three general classes of mineral properties are likely to require valuation:

Case 1—Valuation of new or only partly developed property.

Case 2—Valuation of "going" mine.

Case 3—Valuation for merger—the same valuation procedures used in Cases 1 and 2 may be followed in Case 3 for the purpose of merger.

The important first step in any mine valuation is to assemble all the facts. These should cover not only ore reserves and mining, underground, and surface construction conditions, but items such as capital costs, operating costs, markets, competition, management, raw materials, research, growth, taxes, prices, and policy matters. These factual data should be classified, analyzed, and set up so as to answer readily any question by the buyer, seller, or other interested party. Typical questions generally are:

1. What are the total capital requirements for financing this new project, or placing the property on a profitable basis?

2. What period of time is involved before reaching a profit-making stage?

3. What are the average expected annual profits and what portion of such profits will be available for dividends?

4. How much time will be required to "pay out" the investment under a reasonable financing program?

5. What rate of return will be expected on the investment?

6. What is the possibility of developing new ore reserves?

7. How much should the investor pay for the mineral property?

The objectives of this chapter are to cover the broad methods and basic principles applicable to the step-by-step procedure and organization of a mine valuation and to indicate some of the pitfalls in handling the basic data. A case study is given to show what is considered and how the data are handled. A hypothetical open-pit "M.E. Mine" (the initials refer to Mineral Economics and have no other significance) is used as an example, taking it from the partly developed stage into the active mining stage. All illustrative material is related to and is a part of this hypothetical project. The tables are presented as a useful form; the work sheets are tabulations related to this particular project and show the basic derivations.

It should be pointed out that the valuation of the mineral property is not the same as the valuation of the overall mine and plant operation or business. The mineral property is only part of the overall investment in the operation, and necessarily involves a somewhat different accounting terminology and special treatment, particularly as regards legal applications.

CASE 1—VALUATION OF NEW OR PARTLY DEVELOPED PROPERTIES

Factors for Consideration

The value assigned to new or partly developed property is generally the difference between the total present value, based on discount of all future earnings, and the value assigned to physical assets, working capital, and other costs that are necessary to get the property into operation. A detailed field examination is necessary to ascertain any adverse physical conditions that may be met and corrected before the final decision on methods of mining and milling, the installation of various surface buildings and facilities, and on the transportation needed to extract and prepare a marketable product. Thus, the valuation of a partly developed mineral property is basically an estimate of potential net income from future exploitation of known ore reserves even though mine facilities, plant, or other surface facilities do not exist. Engineering skill and experienced judgment are required because nearly all factors must be estimated. Adequate contingency allowances must be provided. If the ore reserve has not been fully explored, a speculative element must be considered.

Where there are no production experience records to serve as a guide at a new property, the risk factors are of course greater than for a going mine for which the mining and processing methods and costs have been worked out and actually demonstrated. Thus, the investor or buyer is usually justified in demanding a higher rate of return on such new or prospective mines.

The valuator (appraiser or valuation engineer) is usually confronted with the questions of what constitutes ore reserves and how much ore is required to justify a project. He is probably faced with measured ore, partly measured ore, geological ore, or "wildcat" possibilities. Actually, the only firm basis for valuation will be found in the measured ore category, where volumes and grades can be demonstrated. Other ore reserve categories, which must be assumed or considered probable, should definitely be assigned to the speculative category. These speculative ores, however, may serve as a basis of trading between the buyer and seller.

In the ideal situation, a property has sufficient ore blocked out to show clearly that the operation (earning period) will be adequate to amortize the investment from the earnings, provide a return for use of invested funds, and give some indication of continuation of profitable operations. Where the ore reserves are not clearly adequate for a reasonable life of operation for a substantial investment for, say, 15 to 20 years, then it rests with the valuator to point out the speculative element or the chances of eventually developing sufficient additional ores to pay out the investment. Where the margin of profit is high, the payout period may be correspondingly less than the period just mentioned.

Sometimes it may be necessary to value mineral tracts not large enough to make a mine in themselves. Adjoining active mine operations are then usually in the market for such tracts and thus create a market for the limited ore reserves. The valuation can then be considered on the basis or present worth of future royalties (after provision for federal taxes) or present worth of future (deferred) earnings.

Properties having a limited ore showing may be considered in the speculative development stage; the valuator must then use his judgment in advising the investor of the amounts that he might be justified in spending on the prospect to try to bring it to the stage of development for actual valuation of potential production.

Reminder List for Determining Value of New Mine Venture

The following steps are handy reminders for organizing a commercial valuation:

1) Calculate the ore reserves and indicate grade or quality under the following classifications: measurable ore and speculative ore. (This requires a preliminary estimate of costs and determination of mine cutoff grade.)

2) Estimate recoverable ore, taking into consideration such factors as mine dilution, mine losses, and cost of making ore available.

3) From study of flowsheets and metallurgical tests, calculate the treatment losses or metallurgical recovery.

4) Estimate rate of production as determined from the mine potential, and the sales possibilities, as well as limitations, such as availability of power and water.

5) Divide reserves by annual production to obtain life of property or operations.

6) Using recovery and treatment factors, calculate total yield of salable product. Calculate the "smelter settlement value" of the ore or concentrate, or salable products.

7) Estimate average sales price per annum and total average sales volume and total annual gross revenue.

8) Estimate cost of sales (per ton basis), labor, materials and supplies, and overhead.

9) Estimate selling (marketing), administrative, and central office costs.

10) Subtract cost of sales from sales income to get gross profit.

11) Subtract selling and administrative expenses from gross profit to get profit before depletion allowance—also any interest payments.

12) Subtract depletion and depreciation allowances to obtain basis for computing income tax.

13) Determine the income taxes.

14) Estimate the total annual net profit after taxes.

15) Set up work sheet to show estimated cash flow including payments of such items as interest, principal on loans, tax allowances for period of operations, and repayment of initial investment through depreciation and depletion.

16. Consider special risks and hazards to operation and consider a reasonable rate of return on investment, or discount factor to be used.

17. Estimate ultimate speculative tonnage

that may be expected in addition to the measured reserves.

18) Determine the net income before depletion and deduct return on working capital, return on investment in mine, plant and facilities, and return on investment in non-mineral land. This gives the residue earnings applicable to the mineral property.

19) Discount the total residue over the life of operations to get the gross present value of the residue.

20) Adjust for any unrecoverable working capital such as obsolescent spare parts inventory or accounts receivable.

21) Add present value of salvage at end of operations.

22) Adjust for any cost of deferring investment in mineral land and any cost of proving-up mineral reserves.

23) Compare earnings against investment with those current in alternative enterprises.

Factors Bearing Upon Potential Earnings

In the process of developing all the major elements bearing upon earnings, the data should be set up in a useful and timesaving manner (as shown herein) since many of these data are used subsequently for various purposes in the valuation procedure.

Ore Reserves and Life of Project with supporting data must be given both to substantiate the quantity and grade of the recoverable ore reserves and to indicate the development problems met in making them available for mining. To take care of considerable variations in the grade of ore, which would appreciably affect the total earnings, particularly in the early operating stage, the ore reserves should be shown by level (bench or horizon) and by grade in or-der to develop a production schedule for specific sections of the property. Where loan capital is involved it may be desirable to show the possibility of mining high-grade ore during the early payout period so as to speed up payments if necessary. The possibilities for developing additional ore should be adequately covered. All yield and mill recovery factors should be supported by test data and volume factors that demonstrate the tonnage and quality of marketable products. If the physical nature of the deposit indicates a limited productive capacity for the project, this should be clearly presented.

Table 12.1 will be helpful in developing economic factors such as practical mining rates and volume and grade trend with deepening of the mine.

Total ore reserves are then scheduled for extraction so as to coordinate with operating cost estimates and capital costs of equipment.

In this example ore reserves have been scheduled on a 25-full-year basis, largely dictated by market considerations.

Markets and Future Price Levels are keys to the future earnings of any mining project. To substantiate the estimates relative to future markets, analysis of statistical data on production and consumption (consumption-in-use pattern or historical price trends) may be useful. Where the marketing is complex, a careful study and investigation of consumers and competitors may be required before reasonable estimates can be made of the time required and possible share of the market (sales volume) anticipated after launching the project. The market study should also give convincing evidence about the acceptability of the product (grade, quality, etc.);

Table 12.1—Summary Ore Reserves by Levels or Benches

Mine Level or Quarry Bench No.	Tons Ore Recoverable	Grade or Yield	Value per Ton
(A) Total	23,000,000 Average	Average	$5.02 (B)

Note: (A) Data covering each level or bench should be listed here.

(B) Mining staffs nearly always carry out ore reserves estimates to cents. Figures used herein follow the same practice to facilitate checking. Rounding out to requisite accuracy is done with final figures.

this is important not only for metal products but particularly for most nonmetallic products. In highly competitive markets, trade discounts may be important aspects in determination of net sales income.

If long-term contracts can be made, future prices may be projected with considerable accuracy. However, some long-term contracts have escalator clauses tying in prices with fluctuations in labor and materials. The valuator must usually estimate future long-term price levels based on the demand and supply outlook as foreseen at date of valuation. For some large projects, it may be necessary to determine the effect of increased volume of production on the market price of the product. And because transportation may limit the market area and the sales income, a careful study of delivery costs may be essential. Where mineral products having a long record of price stability or uniformity of rising prices are involved, it may be assumed that over a long period changes in costs of operation will be accompanied by corresponding changes in price of products. This may be likely for such commodities as gold and oil. Other mineral commodities, however, have shown rather violent fluctuations in the past, copper for example. Thus, future estimates of prices must be based on the known facts and trends at the time of the report; but at the same time estimates of operating costs must be adjusted in line with any price adjustments likely to result from inflation.

The market potential may limit the size of the operations, but the desired payout period and the physical limitations of the mine may also be factors. If there are no restrictions it may be desirable to schedule the operations for the optimum economic rate of return.

Development of the sales realization will be required, taking into consideration sales discounts or other allowances which tend to reduce income from sale of products. In the example given herein the average sales realization or net sales income is estimated to be $4.78 per ton (compared to the ore-in-place value of $5.02).

Capital Costs can be developed only after fairly complete mine and plant layout plans are made (Table 12.2). Consideration must be given to (1) use of existing physical assets on the property, if any, as well as (2)

Table 12.2—Design Capacities for Mine and Mill (Tons)

Material	Total Ore Reserves	Designed Capacity Annual Basis	Average Daily Capacity, 5-Day Week 2-Shift Basis	Hourly Capacity, 7½ Hr. per-Shift Basis
Run-of-mine ore	23,000,000	1.000,000	4,000	240

new equipment and facilities. A detailed listing can be made under each of the following major headings:

    1) Cost of Property
    2) Preproduction Costs
    3) Mining Buildings, Equipment, and Facilities
    4) Milling Buildings, Equipment, and Facilities
    5) General Buildings, Equipment, and Facilities (includes housing, schools, recreation buildings, hospitals)
    6) Working Capital Requirements

To make the foregoing data useful in the subsequent cost analysis procedures, such as developing a depreciable base, insurable values, and income tax allowances, it is desirable to separate capital costs into Buildings, Building Equipment, and Equipment and Machinery.

The date should be indicated for the estimates covering construction costs involving prices of materials, labor, and other expenses, so that if there is an inflationary trend and the project is delayed, all the figures can be adjusted.

Initial Working Capital requirements may constitute a substantial portion of the total capital or financing necessary for a new project. Sufficient working capital must be assured to sustain the project—that is, to provide funds to "fill the pipeline" or build up operations, including raw material in stockpiles or bins, etc.; inventory stores, usage of materials during tune-up period, semifinished or finished materials en route to market, and payrolls (accounts receivable) and other costs. Table 12.3 is an example or reminder list.

Table 12.3—Estimated Annual Working Capital Requirements (Basic, Tons Annually)

Item	Total Annual Amounts, $
1. Inventories	
Raw materials (months)	
Supplies (months)	
Spare parts (months)	
Work-in-process (months)	
(between usage and monetary return) including	
Payrolls	
Raw materials	
Supplies	
Other operating costs	
2. Preparation and training	
Payroll (months)	
Raw material usage	
Supplies usage	
Other operating costs	
Contingencies	
3. Accounts receivable	
Total initial working capital requirements	

Total capital cost requirements for financing purposes may be conveniently summarized as in Table 12.4 which may be used as a guide for depreciation and depletion and as an aid in other calculations for earnings statement purposes.

Costs of Production incurred in producing a marketable product must reflect actual conditions and difficulties to be experienced in the operation of the property. Each of the major items of operating costs is discussed in the following:

1) Labor costs, including complete manning requirements or development of the payroll, establish a basis for ready calculation of such expenses as Public Liability, Workmen's Compensation, Use and Occupancy, Unemployment, Federal Old Age Benefits, Health and Accident, Holiday and Vacation Pay, Housing, Medical, Recreation, and Outside Transportation. Table 12.5 is a useful form for developing the working force and payroll required for the operations.

2) Materials and Supplies to cover all needs must be estimated on an annual basis. Firsthand knowledge of operations and of

Table 12.4—Preliminary Estimate Capital Costs M. E. Mine

	Mineral Land (Ore Reserves)	Preparation*	Buildings and Facilities*	Building Equipment*	Machinery and Equipment*	Total
Preproduction expenses	$1,900,000†	$198,000	$333,000	$9,000		$2,440,000
Mining			106,000	10,000	$ 710,000	826,000
Milling			994,000	45,000	2,470,000	3,509,000
General			475,000	25,000	301,000	801,000
Total	$1,900,000	$198,000	$1,908,000	$89,000	$3,481,000	$7,576,000
Contingency 5%— except mineral land						284,000
						$7,860,000
Working capital						1,300,000
Total investment requirements for financing						$9,160,000‡

* Includes engineering, supervision, and contractor's fees

† See Work Sheets I-A and I-B for derivation of tentative property valuation.

‡ This figure less cost of mineral property ($9,160,000-$1,900,000 = $7,260,000) represents the capital requirements to get the property into production, exclusive of the cost of the mineral property.

Table 12.5—Estimated Working Force and Annual Payroll (Basic—Tons Annually)

	Number Personnel	Hourly Rate	Total Hours	Straight-Time Earnings	Shift Differential	Payroll
Production						
(a) Mine*						
(b) Mill*						
(c) General surface*						
Engineering*						
Selling, administrative,						
and accounting (1)						
Total						

* Detailed list of jobs or labor categories to be given here.

the equipment is needed here. Table 12.6 gives the general headings to be followed, with detailed listing requiring separate tables.

3) Overhead costs are likely to involve numerous items such as those listed in Table 12.7 for convenience of checking and estimating.

4) Depreciation (not general, ordinary, special maintenance) involves all mine and mill operations. It includes only the replacement cost of major equipment and facilities that wear out or become obsolete before the end of the life period for the project. Such expenditures are necessary to sustain operations but since they do not occur uniformly, a reserve is set up. Such depreciation is an important cost item that has an important bearing upon the earnings of the project. Sound judgment should be used in establishing depreciation rates, taking into considera-

Table 12.6—Detailed Estimate of Materials and Supplies (Basis, Working Day Tons Annually)

Items	Quantity	Unit Price	Total	Freight and Handling	Total Cost On the Job
Raw materials					
usage*					
Supplies					
usage*					
Spare parts					
usage					

* Detailed items may be listed here and allocated to mining, milling, etc., in cost estimates.

tion the life of all equipment and facilities involved. The actual replacement expenditure may not be made during the first three to five years, perhaps, and it may not be practical to make any significant replacements at a period when the mineral reserves offer only a few remaining years of operation, but the depreciation reserve should be adequate to meet these replacements (even with inflated costs) when necessary to maintain the project in a satisfactory operating condition and recover the cost of equipment and facilities. The depreciation cost should also allow for replacement of obsolescent equipment. Table 12.8 is a general check list to ensure that this item is adequately provided for in the operating cost estimate. (Depreciation, as discussed herein, is not straight-line depreciation as applied for tax purposes, which takes into account only the original cost of the item purchased.)

5) Selling and Administrative costs are developed separately as they are likely to be off-the-property costs. In this example, the figure of $105,000 per year or $0.12 per ton of run-of-mine ore is used.

To simplify the calculations at this point, no interest on borrowed money is assumed but if present, this would affect the cash flow and income tax calculations.

The check list in Table 12.9 gives the summary of all the production cost items.

Income Taxes

Income taxes must be carefully determined, particularly making allowances for such tax deductions as depletion, deprecia-

Table 12.7—Estimated Overhead Costs (Annual Basis)

Expense Item	Basis of Calculation: Quantity,	Rate,	Other	Total Annual	Cost per Ton— Run-of-Mine Ore
Telephone and telegraph					
Stationery and printing					
Miscellaneous office supplies					
Traveling expenses					
Employee training					
Engineering					
Research					
Property tax					
Franchise tax					
Fire insurance					
Public liability insurance					
Use and occupancy insurance					
Medical and first aid					
Hospitalization					
Payroll tax					
Workmen's compensation					
Vacation pay provision					
Fringe benefits					
Holiday pay					
Health and accident Insurance					
Group insurance					
Occupational disease insurance					
Pension					
Total					

Table 12.8—Depreciation

Items	Cost, $	Nondepreciable Portion, $	Depreciable Portion, $	Depreciation Rate, %	Annual Depreciation, $
Buildings					
Machinery and equipment					
Service systems					
Land improvements					
Total					

tion, and amortization of preproduction expenses. The impact of these items may well be a critical factor in financing. Income taxes must be estimated and deducted to obtain the amount of net earnings and to establish the actual return on the investment. Current tax rates should be obtained from reliable sources and the tax computed and applied in the prescribed manner, since there is considerable variation according to the mineral involved for each state and for each country. Knowledge of income tax application will allow proper handling of deductible items like depletion and depreciation and permit taking advantage of tax-free periods where these deductions are allowed. Such knowledge will also enable one to secure maximum benefits from fast write-offs, and tax deductions for interest on loans where such are involved. Details of U.S. income taxes are discussed in Chap. 4.17.

In the example of the M.E. Mine cited herein, a Canadian case is used to show how the tax-free period and the other tax allowances for depletion and depreciation create unequal annual earnings. For those who wish to compare United States taxes with Canadi-

Table 12.9—Summary Operating and Fixed Charges (Basis: 265 Working Days, 900,000 Tons Capacity)

Element of Cost	Total Annual Costs, Average Operating Basis, $	Costs per Ton Run-of-Mine 'Ore, $
Labor		
Direct and indirect	902,000	1.00
Raw materials)		
Supplies)	973,000	1.08
Spare parts)		
Overhead	83,000	0.09
Unit depletion	74,000*	0.08
Depreciation (ordinary and replacements)	520,000†	0.58
Selling and administrative	105,000	0.12
Total	2,657,000	2.95

* Unit depletion here provides for the return of all original costs of the mineral land. This is obtained by dividing total original cost by total ore reserves to be produced for sale over life of property. On annual basis $\frac{\$1,900,000}{25.5 \text{ yrs}}$ or per ton basis $\frac{\$1,900,000}{25.5 \div 898,800 \text{ tons}}$ (average annual production).

† This figure can best be derived by first-hand knowledge of the particular industry: knowing the life of the type of equipment used, knowing the amount of replacements and adjustments experienced from obsolescence (this is substantial in some industries where the treatment is not well developed), and making provision for any inflationary trends in prices. In this particular case the unit depreciation figure includes depreciation of original investment in plant and facilities ($5,960,000) as well as investments for replacements and obsolescence ($7,300,000).

an taxes, an example of United States application is presented:

U.S. Income Tax—Percentage Depletion Example—The percentage depletion allowance for purpose of estimating Federal Income Tax is based on a percentage of gross income from the sale of product and has no relation to costs.

For the product involved herein, the percentage allowed is 23% (if the mine were in the United States), not to exceed 50% of taxable income computed with depletion.

The data for calculation of the Federal Income Tax as given in Table 12.10 are from Work Sheet II.

In considering foreign investments the taxes of both the country of operation and the country of residence must be considered.

Time Factors

Time factors are of utmost importance to the investor and are:

1) Time required for preproduction work to reach the first production stage.

2) Time required to get the property up to the designed production stage or rate.

3) Time required to recoup the investment or to pay debt retirements.

The answers to the foregoing will allow the valuator to (1) estimate the interest charges during the preproduction period, and (2) to establish the deferment period before the earnings start and to calculate present worth of future earnings; and (3) to indicate the financing problems or the possibility of financing the project. If the analysis of the cash flow in relation to time shows that the income to the investor is likely to be inadequate during the early period of operations or is too long delayed, the project might

Table 12.10—Calculation of United States Federal Income Tax (Based on Data from Work Sheet II for Average Year)

Gross annual income from sale product	$4,302,000
Royalty (would be subtracted)	(none)
Base for applying percentage depletion	$4,302,000
Allowance: 23% of $4,302,000	989,460
or 50% of taxable income (maximum)	818,500*
The allowable deduction	818,500
Taxable income (before depletion)	1,637,700
Less percentage depletion allowable	818,500
Taxable income	$ 818,500
U.S. Federal income tax:	
Normal 30% on $818,500 = $245,550	
Surtax 22% on ($818,500-25,000) = 174,570	$ 420,120

* $818,500 is 50% of net income before income taxes plus book depletion less state tax.

prove difficult to finance. If the analysis shows that the property can pay for itself in three to ten years, the project probably can be financed.

Management

Management and know-how are too often taken for granted in valuations of mining projects. The term management used herein refers to the entire staff. With severe working conditions where staff turnover is high, the operating costs would reflect higher supervisory and management costs. Where a company is entering a new type of business for the first time, allowance must be made for time and money spent in learning the business. Sometimes this can be shown by providing specific capital allowance to cover start-up costs, a longer tune-up, and a longer training period for the personnel. An estimated lower sales volume for the first five-year period may be desirable to allow for breaking into new markets. In mines employing a high proportion of untrained or unskilled labor where more supervision is required, it may be necessary to apply a lower labor efficiency figure (lower output per man shift). All the foregoing factors relate to the ability of management to plan and carry out its task. If it is assured that the property will have capable management with experience in the industry, the efficiency of operation can be assumed high or equal to that of active competitors. In "going" operations the success of high quality management is often reflected in the public's demand for stock, particularly oil and gas projects.

Unusual Risk Factors

In reality, risk factors (discount factors) as applied by the valuator are so used as to reflect properly the relationship of the hazards to the earning rate. The risks at the early stages of mine development are likely to be high. The engineer should have a clear understanding of the risks because of their effect on the source and cost of financing the project. Such risks might include the natural hazards of excessive underground water, "heavy" ground, and long periods of severe weather, also social instability involving political unrest, and labor difficulties.

Financing

Financing of the typical M.E. mining project could be done through issuing stock, borrowing, or reinvestment of earnings. If part of the capital is to be raised by long-term borrowing, such as through sinking fund mortgage bonds or serial debentures, these obligations would affect the amount available for dividend payments.

For the sake of simplicity, debt financing is not considered here. However, debt financing is common and usually improves the profitability of a property, partly because interest rates are tax deductible and partly because borrowing reduces the requirement for equity capital and calls for a lower interest rate than is usually demanded for equity capital. In this case, the equity capital rate is considered to be 12%.

Various other methods of financing involving consideration of special income tax aspects have been used in the mining industry. An example would be ABC financing wherein a buyer, a seller, and a banking institution are involved. Since these methods are generally not applied to high-risk ventures and since they are confined to short-term situations, they are not discussed here.

For simplification it is assumed that the M.E. project will be financed out of earnings and hence no public financing costs as such are provided. If present, however, these might add 7% or more to the total of the project.

Effects of Trading Deals on Appraisal of Mineral Property

Trading between buyer and seller may result in a figure differing from the fair market value for the reason that special conditions may exist which would favor one party or the other. As an example of a buyer's special viewpoint, it is not unusual to find a prospective buyer committed, say, to expansion and diversification who may attach some additional value to the mineral property. The property may be favorably located to his own operations, the beneficiation process involved may produce a higher quality product, or he may wish to short-cut the time required to explore and develop much needed reserves.

The determination of various possible equities is a matter of trading but some of the arrangements generally result in advantages to one side or the other, particularly as regards income taxes. If there is a tax saving and other savings to the operations their effect on the appraised value should be calculated. The various provisions in these deals affect the amount of money available for dividend payments, and thus sometimes have a bearing on the value of the property. Table 12.11 gives a comparison of the effects of two different deals.

Table 12.11—Comparison of Effects of Method of Financing on Value of Mineral Property (United States Conditions)

	Indicated Equivalent Purchase Price of Mineral Property, $
1. Outright purchase	990,400
2. Royalty basis—5% of net sales	1,224,100

* Present value of royalties deferred three years.

Setting Up Data for Measuring Earnings

To best illustrate the process and problems inherent in mine valuations, an example has been selected wherein the ore reserves are sufficient for a long potential profit period, a good possibility exists for finding additional ore, but the profit margin is on the low side and there is some risk as regards competition and its effect on future prices.

Most mining companies follow standard accounting practice, hence the valuator should recognize certain basic accounting procedures in setting up and compiling the data in accordance with sound accounting methods and practices. A cash-flow sheet showing cash and noncash items just as they would be experienced in actual practice is desirable. (See Work Sheet II.)

Assuming that the valuator selects a 12% return on the investment and that the future market conditions are based on a continuing conservative level of business activity (as compared to a continuing or increasingly high level of prosperity), he might set up a preliminary test figure for the value of a new mineral property as shown in Work Sheets I–A and I–B. This step is necessary to estab-

Work Sheet I-A—Tentative Estimate Average Net Income

	$ Per Ton
Net Sales	4.78
Cost of sales, exclusive of provision for depreciation and depletion	2.17
Gross profit before provision for depreciation and depletion	2.61
Provision for depreciation and depletion	0.66*
Gross profit	1.95
Selling and administrative expenses	0.12
Net profit before provision for taxes on income	1.83
Provision for taxes on income	0.61
Net income (average annual) 898,000 tons @	1.22—
	1,099,000

* The depletion portion of this figure ($.08) is derived through trial, checked by the procedure in Work Sheet I-B.

Work Sheet I-B—Capitalization Test* To Determine Approximate Value of Mineral Property to be Used for a Tentative Depletion Figure

Calculated gross investment based on capitalization rate of 12% (Factor 8-⅓ X $1,099,000) rounded to	($9,158,333) $9,160,000
Estimated costs of physical property and working capital	7,260,000
Tentative amount assigned to mineral property (including any other assets on property when purchased and any interest until start of earnings)	$1,900,000
Less estimated exploration drilling costs before purchase of property	188,900
Tentative price to be paid for mineral property	$1,711,100

* Capitalization of earnings used here is an accounting procedure to obtain the value of a business as a going concern on a given date. Its key factor is the estimated average annual earning. See formula, page 452.

lish a tentative book depletion allowance to be used in development of the important operating statement (Sheet II), which shows the estimated net earnings. Work Sheets I–A and I–B provide a preliminary estimate of the depletion figure to cover the cost of the mineral property.

Work Sheet II is a basic step-by-step method of showing potential earnings and cash flow for a project. and can be applied to all three commercial valuation examples. In applying Work Sheet II to Case 1, the determination of the value of the mineral land, it is assumed that the price of $1,900,000 (as developed in Work Sheet I–B) has been selected after careful analysis of all the factors involved as a possible figure that will meet the profit potential demanded by the purchaser, and is now to be tested for its validity on the detailed Work Sheet II.

As indicated on line 12 of Work Sheet II, total net earnings for the 25 full.years of operation and the tune-up period amount to $28,025,000. In addition. the original capital investment of $5.960.000 and deferred capital investments of $7,300,000 are recovered through annual provisions for depreciation, and the initial value of $1,900,000 for mineral land is recovered through annual allowances for depletion. Because a stockholder or investor may expect a reasonably quick return of the investment, there is shown on line 11 of Work Sheet II a theoretical minimum payout time, which indicates the possibility of a cash accumulation in the sixth year of $7,860,000, sufficient to return the original investment in buildings, machinery, equipment, and the initial investment in mineral land. However, Work Sheet III shows that the cash throw-off is not sufficient in the early years to meet this theoretical condition. Working capital may be assumed to be, at this minimum payout period, principally in a form readily convertible into cash on a dollar-for-dollar basis. The operations also yield the 12% on the original investment as intended.

In actual practice, early dividends seldom account for the full return of the investment. This deferment, caused by other expenditures, as for a search for new properties, tends to lower the present worth of the return-of-investment portion of the cash flow.

Discounting Future Earnings to Present Worth— Straight Line Method

The mining project is now set up as a business wherein capital has been put into (1) the mineral property, (2) mine equipment and facilities, (3) plant equipment and facilities, and (4) working capital. Since the capital has been put in, it can now be taken out over the life of operations through depreciation, depletion as shown in Work Sheet II, and salvage at the end of operations.

The object here being to prove the value assignable to the mineral property, provision must be made to discount the remaining cash flow (DCF) so as to provide the return *on* the investment (as the return *of* the investment is already provided).

Shortcut methods of discounting cash flow from the preliminary work sheet often result in a higher value than actually experienced in the operation. The reason for this is the shortage of income in the early period to cover the full return of and on the investment and because of uneven capital replacement requirements. Thus the net income and the net book investment should be worked out year by year.

Proof of the validity of the value of the mineral property is shown in Work Sheet III, wherein the net income is discounted to provide the designated return *on* the investment in each segment of the overall business.

It is to be noted from Work Sheet III that the discounted cash flow, adjusted for any unrecouped working capital and salvage of land at the end of operations shows a value of $1,900,000, hence proving the tentative figure used in the operating statement to cover the investment in the mineral property.

It has been common practice to use an average discount rate or risk rate on higher risk mining ventures on the premise that the investment in one segment of the project is no more risky than the other segments. With capital requirements increasing, it is now necessary to analyze and support rates used for working capital, plant, mine equipment, and reserves. For this reason, the example used here in Work Sheet III provides for varying risk rates so as to arrive at a composite rate that more or less fits the experience in the industry. The working capital requirements for an isolated property that en-

Work Sheet II—Showing Development of Profit, Cash Earnings, and Returns M. E. Mines; Project Based on 898,800 Tons Annual Production over Life of Mine (Dollar Figures in Thousands Where Applicable)

Line No.	Start-Up Period	1st	2d	3d	4th	5th	6th	7th	25th	Average Year	Unit Basis per Ton
1. Net sales	$1,311	$3,496	$4,371	4,371	$4,371	$4,371	$4,371	$4,371	$4,371	$4,302	$4.78
Mining and milling costs:											
Labor, direct and indirect	$ 451	$ 902	$ 902	$ 902	$ 902	$ 902	$ 902	$ 902	$ 902	$ 902	$1.00
Materials and supplies	486	973	973	973	973	973	973	973	973	973	1.08
Overhead and general expense	42	83	83	83	83	83	83	83	83	83	0.09
Total mining and milling costs before provision for depletion and depreciation	$ 979	$1,958	$1,958	$1,958	$1,958	$1,958	$1,958	$1,958	$1,958	$1,958	$2.17
Provision for depletion	37	74	74	74	74	74	74	74	74	74	0.08
Provision for depreciation	260	520	520	520	520	520	520	520	520	520	0.58
2. Total mining and milling costs	$1,276	$2,552	$2,552	$2,552	$2,552	$2,552	$2,552	$2,552	$2,552	$2,552	$2.83
3. Gross profit	$ 35	$ 944	$1,819	$1,819	$1,819	$1,819	$1,819	$1,819	$1,819	$1,750	$1.95
4. Gross profit as % of net sales									40.7		
Selling and administrative expenses:											
Selling	20	52	65	65	65	65	65	65	65	64	.07
Administrative	20	40	40	40	40	40	40	40	40	40	.05
5. Total selling and administrative expenses	$ 40	$ 92	$ 105	$ 105	$ 105	$ 105	$ 105	$ 105	$ 105	$ 104	$ .12
6. Net profit before provision for income taxes	$ -5	$ 852	$1,714	$1,714	$1,714	$1,714	$1,714	$1,714	$1,714	$1,646	$1.83
Provision for provincial tax (State, if U.S.)	—	43	86	86	86	86	86	86	86	83	0.09
Provision for dominion tax (Fed. tax if U.S.)*	Tax-free first 3½ years				$21	589	628	645	568	464	0.52
7. Total income taxes	—	$ 43	$ 86	$ 86	$ 607	$ 684	$ 714	$ 731	$ 654	$ 547	$0.61
8. Net income	$ -5	$ 809	$1,628	$1,628	$1,107	$1,030	$1,000	$ 983	$1,060	$1,099	$1.22
9. Net income as % of net sales										25.5	
10. Percent return on investment of $9,160,000 (includes working capital)										12.0	
11. Recovery of investment—cumulative net income plus provision for depletion and depreciation	292	1,695	3,917	6,139	7,840	7,860	(Return of investment)				
12. Cumulative net earnings from line 8 (Total $28,025)	-5	804	2,432	4,060	5,167	6,197	7,197	8,025	(Return on investment)		

* Because taxes are constantly changing and alternate choice features are involved, no attempt is made to clarify all details.

73

Work Sheet III—Discounting Approach to Fair Market Value Mineral Land (2.5-½ Year Life)

Net Income After Taxes and Depreciation but Before Book Depletion, $	Return on Working Capital (8%), $	Return on Equipment and Facilities (12%), $	Total, $	Residue Income for Mineral Property, $	Discount Factor (12% Return on Investment in Mineral Property)	Present Value Mineral Property, $
32,000	52,000	357,600	409,600	(377,600)	0.89286	(337,144)
883,000	104,000	718,353	822,353	60,647	0.79719	48,347
1,702,000	104,000	724,659	828,659	873,341	0.71178	621,627
1,702,000	104,000	730,965	834,965	867,035	0.63552	551,018
1,181,000	104,000	737,270	841,270	339,730	0.56743	192,773
1,104,000	104,000	743,577	847,577	256,423	0.50663	129,912
1,074,000	104,000	749,883	853,883	220,117	0.45235	99,570
1,057,000	104,000	756,188	860,188	196,812	0.40388	79,488
1,179,760	104,000	762,494	866,494	313,266	0.36061	112,967
1,179,760	104,000	768,800	872,800	306,960	0.32197	98,832
1,179,760	104,000	775,106	879,106	300,654	0.28748	86,432
1,179,760	104,000	781,412	885,412	294,348	0.25668	75,553
1,179,760	104,000	787,718	891,718	288,042	0.22917	66,011
1,179,760	104,000	794,024	898,024	281,736	0.20462	57,649
1,179,760	104,000	800,330	904,330	275,430	0.18270	50,321
1,179,760	104,000	806,636	910,636	269,124	0.16312	43,900
1,179,760	104,000	812,942	916,942	262,818	0.14564	38,277
1,179,760	104,000	819,248	923,248	256,512	0.13004	33,357
1,179,760	104,000	825,553	929,553	250,207	0.11611	29,052
1,179,760	104,000	831,859	935,859	243,901	0.10367	25,285
1,179,760	104,000	838,165	942,165	237,595	0.09256	21,992
1,179,760	104,000	844,471	948,471	231,289	0.08264	19,114
1,179,760	104,000	850,777	954,777	224,983	0.07379	16,601
1,179,760	104,000	857,083	961,083	218,677	0.06588	14,406
1,179,760	104,000	863,388	967,388	212,372	0.05882	12,492
1,134,080	104,000	869,694	973,694	160,386	0.05252	8,423
Total 29,925,000*	2,652,000	20,208,195	22,860,195	7,064,805		2,196,255

Less present value unrecouped working capital   -42,504

   2,153,751

Add present salvage value at end of operations   +44,700

Gross present worth $2,198,451
Cost 2-½ year—carrying charge on investment in mineral property   -298,451

Value mineral property on books $1,900,000
Cost proving up drilling   -188,900

Price to be paid for mineral property $1,711,100

* Total net earnings plus total book depletion of $1,900,000 from Work Sheet II.

74

joys good earnings are likely to be much greater than a similar property close by, yet the risk is the same. Also, in many cases plant costs and equipment costs (on turnkey jobs, for instance) can be estimated with a fair degree of accuracy, hence the risk rate on plant may well be considered lower than the risk on the reserves where the same degree of measurement is not possible. Such varying risk rates are particularly applicable in cases wherein the project is a going concern with proven operating costs, and detailed capital cost data are available but the reserves are less defined.

Determining Present Worth of Mineral Land

The determination of the fair market value of, or purchase price to be paid for, the mineral land is summarized in Table 12.12.

Discounting by Hoskold Method

Some appraisers of mineral property still use the Hoskold formula for purposes of comparison and checking. In using this formula, caution must be exercised to avoid errors in handling depletion and depreciation, as this formula provides for the replacement of the original capital investment. An example is given in Work Sheet IV using the same basic data but assuming uniform net income (which is not the case here as is shown in Work Sheet II).

The basic Hoskold formula is as follows:

$$V_P = \frac{A}{\dfrac{r}{R^n - 1} + r'}$$

where A is annual earnings ($1,383,700), r is "safe" interest rate on reinvested capital redemption (4%), r' is risk rate of return on invested capital (12%), n is life of operations (25 years). V_P is present worth or present value of future income ($9,608,100), and R is amount of $1 with one year's interest (1 + r or 1.04).

Table 12.12—Summary Fair Market Value of Mineral Property

Production, tons		23,000,000
Total net Sales		$109,701,000
Total cost sales:		
Fixed and variable costs	$60,654,000	
Depreciation	7,290,000	67,944,000
Income before taxes and depletion		$ 41,757,000
Income tax provision		11,832,000
Net income before cost depletion		$ 29,925,000
Less return on investment in:		
Working capital (8%)	$ 2,652,000	
Equipment and facilities (12%)	20,208,195	22,860,195
Remainder income for mineral property		$ 7,064,805
Present value of remainder income		$ 2,196,255
After adjustment for unrecouped working capital		$ 2,153,751
Salvage land at end operations		+44,700
Gross present value		$ 2,198,451
Carrying charge for investment in mineral land, 2½ years @ 8%		298,451
Value of mineral land and reserve Prove up costs— book depletion		$ 1,900,000
Cost proving up drilling		-188,900
Price to be paid for mineral land only		$ 1,711,100

Work Sheet IV—Method of Estimating Future Annual Net Income for Purpose of Capitalization, by Hoskold

	$ Per Ton	Average Annual Amount, $
Net sales	4.78	4,302,000
Cost of sales (exclusive of provision for depletion and depreciation)	2.17	1,958,000
Gross profit before provision for depletion and depreciation	2.61	2,344,000
Provision for net depletion and depreciation	0.66	594,000
Gross profit	1.95	1,750,000
selling and administrative	0.12	104,000
Net profit before provision for taxes on income	1.83	1,646,000
Provision for taxes on income	0.61	547,000
Net income	1.22	1,099,000
Add back unit depletion and amortization of plant over life	0.32	284,700
Basis for discounting by Hoskold (assumes uniform earning rate)	1.54	1,383,700
Present worth factor 25 years @ 4% and 12% = 6.9438		
Value of total operating profit at start of operations $1,383,700 × 6.9438		9,608,100
Add present worth of salvage value at end of operations		44,700
Total value of project at start of operations		9,652,800
Less capital cost to get into production (exclusive of cost of mineral property but includes $188,900 for prove-up drill)		7,448,900
Value of mineral land at start of operations		2,203,900
Value of mineral property at date of purchase (includes any other assets involved at date of purchase)		
Deferred 2½ years before profit starts		1,819,496

EARNINGS ANALYSIS

Yardsticks For Earning Power and Value Comparisons

The valuator may be interested in comparing the estimated earnings of the project with actual earnings of similar operating companies or with earnings of current or active ventures on the security market. The purpose is to show whether or not the project is a reasonable one justifying the risks—whether or not the earnings fall within a reasonable range of return as might be expected or demanded by the public, particularly if the venture is to be publicly financed. The basis for comparing the investment is assumed to be the average investment over the life of the operations. Some of these yardsticks (operating ratios and percentages) can be easily derived from Work Sheet II.

1. *Gross profit* as a percentage of Net Sales, averages about 40% during the life of the operations.

2) *Net profit* as a percentage of Net Sales, averages about 25.5% during the life of the operations.

3) *Average annual net earnings* after taxes when compared to total investment gives a rate of 12.0%.

4) *The payout period* is in the sixth year after start-up.

Table 12.13 shows a comparison of "yardsticks" of earning power for various mining companies.

Any comparisons must be applied with considerable judgment or knowledge of the operations concerned. One factor making comparisons difficult is that most of the mining companies available for comparison are integrated with smelting and fabricating, hence are not strictly comparable with a project wherein only mining is involved.

Other factors affecting "yardstick" or earning power comparisons are: (1) sudden changes in prices of metals, (2) sudden increase in labor costs, (3) expansion or increase in output, and (4) production restrictions (as with oil and gas).

Capitalizing as Test of Value

In this method an earning rate corresponding to the amount an investor would

Table 12.13—Yardsticks or Comparisons of Earning Power (Mining Companies) Based on Data Available for 1967

Company Listed on Exchange	Price Range, $	Earnings per Share, $	Dividends, $	Dividend % of Earnings	Yield, %	Gross Profit % Net Sales	Price Earnings Ratio	% Earnings of Price
A	53–39	4.31	2.38	55	5.1	21.2	10.8	7.8
B	56–43	3.87	2.25	58	4.1	6.1	12.9	7.8
C	7–2	0.05	nil	—	—	2.6	—	—
D	20–12	1.94	0.70	36	4.2	8.3	8.6	13.1
E	19–9	0.24	3% stock	—	—	—	—	—
F	45–55	4.75	1.52	32	3.8	12.2	8.4	12.2
G	37–25	2.31	1.70	74	5.4	32.5	13.6	7.5
H	56–40	0.30	0.50	—	1.0	4.1	—	0.6
I	58–37	1.64	0.80	49	1.7	25.8	29.4	3.5
J	70–44	5.53	2.20	41	3.8	—	13.0	9.4
K	80–62	5.04	3.40	67	4.8	12.5	14.1	7.2
Compared to M.E. Mine example see Table 12.17	78–5/8	9.90	6.00	60.0	7.6	40.7	7.9	12.6

expect from a similar going concern is selected using the following formula:

$$\text{Value} = \frac{\text{Annual Earnings}}{\text{Rate of Return, \%}}$$

An example has been worked out on Work Sheet I-B (page 444) for the purpose of obtaining a test figure for value of the mineral property.

Capitalization is a method of valuing net earnings on a business that has no fixed life, hence is not used in mining valuation except as a shortcut check method. It does not discount cash flow.

CASE 2—VALUATION OF A GOING MINE

Factors for Consideration

The fair market value of a going concern essentially consists of discounting future annual earnings, adding value of salvage at end of operations and net liquid assets, and subtracting any additional capital costs plus interest charges, if any.

Usually, operating records are available on which to base the future performance of the operations and more of the unknowns are eliminated because of actual operating rec-

ords and experience. Also, the property will have tangible assets such as buildings, mine and mill equipment, and facilities which must be appraised as to their importance and use in future operations to determine adjustments necessary to provide adequate depreciation and to determine new capital requirements. Other facilities on the property will likely not be usable but may be considered as having salvage value. Other nonoperating assets may exist which contribute to the earnings.

A thorough examination and study of the mine and plant operations are necessary to determine possible future improvements in mining, milling, or other operating factors, including adverse factors to be met. The following reminder list may be useful:

Reminder List for Determining Value of a Going Mine

1). Thoroughly examine the property, study the history and experience of the organization, and analyse the property, operating, and financial records.
2). Determine the special conditions likely to be involved in the valuation.
3.) Calculate the remaining mine recoverable ore reserves and ex-

amine the possibility of developing additional reserves.

4.) Determine the relation of the ore reserves in place to the volume of recoverable and salable product, taking into consideration the actual experience record in mining, handling, and treatment losses.

5.) Using foregoing data calculate the total salable product to be recovered from the reserves.

6.) Determine the annual market potential or the desirable rate of production indicating physical or economic limitations and thence determine the life of operations based on existing ore reserves.

7.) Study production costs—the trend for the past five-year period considering the direct labor and materials costs and the noncontrollable and overhead costs.

8.) Study sales for five-year period or more—by grades, customers, tonnage and dollars, transfer to departments, sales by areas; determine if sales include freight to destination.

9.) Study income statement of operation for other data such as inventory increases or decreases, cost of sales, sales expense, general administrative expense, other income less other charges, net profit before income taxes, federal and state income taxes, net income.

10.) Study annual depreciation rates for each item of physical property or by groups.

11.) Set up future average annual sales potential by tonnage and dollars.

12.) Set up future average annual net earnings (after income taxes, taking into consideration tax allowances) based on future trend of sales revenue and operating costs.

13.) Determine all risk factors and discount annual profits to obtain present value of all future annual earnings to be derived from the ore reserves.

14.) Make a detailed inventory of the mine, plant equipment, and facilities, indicating their condition and future usefulness to the operations. Check these physical assets against life of reserves.

15.) Estimate the reproduction or replacement cost of all usable fixed capital physical property (excluding land) and adjust for depreciation and future value. Estimate cost of new capital requirements to place in good operating condition.

16.) Estimate salvage value of fixed assets at end of operation.

17.) Determine the value of all the nonmineral land.

18.) Estimate the value of all working capital and warehouse stocks.

19.) Estimate the present value of any intangible assets.

20.) Consider the capability of management and operating staff.

21.) Determine present value by discounting adjusted cash flows. To this add net current assets and the present value of any salvage residue at end of operations.

22.) Add present value of nonoperating assets.

23.) Subtract any new capital costs and interest charges, if any.

Appraisal of Physical Assets

A going mine will have fixed assets, and the value of their future use in the operation must be determined. These assets are likely to include (1) land or real estate, (2) buildings and facilities, and (3) inventory and supplies. The real estate value of land can be determined by comparing with going rates or past sales of property of similar type. The value of inventory and supplies can be priced on their value in use or cost of placing them at the property. The value of buildings and facilities, however, requires that the cost of duplication be established and thence discounted for depreciation or remaining life and usefulness. This requires careful inspection of the buildings, equipment, and facilities to determine the state of wear and tear or their obsolescence. Only after such inspection and reproduction cost appraisal can the engineer estimate the new capital to be added for future operations and the proper depreciation charges against future operations. This process usually requires experienced engineers and construction cost estimators. A summary of their detailed calculations would likely take the form shown in Table 12.14.

Usefulness of Facilities

The purchaser of a going mine is not interested in investing money in equipment and facilities that have no useful purpose in the future operations of the property. Most mines that have operated for any length of time may have considerable equipment and facilities of this nature. Those items having no future use to the operations may be appraised at salvage value.

Nonoperating Assets

A mining company may realize profits on various extraneous activities outside its mine operations as follows:
1) Sales of securities (nonrecurring).
2) Lease of land for farming.
3) Operations of railroad.

Table 12.14—Summary of Physical Property

| Classification | Estimated Cost of Reproduction on | | Value to Future Project $ |
	New $	Less Depreciation $	
Land			
Development			
Buildings			
Furniture and equipment			
Machinery and equipment			
Service facilities			
Inventory			
Subtotal			
Miscellaneous expenses for engineering, construction, taxes, insurance, etc., during period of construction or installation of any new facilities or rehabilitation			
Total			

4) Sale of power.
5) Sale of water.
6) Sale of tailings.
7) Earnings on stock in other concerns.

The problem of the valuator is then to determine the nonrecurring income as distinguished from that which is likely to continue in the future.

Once the annual future earnings can be established on any source of "outside" income, these can be reduced to present worth by a discount method as covered previously.

Intangible Values (Good Will, Brand Name, Etc.

Sometimes intangible values must be considered and given a value. An example is a vertically integrated mining company which produces a pig metal with a brand name that is well established in the trade. Anyone contemplating entering the metal business as a new venture would likely find much sales resistance in trying to establish the same position in the trade and would have to resort to an aggressive marketing and sales campaign over a period of time to get established and obtain a fair share of the market. The purchase of a company with an established market would have a definite value in terms of avoiding the cost of a sales campaign and of added value by avoiding delays in securing a share of the existing market.

The value of intangible ore discovery prospects may be a significant factor in the opinion of either the buyer or seller of a mineral property and is often a point involved in the "trading."

Management and know-how also have an important bearing on the value of a going concern and should definitely be considered. This is particularly shown by favorable market reactions on properties having good management.

Summarizing Value of a Going Mine

Assume that the M.E. Mine was appraised for sale 15 years after the mine had started operations and during this time five additional years of ore reserves had been found. Also, the company had been able to maintain its profit margin although a careful appraisal of the physical assets indicated that a buyer would have to invest $120,000 in new capital to correct the condition of deferred maintenance, and there would be a salvage residue recoverable at end of operations of $800,000. Engineers estimated the value of the ore reserves, costs of future operation and total future earnings, as shown in Table 12.15.

During the examination, the M.E. Mine was found to have a nonoperating income item in the form of a contract to sell power at a rate which allows an annual net income of $10,000 for ten years, or a total of $100,000.

Figures in the company's balance sheet were analyzed to determine the amount of current assets that can be recovered and the necessary working capital required to keep the business going since the buyer must pay the seller for the net current assets. Table 12.16 shows the company's balance sheet and the estimated value of liquid assets.

Determining Present Value and Price to be Paid for Going Concern

The valuator can now assemble all the basic elements of value as shown in Work Sheet V.

The purchaser could invest $8,645,200 plus the new capital requirements or a total of $8,765,200. He could expect to earn 12% on his investment and also recover that investment by the end of the operations.

Indicated Intrinsic Value of Common Shares

A share of common stock is an equity in the business and if a good fair market value is determined, the total value of the shares on the average should theoretically represent the present value of all future dividends after adding net current assets and subtracting all priorities ahead of common shares. However, the market quotation is a value of common stock set by the public at a given time, hence may be higher or lower than the fair value as mentioned previously. Also, dividends generally do not equal total earnings as shown in Table 12.13.

Stock quotations may vary because of such items as (1) fluctuation in price of commodity sold, (2) recent exploration success, (3) change in methods of management affecting operating costs, and (4) unexpected adverse hazards.

Although the province of valuing securi-

Table 12.15—Estimation of Future Earnings M. E. Mine—15 Years after Original Appraisal

Mine Recoverable Reserves	Tons	Estimated Recoverable Value per Ton, $
Broken ore inventory	295,000	5.85
Developed ore	13,205,000	5.39
Total	13,500,000	5.40

Total value of ore reserves 13,500,000 × $5.40 = $72,900,000
Cost as per annual report

Development	$0.10
Mining	0.88
Milling	1.64
Overhead	0.10
Depletion and depreciation	
Total cost of manufacturing	$3.44
Selling and administrative	0.13
Total operating costs	$3.57
Estimated income taxes	0.61

Estimated cost of operation and taxes per ton $4.18
Estimated total cost of sales 13,500,000 × $4.18 = $56,430,000

Estimated total future earnings $16,470,000

Table 12.16—Value of Current Asset as per Balance Sheet M. E. Mine, Annual Report

Assets	As per Balance Sheet, $	Estimated Recoverable Value, $
Cash on hand, in banks	270,000	270,000
Salable products on hand and in transit	640,300	640,300
Accounts receivable from sale of product	53,800	43,000
Accrued interest receivable	200	200
Inventory supplies	340,000	225,000
Investments	10,000	14,000
Prepaid expenses	25,000	25,000
Fixed assets less depreciation	1,805,000	Considered in salvage figure
Preproduction expenses	25,000	—
Advances and miscellaneous	35,000	30,000
Total assets	3,204,300	1,247,500
Liabilities		
Current liabilities	163,400	163,400
General reserve	496,000	—
Capital stock less discount	2,544,400	—
Surplus	500	—
Total	3,204,300	163,400
Total net recoverable current assets		1,084,100

ties is in another field, the mining engineer should have some idea as to the indicated intrinsic value of a common share in the project he is valuating, so as to compare with shares of established mineral operations. Table 12.17 shows the derivation of the market value of a share in M.E. Mine based on the appraised value.

It is assumed in this project that a syndicate originally supplied the necessary capital ($9,160,000, Table 12.4) to get the property into production; they formed a company and issued to themselves 110,000 shares of stock. Table 12.17 shows the derivation of the market value of a share in M.E. Mine based on the appraised value. The figures appear in Table 12.13 for purpose of comparison with other mining companies.

CASE 3—VALUATION FOR MERGER

The mining engineer may have occasion to assist in valuation work involving a merger. For this reason he should have some idea as to the approach to the problem and factors to be studied.

A merger of two mining concerns or a mining concern with some other industrial

Work Sheet V—Summary Fair Market Value M. E. Mine as Going Operation

Present value future earnings 13,500,000 tons × $1.22 = $16,470,000 total profit discounted at 12% over 15 years × 6..8109	$7,478,400
$\overline{15}$	
Add salvage at end of operation— $800, 200 discounted at 12% after 15 years (× 0.18270)	146,200
Add net current assets (from Table 12.16)	1,084,100
Total	$8,708,700
Add present value of income from nonoperating assets $100,000 discounted at 12% over 10 years $(\times \dfrac{5.6502}{10})$	56,500
Total present value future income	$8,765,200
Subtract new capital requirements to take care of the deferred maint.	120,000
Fair market value as going concern	$8,645,200

concern usually implies that the effect of such action will result in a mutual advantage. Thus the advantages of the merger must be weighed against the disadvantages.

Table 12.17—Market Value of Common Share M. E. Mine as Going Operation

Total present value as going concern (from Work Sheet V)	$8,645,000
Market value per share $8,645,000 ÷ 110,000 shares	78.60
Annual earnings $\dfrac{\$16{,}470{,}000}{15}$	$1,099,000
Annual earnings per share (÷ 110,000 shares)	$9.90
Price-earnings ratio $\dfrac{78.60}{9.90}$	7.9
Dividend assumed, 60% of annual earnings	$6.00

The valuation process involves the commercial valuation of each company and, where required, the fair ratio of exchange of securities of the two companies. The valuation procedure is essentially the same as in the example of the going concern, except that the specific advantages and disadvantages of the merger must be weighed.

Factors Affecting the Exchange Ratio of Common Stock

Factors that may be considered include the nature of each business, historical operating records of each company, the current earnings, comparative book net worth, and the future potential earnings of each company, the latter being the most important factor. The following is a list of items that might be involved in a study and determination of the commercial value of each company:

1) Inspection of properties and installations.

2) Analysis of ore reserves of both companies, their availability and effect on future earnings.

3) Analysis of operations, showing degree of obsolescence, general operating condition, maintenance practice; also consideration should be given to source of raw materials and supplies.

4) Analysis of financial records:

(a) Corporate history.

(b) Financial position. A review of total assets will indicate the percentage of total assets that are current assets and the percentage that are fixed assets, such as land, buildings, machinery and equipment, adequacy of reserves for depreciation, and amortization of buildings and equipment. Compare past capital values and indicate changes in value of properties.

(c) Working capital. The difference between the total current assets and the total current liabilities gives the amount of working capital. The amount should be sufficient to cover operating expenses and usage over intervals required between production and sales realization. A fairly uniform ratio should exist between the inventory figure and total current asset figure.

(d) Inventory records of raw materials, work in progress, and finished products should be examined.

(e) Loans should be analyzed as to date of maturity, and any other restrictive feature which might affect payment of dividends, working capital, future credit, and cash flow. Indication of indebtedness on future earnings.

(f) Capital structure factors, such as the amount of unissued stock and recent increases in number of shares, may be indicative of possible management problems.

(g) Dividend records, especially the amount of cash dividends in relation to the total net income, may be revealing if compared on an annual basis and related to the number of shares. Comparison of earnings of each company in recent years.

(h) Market analysis and sales records, distribution methods, competitive position, and sales contracts may involve advantages or disadvantages or disclose future restrictions.

(i) Management policies such as organization, controls, labor and executive employment contracts, retirement or pension fund, and bonus plans are likely to have an important bearing on any merger as well as indicate future financial obligations of each company and the effect of the merger on employe relationships.

(j) Gross profit records may be a useful guide to estimating future profit potential.

(k) General, administrative, and selling costs may indicate the source of savings through merger.

(1) Other income or other deductions from income items should be studied as to their future effect on the business.

5) Analysis of trading in common stocks of each company and the valuation of the stocks by the public may be a useful guide or offer supporting evidence as to the final ratio derived.

6) Placing of all economic indices or factors on a comparable basis, giving proper weight to each factor.

7) Determination of the earning and commercial value of each company and thence ratio of prospective earnings.

8) Consider tax applications.

A consideration of the foregoing should indicate whether or not the companies to be merged can complement each other's existing facilities or whether duplications are involved. Also the data will form the basis of estimating economies through combined operations, and any advantages through increased capacity in the light of opportunity for improved buying, marketing, product diversification, and tax advantages. Any advantages would likely improve the marketability of the stock of the merged company.

From the foregoing the valuator can determine (1) the advantages, (2) the disadvantages, (3) the desirability, and (4) the reasonable ratio for exchange of stock or equity in the earnings.

CASE 4—VALUATION IN EMINENT DOMAIN CASES

In cases of eminent domain wherein the federal, state, or local authorities have condemned mineral land for public use, the courts have handed down a most interesting record of awards or settlements, many of which have little relation to values derived through valuation procedures accepted in the market place of the mining industry. The court records often clearly show a profound struggle by judges, lawyers, and jury members to comprehend the subject of valuation where mineral land is involved.

As every valuator knows, when he fixes the value on a mineral property it is neces-

sary to calculate the future profits from the extraction of the mineral. Such evidence, however, is always improper in many of the courts under the current established rules of evidence. Hence it is difficult for the engineer to understand why it is not possible to present all the evidence to the court. A major portion of the voluminous transcripts covering eminent domain cases consists of involved discussions which point out that "value" is not gross value (amount paid by the buyer for the mineral processed) nor the naked value of the land as though it contained no mineral, but something in between. This search for the "something in between" has taken some fantastic paths by judges and juries in their attempts to understand, in a short time in the court, what is frequently a fairly complex valuation problem.

So it is not surprising to find a mineral property owner confronted with a "taking" and later find "the court's reluctance to accept evidence relating to anticipated future earnings, profits, royalties, and costs in the valuation of mineral properties." Even the legal profession has recognized the general confusion. For example, in Orgel's *Valuation Under Eminent Domain,* Vol. I, Sec. 165, is to be found the following quotation:

"The courts have generally rejected . . . attempts to introduce into evidence the various forms of earnings and profits data. They have usually stated that the measure of compensation in the mine and quarry cases, as in other eminent domain cases, is the market value of the land, but that the stone or mineral deposits may be considered as bearing on the market value of the land. Accordingly, it is proper to admit evidence that the land contains valuable mineral deposits, but the award may not be reached by separately valuating the land and the deposits.

"Evidence of realized profits derived from quarry operation receives the same treatment as general business profits, and similar reasons are advanced for the exclusion of the testimony. . . .

"A fortiori, it is error to consider testimony based on assumptions as to annual productivity over a long period of years, fixed future costs of production and sales prices of the quarry products."

Orgel further states:

". . . the facts that the courts have failed to set any real standards of 'value of the part taken' and that juries must therefore guess almost blindly as to what the concept means."

The law involving condemnation assumes that the owner is to be given "just compensation" for the injury suffered or the land denied by the taking and damage, if any, to the remainder in the case of a partial taking. Since the law is generally based upon precedent, it is disconcerting to review the hundreds of early cases of eminent domain and to note the differences in the bases of awards of the various juries by geographic areas.

Recently the many large public works projects have eliminated tremendous reserves of minerals from future use. Such cases of necessity have required careful consideration of facts presented by witnesses or specialists in mineral land appraisals as contrasted with real estate appraisers.

In the court, the valuator has the role of "expert witness" and thus is to serve as a guide in the determination of the "value" (the "something in between" mentioned previously). His objective, without infringing on the rules of evidence, is to clearly develop the value attached to the mineral property as separated from the total value of the enterprise (actual or theoretical mining enterprise involved in the extraction of the mineral). Under the present methods, jurymen who have no independent knowledge of valuation are asked to pass their opinion on the market value of the mineral land after hearing the views of expert witnesses, and forming their own impressions on data presented that has been much restricted by the rules of evidence. In theory, this may appear to be a just method of settlement, but in practice many of the records do not support the theory.

The courts appear inclined to approach the method of proof of just compensation, where severance damage is involved, in their own legalistic ways. Usually, with supposed severance damage the value concept of "Before and After" is generally followed. This method tries to prove the fair market value of the mineral land before the taking, subtracts the fair market value remaining after

the taking, and the remainder is the total loss (value of land taken plus damage to remainder not taken) to the landowner. In both instances the land is the issue and not the profits that can be made or the value of the mineral in place on the parcel. Unfortunately, many factors are involved that make it difficult to follow accepted valuation practices in the marketplace and yet satisfy the phraseology used in precedent cases.

Due to the desire of both the bench and the bar to adhere to precedent cases, and to their fondness for attaining "well settled" cases despite the undesirable consequences, little judicial criticism of the present rules has been made and hence no move toward reform.

The cases show that three general approaches are used in cases involving mineral property. One involves sales prices of comparable properties. Since sales of the same type of mineral property are not frequent and rarely would they involve the same conditions, this method is unlikely to be adaptable to most condemnation cases involving valuable mineral. The second is the cost approach. If by chance the property is new and improvements reflected its highest and most profitable use (which is seldom true when the property is subject to condemnation), the total cost may be synonymous with value. The third approach involves capitalization of income from the property (not income from the business). In other words, the law appears to allow compensation for loss of the value of the mineral land exclusive of any loss of income from investment in plant and facilities even though these are essential to recover any investment in the mineral property.

In cases of eminent domain, therefore, it is necessary and desirable that accepted sound valuation be harmonized with the legal requirements through guidance by members of the legal profession. Additional practical aspects involve human factors including the proficiency of the judge to properly instruct the jury, the background of the jury members, and the skill of the jurists. There is genuine need for the legal profession to bring more uniformity of valuation procedure into the courts and to clarify some of the existing confusion regarding use of basic valuation procedures.

Bibliography

1. Baxter, C. H., and Parks, R. D., *Examination and Valuation of Mineral Property*, 3rd ed., Michigan College of Mining & Technology, Houghton, 1952.

2. Campbell, J. M., *Oil Property Valuation*, Prentice-Hall, Englewood Cliffs, N.J., 1959.

3. Dewing, A. S., *Financial Policy of Corporations*, 5th ed., Vol. 2, Ronald Press, New York, 1941.

4. Glanville, J. W., "Rate of Return Calculations as a Measure of Investment Opportunities." *Journal of Petroleum Technology*, Vol. 9, No. 6, June 1957.

5. Kennedy, E. J., "Valuation Reports and Security Analysis," *Journal of Petroleum Technology*, Vol. 5, No. 7, 1953.

6. Matthews, T. K., "Tax Planning: A Guide to Financing in the Mining Industry," *Mining Engineering*, Vol. 21, No. 1, 1969.

7. Paine, P., *Oil Property Valuation*, John Wiley, New York, 1942.

8. Sheldon, D. H., "Valuation of Oil Properties," *Journal of Petroleum Technology*, Vol. 5, No. 7, 1953.

9. Weller, P., "Put Policy First in DCF Analysis," *Harvard Business Review*, Jan.-Feb. 1970.

10. Wilcox, F., *Mine Accounting and Financial Administration*, Pitman & Sons, New York, 1949.

[*Editor's Note:* In the original, Worksheets II and III appear on pages 446–449.]

6

Copyright ©1980 by the Canadian Institute of Mining and Metallurgy

Reprinted from Canadian Mining and Metall. Bull. **73**(814):87–99 (1980)

QUICK GUIDES TO THE EVALUATION OF OREBODIES

T. A. O'Hara

ABSTRACT

Annual mining revenue can be computed by formulae relating metal prices, smelting schedules, concentrate grade and recovery to known ore grade and an assessment of core samples. Operating costs at different daily tonnages are related to orebody shape, mining method, milling process and general plant services. Capital costs are related to sizing of mining equipment, mine development, plant-site topography, climate and accessibility, plant services and personnel housing.

Introduction

A crude guide to the average 1978 cost of mine projects is shown in Figure 1, but the costs of many mine projects have differed widely from the graphed average costs of $800,000 $T^{0.6}$ for underground mine/mill projects and $400,000 $T^{0.6}$ for low-grade open-pit mine/mill projects.

High mine project costs and high operating costs are characteristic of open-pit mines with high stripping ratios, underground mines with thin orebodies, mills with complex metallurgy, and projects located in isolated regions with severe climate, mountainous topography, and lacking access to existing roads, towns and electric power. When none of these adverse conditions prevailed, the mine project cost was typically much less than the graphed average cost. Virtually all the conditions likely to cause high capital and operating costs are known from knowledge of the local topography, climate or accessibility, or can be assessed by tests on core samples, as soon as the orebody has been outlined by drilling.

A more accurate estimate of over-all project capital cost and operating cost can be made from a summation of items of cost after judging the effect of specific local conditions on each item of capital and operating cost. This paper offers guides in judging the effect of local conditions on the sizing and 1978 cost of specific items of capital cost and operating cost and operating cost of mine projects. When the specific items of capital cost and operating cost are totalled and measured against expected net revenue and taxes over the life of the mine using the net-present-value or discount-cash-flow methods, the result should show whether the mine project development is clearly feasible, doubtfully feasible or clearly uneconomic. If the development of the mine project is clearly feasible, or probably feasible, a detailed feasibility study by experienced consulting engineers should be commissioned before any financial commitments are made; but if project development appears to be uneconomic, the time and cost of a detailed feasibility study is not warranted.

The chief variable affecting all items of capital cost, revenue and operating cost is the daily tonnage of ore that is treated by the process plant. The guides to capital and operating cost offered in this paper are based on the assumptions that the process plant will be operated continuously for seven days per week, but mining operations and crushing plants will operate for only five days per week, with only necessary maintenance services on weekends. Thus, the daily ore tonnage mined and crushed will be 40% greater than the daily ore tonnage milled.

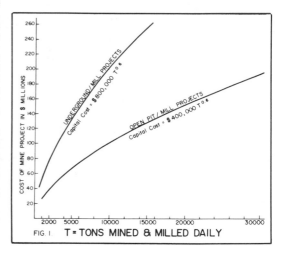

FIG. I   T = TONS MINED & MILLED DAILY

Estimates of the number of employees required, the operating costs per ton and some items of capital costs will require adjustment if the plant operates with seasonal shutdowns or if the mine and the mill operate on the same shifts per week.

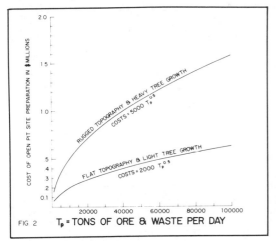

FIG 2 T_p = TONS OF ORE & WASTE PER DAY

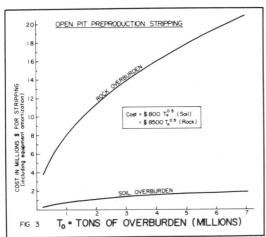

FIG. 3 T_o = TONS OF OVERBURDEN (MILLIONS)

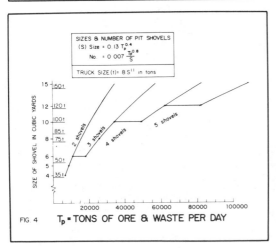

FIG. 4 T_p = TONS OF ORE & WASTE PER DAY

Capital Cost Estimation—Open Pits

If the ore deposit is judged to be amenable to open-pit mining methods, an estimate must be made of the amount of soil and rock overburden that must be stripped before ore mining can be started, and an estimate must be made of the average ratio of waste tonnage to ore tonnage when the open pit is producing ore at the selected daily milling rate.

Site Preparation, Plant and Pit Roads

The cost of site preparation, as shown in Figure 2, will depend on the area, the topography and tree growth on the site of the proposed open pit, waste dumps and process plant. The area of the site and length of the access road to be prepared will typically vary with the square root of the size of the open-pit mine as expressed in terms of the tonnage of ore and waste (T_p) to be mined daily.

Pre-Production Stripping Costs

Figure 3 shows the cost of stripping soil and rock overburden, assuming that soil will be stripped by contractors using scrapers, and rock overburden will be removed by drilling, blasting, loading and haulage of rock by readily available sizes of drills, loaders and trucks. Because the size and condition of equipment readily available for rock overburden stripping is rarely suitable for production mining, the cost of this equipment must be amortized over the pre-production period.

Open-Pit Equipment—Sizing and Cost

The most important items of pit equipment are usually the pit shovels and haulage trucks, and because the size and number of haulage trucks depends on the size and number of shovels, selection of shovel size affects all other items of pit equipment. Figure 4 shows the typical selection of shovel size and number of shovels for different daily pit tonnages. Although some pits may have conditions suitable for partial replacement of shovels by front-end loaders at lower capital cost and improved flexibility, this study assumes that only shovels will be used for loading rock.

The size of haulage trucks should be matched against the shovel size so that a unit number of shovel loads fills the waiting truck. In general, the truck size in tons is about nine times the shovel size (in cubic yards) for small shovels, and about ten times the shovel size for large shovels as shown by the formula $t = 8\, S^{1.1}$.

The cost of open-pit equipment is shown in Figure 5 as a function of tons of ore and waste mined daily (T_p)

Figure 5 shows that the two major items in pit equipment cost are the fleet of shovels, each of which costs about

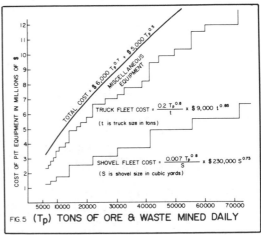

FIG. 5 (T_p) TONS OF ORE & WASTE MINED DAILY

$230,000 S$^{0.73}$, and the fleet of trucks, each of which costs about $9,000 t$^{0.85}$.

Open-Pit Maintenance Facilities

The cost of building and equipping facilities for the garaging and maintenance of open-pit equipment will be dependent chiefly on the sizing and numbers of pit haulage trucks. The haulage truck size varies directly with $T_p^{0.44}$, and the number of haulage trucks varies directly with $T_p^{0.36}$. The cost of maintenance facilities will vary with the square of haulage truck size and directly with the number of trucks required, and consequently the cost of maintenance facilities will vary approximately with the 0.3 power of tons of ore and waste mined daily, as shown in Figure 15.

Capital Cost Estimates— Underground Mines

If the orebody is amenable to underground mining, knowledge of the shape and attitude of the orebody will enable a judgment to be made on: the hoisting distance necessary to extract the lowest ore, and the average stoping width of the mineable ore.

If the nature of the ore boundary and the competence of the wall rock are known from drill core samples, a judgment can be made on the appropriate stoping method.

It is assumed that almost all underground orebodies can be developed by vertical shaft access and lateral levels, but it is recognized that under some conditions development by inclined ramps may be more appropriate. Nevertheless, the capital costs estimated for development by a vertical shaft could be used as a measure of whether or not alternative development methods would be preferable. Generally these alternative methods would be appropriate only when they are both physically feasible and less costly than shaft development.

The capital cost estimates and operating costs for underground mines are based on the higher productivity attained by mechanized equipment for drilling, drifting, raising, stope mucking, drawpoint mucking and trackless haulage. This equipment would be suitable for room-and-pillar stopes over 8 ft high, and for blasthole stopes and cut-and-fill stopes over 15 ft wide. A partially mechanized mine would attain lower productivity for shrinkage stopes, for blasthole stopes and cut-and-fill stopes less than 15 ft wide, and for room-and-pillar stopes less than 8 ft high, and consequently manpower requirements and operating costs are substantially higher in such mines.

Cost of Shaft Sinking

If the ground through which the shaft is to be sunk is known to be fractured, waterbearing, or weak and incompetent, the shaft will probably need to be a circular concrete-lined shaft. If the shaft sinking will be in strong competent rock, a rectangular shaft with multi-compartment shaft sets 8 feet apart and blocked to the bare rock excavation will probably be most appropriate. The size of the rectangular shaft, or the diameter of concrete shaft, required to hoist the daily tonnage of ore and service the mine is shown in Figure 6.

The cost of shaft sinking, as shown in Figure 7, consists of mobilization and demobilization costs (which will be virtually independent of shaft sinking distance) as well as unit shaft sinking costs, including shaft supplies per foot of shaft.

Cost of Mine Development

Mine development for mechanized mines will consist of ramps as well as shaft crosscuts, ore and waste passes, raises, and miscellaneous excavations for pump stations, loading pockets, lunchrooms, etc. Because of the larger size of opening necessary for mechanized mines, the development cost per foot will be higher than for conventionally equipped mines. It is assumed that the initial mine development will develop ore reserve tons equal to 2000 times the daily mill tonnage. The amount of mine development is expressed in terms of footage

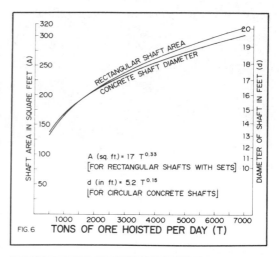

FIG. 6   **TONS OF ORE HOISTED PER DAY (T)**

A (sq. ft.) = 17 T$^{0.33}$
[FOR RECTANGULAR SHAFTS WITH SETS]

d (in ft.) = 5.2 T$^{0.15}$
[FOR CIRCULAR CONCRETE SHAFTS]

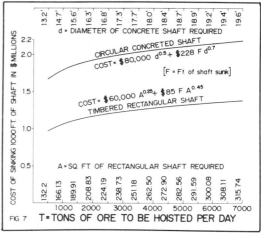

FIG. 7   **T = TONS OF ORE TO BE HOISTED PER DAY**

CIRCULAR CONCRETED SHAFT
COST = $80,000 d$^{0.5}$ + $228 F d$^{0.7}$
[F = Ft. of shaft sunk]

COST = $60,000 A$^{0.25}$ + $85 F A$^{0.45}$
TIMBERED RECTANGULAR SHAFT

A = SQ. FT. OF RECTANGULAR SHAFT REQUIRED

(Fd) of 8- by 8-ft drift or the equivalent in costs for ramps, raises and miscellaneous excavations.

Fd (in feet of equivalent 8x8 drift) = 270 T/W$^{0.8}$, where T is tons of ore milled daily and W is average stoping width in feet.

The average cost per foot of 8x8 drift, as of January 1979, is estimated to be about $150, including all direct labour and material costs, as well as proportionate amount of general expenses (administration, supervision, general supplies, fringe benefits, etc.) incurred during mine development.

Capital cost of mine development = $40,000 T/W$^{0.8}$ (Fig. 8).

It will be noted that "mine development" does not include the driving of stope development drifts, crosscuts and raises, because this is essentially an operating cost associated with the need to develop future new stopes to replace those stopes currently in production.

Hoisting Plant Sizing and Cost

The guides for estimating size and cost of hoisting facilities for underground mines are based on the assumption that a double-drum hoist will be used. It is recognized that for many large mines hoisting ore from great depths, friction hoists may be more suitable than double-drum hoists. Nevertheless, double-drum hoists have a technically wider application to shaft sinking, to multiple-level service requirements and to all sizes of mines.

The hoisting plant size and cost shown graphically is based

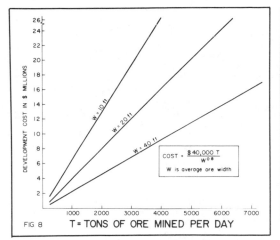

FIG 8   T = TONS OF ORE MINED PER DAY

$$\text{COST} = \frac{\$\ 40,000\ T}{W^{0.8}}$$

W is average ore width

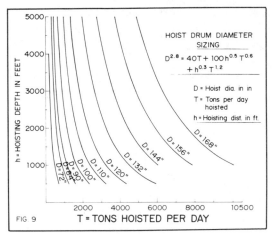

HOIST DRUM DIAMETER SIZING

$$D^{2.8} = 40T + 100h^{0.5}T^{0.6} + h^{0.3}T^{1.2}$$

D = Hoist dia. in in

T = Tons per day hoisted

h = Hoisting dist. in ft.

FIG 9   T = TONS HOISTED PER DAY

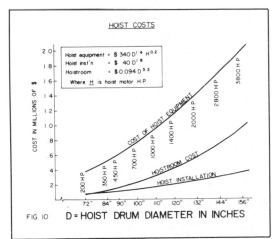

HOIST COSTS

Hoist equipment = $ 340 D^{1.4} H^{0.2}

Hoist inst'n   = $ 40 D^{1.8}

Hoistroom   = $ 0.094 D^{3.2}

Where H is hoist motor H P

FIG 10   D = HOIST DRUM DIAMETER IN INCHES

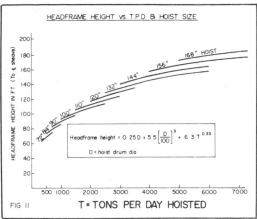

HEADFRAME HEIGHT vs. T.P.D. & HOIST SIZE

Headframe height = 0.25 D + 5.5 $\left[\dfrac{D}{100}\right]^{3}$ + 6.3 $T^{0.33}$

D = hoist drum dia.

FIG II   T = TONS PER DAY HOISTED

large mines will use one hoist for ore hoisting only and a second hoist for cage service.

The drum size of a double-drum hoist required for hoisting T tons of ore daily from a hoisting depth of h feet is shown in Figure 9. The rated horsepower of the motor installed on the hoist will be dependent on the hoisting rope speed S (in feet per minute), but the formula shown in Figure 9 assumes that the appropriate hoisting speed for hoisting T tons of ore over a distance of h feet will be about $1.6\ h^{0.5}T^{0.4}$.

A hoisting speed slower (or faster) than this will require a hoist drum diameter slightly larger (or smaller) than D as computed by the formula in Figure 9, but the motor horsepower required would be somewhat smaller (or larger) than the size typically used on hoists of diameter D.

The motor horsepower suitable for driving a hoist of diameter D at a rope speed of S f.p.m. is:

Motor horsepower $(H_1) = 0.5\ (D/100)^{2.4}S$

Figure 10 shows the capital cost of a fully equipped double-drum hoist of diameter D and H_1 motor horsepower, including all electric equipment. The cost of installing hoists, and the cost of constructing a hoistroom to house the hoisting equipment, is also shown in Figure 10. The hoistroom cost is based on the assumption that the hoistroom will have an area of $0.125\ D^{2.2}$ square feet, which will be sufficient for the installation of mine compressors in addition to hoisting equipment.

The height of the headframe must be sufficient to permit the skips to dump into orebins sized in relation to daily tonnage hoisted in addition to a safe distance for skip overtravel and hoist braking distance. Headframe height for different daily tonnages and hoist sizes is shown in Figure 11. The weight of structural steel (in pounds) in a headframe of height h feet and safely designed for hoisting ropes of 1/80 of the hoist drum diameter D will be approximately $0.12\ h^3\ (D/100)^2$. The cost of erecting the structural steel, headframe sheathing, piping and electrical equipment, deckhouse, sheaves, cages, skip dumps and orebins is shown in Figure 12.

Mine Compressor Plant

The size and cost of the mine compressors installed in the hoistroom building complex is shown in Figure 13. The size and consequently the cost of the compressor may be somewhat higher if the mine expects to utilize air-powered slushers and mucking machines instead of diesel-powered LHD equipment.

Underground Mining Equipment

The cost of equipment and equipment installation for underground development, stoping, loading, haulage, pumping, ventilation, crushing and miscellaneous mine services will increase as the daily tonnage increases, as shown in Figure 14. In the older conventionally equipped mines using portable

on the use of one double-drum hoist for dual duty in hoisting ore and providing service facilities to all mine levels 24 hours per day. In practice, however, the larger mines, hoisting over 4000 tons of ore daily, will rarely depend on one large hoist for all skip hoisting and cage service requirements. Normally,

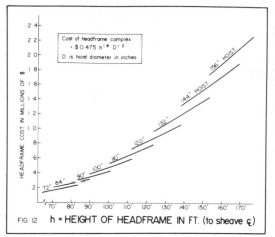

FIG 12   h = HEIGHT OF HEADFRAME IN FT. (to sheave ℄)

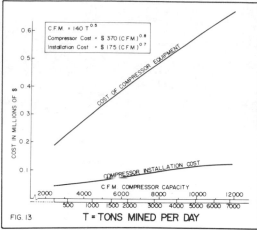

FIG. 13   T = TONS MINED PER DAY

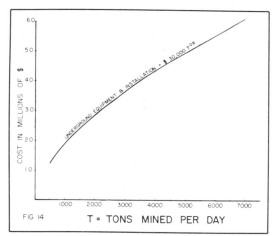

FIG 14   T = TONS MINED PER DAY

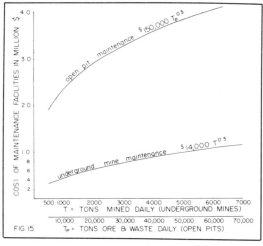

FIG 15   T = TONS MINED DAILY (UNDERGROUND MINES)
   Tp = TONS ORE & WASTE DAILY (OPEN PITS)

jacklegs and stopers for drilling, slushers for stope mucking, mucking machines for drawpoint loading, and rail haulage, somewhat more equipment was required for mines with narrow stopes. If, however, modern mechanized equipment is employed whenever the stope width is adequate for its use, the cost of equipping a mine will depend primarily on the daily tonnage required.

Underground Mine Maintenance Facilities

The cost of constructing and equipping facilities for repair and maintenance of equipment drilling, loading, haulage and general mine services is shown in Figure 15.

It should be noted that the mine maintenance facilities would not include facilities required for maintenance of process plant equipment or general surface equipment. Most of the larger process plant equipment would be repaired within the mill repair bay, and only the smaller electrical items would be removed from the mill for repair in the general services maintenance facilities.

Capital Cost Estimation—Process Plants

Plant-Site Clearing and Mass Excavation

Clearing and excavation costs will vary widely depending on the topography and type of ground to be excavated, but for similar site conditions costs will vary with the plant-site area, which will be approximately proportional to the 0.3 power of the plant tonnage rate, as shown in Figure 16.

As compared with a flat plant site with less than 10 feet of soil overburden, excavation costs will be increased by 50% when the plant site is moderately sloping and some rock blasting is necessary to attain a level excavated area; excavation costs will be increased by 150% when the plant site is steeply sloping and extensive rock blasting is necessary to attain a terraced excavation area for the plant.

Concrete Foundations and Detailed Excavation

The cost of detailed excavation, compacted backfill, formwork, reinforcing steel, and concrete foundations for the plant and process equipment will be primarily dependent on the volume of concrete required. Mass excavation will have modified the localized site topography to levelled areas suitable for plant construction, and concrete volume does not depend on site topography.

The volume of concrete will, however, depend on the bearing capacity of the excavated ground; if this excavated ground is all in sound rock, concrete volume will be minimized, but if this excavated ground consists of compressible soil, the concrete volume and cost required will be substantially increased by the need for piled foundations and massive concrete slabs

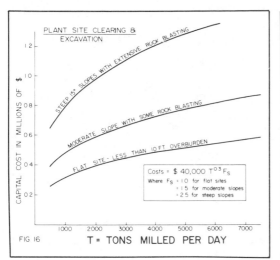

FIG 16

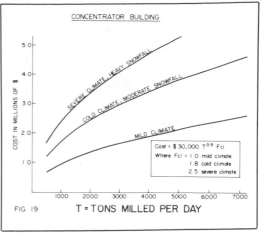

FIG 19

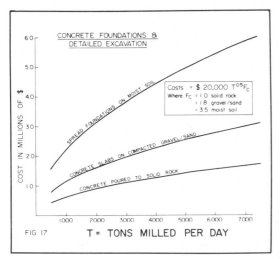

FIG 17

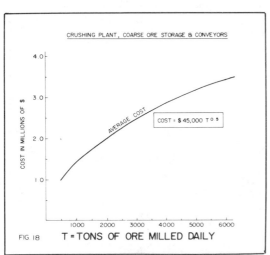

FIG 18

to support the process equipment, as shown in Figure 17.

Crushing Plant, Coarse Ore Storage and Conveyors

For any specific tonnage rate, there could be many different design configurations of primary, secondary and tertiary crushing units linked by conveyors and ore storage to take advantage of site topography. In general, however, there will be an optimum design that minimizes capital cost in relation to daily tonnage capacity without compromising the availability of fine crushed ore for continuous 24-hour-per-day milling. This optimized capital cost will be approximately as shown in Figure 18.

Concentrator Building

Assuming that concrete foundations have been completed for the plant, the cost of constructing, sheathing, insulating and heating the concentrator building to house the fine ore bins, grinding section, flotation or cyanidation leaching section, thickening and filtration section, as well as offices, laboratories and supplies storage, will be approximately as shown in Figure 19.

The cost of the concentrator building will be substantially affected by the regional climate and winter snowload. In increasingly severe climates, building costs will be increased by the need for insulated sheathing and roofdeck, provision of heating equipment, structural strength to support snow loading or wind loading, and increased construction cost under severe climatic conditions.

Grinding Section and Fine Ore Storage

The capital cost of installing grinding mills, classifiers, fine ore bins and all accessory equipment for storage reclaim and grinding on previously prepared concrete foundations will be approximately as shown in Figure 20.

The cost of equipment and equipment installation will be increased when the ore needs fine grinding or is difficult to grind (i.e., has a high work index). Soft ores may be considered as ores which have a work index below 12, medium ores have a work index of about 15 and hard ores have a work index of 17 or greater. The fineness of grain of the ore and the degree of intergrowth with other sulphide minerals will determine the fineness of grind that will be necessary prior to flotation separation of the valuable minerals into several saleable concentrates. The fineness of grind is usually determined by metallurgical testwork on drill core samples, but a microscopic examination of core samples will normally suggest whether or not fine grinding will be necessary to separate several valuable minerals from the gangue.

It should be noted that the costs graphically shown in Figure 20 allow for sufficient fine ore storage to permit the mill to operate seven days per week on ore which has been crushed for five days per week. If the crushing plant will be operated each day that the mill is operating, less fine ore storage will be required and the cost of the fine ore storage and grinding section will be reduced by 25%.

Flotation and/or Processing Section

The capital cost of installing the processing equipment to concentrate the valuable minerals from the waste in the finely ground ore will depend on the type of separation process and the degree of complexity in the flowsheet. Figure 21 illustrates the cost of the process section for several different types of process and flowsheet complexity.

Thickening and Filtering Section

The capital cost of the thickening and filtration section depends on the volume of concentrates to be thickened and filtered, which in turn is a function of the process plant tonnage multiplied by the grade of ore being processed.
Capital cost = $5,000 $F_f T^{0.5}$, where the filtering factor:
F_f = 1.0 for low-grade straight Cu ores
F_f = 1.6 for high-grade Cu ores with recoverable zinc values
F_f = 2.0 for complex Pb-Zn-Ag ores or Cu-Zn-Pb ores
F_f = 3.0 for cyanided gold ores

Concentrate Storage and Loading

The capital cost of concentrate storage depends on the daily tonnage of concentrate T_c produced by the process plant. This can be computed by multiplying the daily ore tonnage processed by the mill recovery and by the ratio of the ore grade (or ore grades) to the concentrate grade (or concentrate grades).
Cost of concentrate storage and loadout = $4,000 $T_c^{0.8}$.

This cost would be appropriate for concentrate storage at the mill site from which concentrates will be transported to the smelter at frequent intervals by truck or rail. If, however, ocean shipment of concentrates will be envisaged, a tidewater storage and shiploading facility will be rquired. It may be possible to lease suitable facilities at an existing port; if leased facilities are not available, they must be constructed at a cost which may vary from several hundred thousand dollars for a small terminal at an existing port to several million dollars for a major terminal and port in a new location. It is not possible to suggest a rough cost for a tidewater concentrate terminal, because this cost will be greatly dependent on localized conditions of land access, tidal range, water depth, shipping access, etc.

Capital Cost Estimates for Plant Utilities and General Services

Electric Power Supply and Distribution

The peak load for mine/mill plants, as shown on Figure 22, increases with increased daily tonnage, but the rate of increase is slower for open-pit mines mining lower-grade ore than it is for underground mines which typically mine ore of higher grade requiring more complete processing.

The cost of electric power supply depends on the peak load requirements and also on whether the power is generated by a coal-fired plant (for large isolated mines) or by diesel generators (for small isolated mines), or is supplied by an existing utility through an extended transmission line and transformer station. The cost of each of these sources of electrical power is shown in Figure 22.

If utility power is available, an additional cost may be required to extend the transmission line to the plant site. This may cost about $60,000 per mile of line. For large mining projects consuming a large amount of electric power, the electric utility will typically install the transmission line and stepdown transformer station on the assured basis that future revenue from mine plant consumption of electric power will repay the

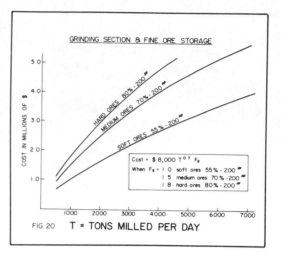

GRINDING SECTION & FINE ORE STORAGE

Cost = $ 8,000 $T^{0.7}$ F_g
When F_g = 1.0 soft ores 55% - 200 #
1.5 medium ores 70% - 200 #
1.8 hard ores 80% - 200 #

FIG 20   T = TONS MILLED PER DAY

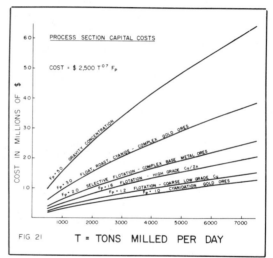

PROCESS SECTION CAPITAL COSTS

COST = $ 2,500 $T^{0.7}$ F_p

FIG 21   T = TONS MILLED PER DAY

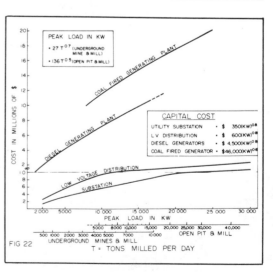

PEAK LOAD IN KW
= 27 $T^{0.7}$ (UNDERGROUND MINE & MILL)
= 136 $T^{0.5}$ (OPEN PIT & MILL)

CAPITAL COST
UTILITY SUBSTATION = $ 350(KW)$^{0.8}$
L V DISTRIBUTION = $ 600(KW)$^{0.8}$
DIESEL GENERATORS = $ 4,500(KW)$^{0.8}$
COAL FIRED GENERATOR = $46,000(KW)$^{0.6}$

FIG 22   T = TONS MILLED PER DAY

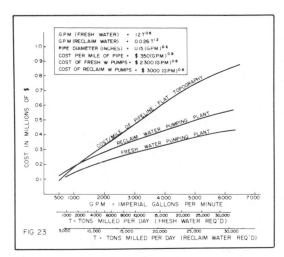

FIG 23

capital costs of the transmission line and transformer substation. For small mines, however, the mine may have to bear the cost of the extension of the transmission line and substation, either directly or effectively by payment of higher operating costs for electric power consumption during the initial years of operation.

The capital costs of the low-voltage electrical distribution from the main substation to the mine, mill and plant utilities will be approximately as shown in Figure 22. This cost will be common to all types of electric power sources: utility power, diesel or coal-fired generators.

Tailings Storage

The capital cost of pipelines and pumps to deliver mill tailings to the tailing storage area, and the cost of damming the storage area and preparation of settling and treatment ponds, is always difficult to estimate even when there is complete topographic data on the area selected for tailings storage. The cost of tailings storage can be drastically affected by environmental constraints and by stringent requirements imposed by regulatory agencies.

If it is judged that an environmentally acceptable tailings basin in moderate topography can be developed by damming with local material within 2 miles of the mill site, the cost of tailings storage for a mill treating T tons of ore daily will be about:

Cost = \$8,000 $T^{0.5}$.

This cost represents the cost for which suitable tailings storage can be developed under favourable conditions of topography, accessibility and minimum environmental effects. Capital costs may, however, be many times larger when local topography is steep and rugged, when the selected tailings basin is many miles from the millsite and uphill pumping is required, or when the nature of the tailings imposes extremely stringent requirements for tailings treatment and containment.

The capital costs may be effectively increased by the administrative and technical costs of preparation of studies of environmental impact, planning of several alternative tailings storage facilities, and delay and uncertainty in obtaining approval of tailings storage planning.

Water Supply

Mines use large quantities of water for mining and milling and most of this water flows to the tailings storage, from which water reclamation may be necessary to avoid downstream pollution. The reclaimed water may be recycled in a portion of the mill circuit that is not adversely affected by the reagent content of the reclaimed water. The smaller mines frequently

operate with fresh water only, whereas the larger mines, which could have much more effect on the regional stream drainage, are usually required to minimize their consumption of fresh water by recycling clarified water from the tailings pond.

In the Precambrian Shield area of Canada, where there is typically a plentiful supply of fresh water within a mile of the process plant, the volume of fresh water required for a mine which mills T tons of ore daily is about 12 $T^{0.6}$ gpm.

In the drier areas of British Columbia, where large-tonnage low-grade open-pit mines have been developed, water conservation is critically important and fresh water usage is typically only 2.5 $T^{0.6}$ gpm.

Figure 23 shows the typical requirements for fresh water and reclaim water for mines and mills of varying tonnages. The cost of the pump stations and pipeline cost per mile of pipe are also shown in the chart in Figure 23. These costs would be appropriate to the low-relief topography typical of the Precambrian Shield area of Canada, but in mountainous country, where high pumping heads are frequent, the cost of pump stations and pipelines would be much higher.

Capital Costs of General Plant Services

These costs include the costs of constructing and equipping the general administration offices, general warehouse, maintenance shops for the mill and surface facilities, vehicle garages, security stations and fencing, employee parking lots, changehouses and the cost of general-purpose vehicles. In general, these costs will vary as a function of the total number of company employees, which is not necessarily uniformly related to the plant capacity. The number of employees required for differing types of mining, milling and general services can be computed as described under operating costs, and, if this number is N employees, the total capital costs of general plant services can be computed as a function of N:

Costs of general plant services = \$8,000 $N^{0.8}$.

Capital Cost of Access Road

If the mine is not readily accessible by an all-weather road, a road suitable for subsequent truck haulage of concentrates must be constructed before major project construction can get underway. A 30-ft-wide gravelled road with adequate drainage, curvature and gradients to permit satisfactory concentrate haulage conditions will cost roughly \$200,000 per mile in regions with moderate topography, some rock outcrops and local sources of gravel. Under ideal conditions of flat topography, absence of exposed rock and well-drained gravelly soil, the cost per mile would be less; alternatively, in steep topography requiring heavy rock cuts, the cost of road construction could be much greater.

Bridges to span creeks or rivers where the total bridge length is L_b feet will cost approximately \$130 $L_b^{1.5}$ (for each bridge).

Townsite and Housing Cost

If the mine site is within commuting distance of an existing town which has available housing and acceptable community facilities for N mine employees, the cost to the mining company may be limited to provision of housing for key staff only. This cost may be:

Existing townsite cost = \$4,000 N.

Whenever a mining project is to be developed in an isolated region, a decision must be made as to whether the mine townsite will be developed as a bunkhouse camp (lower capital cost, high operating cost and transient work force) or as a developed townsite for family living (high capital cost, more attractive lifestyle, more permanent work force). Because a townsite must be of a certain minimum size before it can service the schooling, medical and recreational needs of families, the family townsite will be preferred only for mines employing in excess of 100.

When the number of employees (N) is less than 100, the bunkhouse camp will consist of single accommodation, mess facilities and leisure facilities for 95% of the employees at $12,000 each as well as family accomodation for the remaining 5%, consisting of senior staff at $50,000 each, and minimum townsite facilities at 20% of housing cost:

Bunkhouse townsite cost = $20,000 N.

A family townsite accommodating 65% of the married employees in detached or apartment housing with facilities for schools, commercial, medical and recreational needs of the community as well as single accommodation for 35% of the employees and housing for townsite service employees will cost:

Family townsite cost = $55,000 N.

Project Overhead Costs

A. Feasibility studies, design engineering and technical planning: 4% to 6% of all mining pre-production costs, site preparation, excavation and road building costs, as well as 6% to 8% of all other project costs.

B. Project supervision, contract management, expediting and general construction facilities, including camp costs: 8% to 10% of all direct project costs.

C. Administration, accounting, legal and pre-production employment of key operating staff: 4% to 7% of all direct project costs.

D. Working capital for capital spares, supplies inventory and operating costs for the period between plant start-up and receipt of smelter payment for initial concentrate shipment; typically 4 months of operating costs on full production basis.

In general, the higher percentages for design engineering, project supervision and administration would be applicable to smaller mine projects, whereas the lower percentages would be applicable to major mine projects costing $100 million or more.

Estimation of Revenue

The revenue produced from mining and milling ore from an orebody that contains geologically estimated reserves of T_R tons grading G_R% metal is adversely affected by four factors:

1. Not all the ore reserve tonnage will be recoverable in practice by the planned type of open pit, or the practical shapes of underground stopes. It is difficult to offer any guide to the percentage of ore reserves that can be recovered by practical open-pit or underground stoping methods, because this percentage will vary greatly depending on the irregularity of the shape of the ore reserve blocks.

In general, if the ore reserve blocks are reasonably regular in shape and dip and are above the planned open-pit limits or the lowest underground level, a mineable recovery of over 95% of the ore reserve blocks could be expected.

2. Dilution of waste rock off the underground stope walls will vary with the type of stoping employed, the width of the stope in feet, the angle (A°) at which the stope is dipping and the competence of the stope wall rock. For stope wall rocks of average competence in relation to the type of stoping method employed, the dilution (D%) expressed as a percentage of waste rock in the mined ore will be approximately as shown in Figure 24.

When the stope wall rock is regular and competent, dilution may be only 0.7 of that shown in Figure 24, but if it is unusually weak and incompetent the dilution could be as much as 1.5 times the dilution shown.

Dilution has an adverse effect on operating costs because more tons must be mined to yield the same metal content as the undiluted ore, but it also has an adverse effect on revenue because the metal grade of each ton of diluted ore is reduced and mill recovery will consequently be somewhat lower than for undiluted ore.

Mill Recovery

The recovery of metallic minerals by flotation, cyanidation, leaching or gravity methods is affected by the fineness of grind

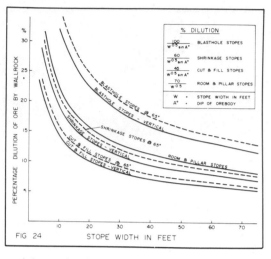

FIG 24   STOPE WIDTH IN FEET

needed to attain mineral unlocking, by the grade of the ore being milled and by the response of the mineral to the specific concentration process. Assuming that the valuable minerals can be fully separated from associated waste minerals at an economic fineness of grind, the recoveries and grades of metallic concentrates and precious metal ores will be about as shown by the formulae shown in Table 1. Somewhat higher recoveries will be attainable if the metallic ores are coarse grained, and lower recoveries will be inevitable if fine grinding does not fully unlock closely associated minerals, or if the valuable sulphide minerals are coated with oxidized minerals or slimes.

Net Smelter Revenue for Base Metals (as of January 1979)

The net smelter revenue (at the mine), in $ per ton of base metal concentrates grading G% metal, is a function of the metal content of the concentrates (either a percentage P or a fixed unit deduction U) multiplied by a standard published metal price (M) less a specific deduction d in cents per lb (to allow for cost of refining and selling the smelted product), less a smelter charge S in $ per ton, less a freight cost F in $ per ton of concentrates transported from the mine to the smelter.

$ per ton = (P/100) $20 (G-U) (M-d)/100 - S - F

Typical Values:

Where P = 100% for Cu, 95% for Pb, 85% for Zn

U = 1.3 for Cu, 3 for Pb, O for Zn

M = 90¢ for Cu, 52¢ for Pb, 39¢ for Zn (as of January 1979)

d = 12.0¢ for Cu, 12.0¢ for Pb, 2.5¢ for Zn (as of January 1979)

S = $60 for Cu, $50 for Pb, $145 for Zn (approx.) (as of January 1979)

F = freight cost per ton of moist concentrates from mine to smelter, assuming that concentrates are trucked T_m miles by road, R_m miles by railroad and (if applicable) shipped by 15,000-ton freighter for D_o days of loading, ocean travel and unloading.

Approximate freight cost F in $ per ton of concentrates = $0.17 T_m^{0.9}$ + $0.26 R_M^{0.7}$ + $0.80 D_o$.

Estimation of Operating Costs

Operating Personnel—Open-Pit Mines

The number of men employed in open-pit operations, as shown in Figure 25, typically varies as the 0.5 power of the tons of ore and waste (T_p) mined daily, but the number of men required on truck haulage and road maintenance tends to vary with the 0.7 power of daily tonnage chiefly because, in the larger-tonnage pits, haulage distances are generally longer, and

TABLE 1. Base metal ores milled by flotation (head grades in %)

		Recovery Formulae	Typical Conc. Grade
Cu	in straight chalcopyrite ores	$R = 100\% (1-0.07 Cu^{-0.8})$	28.5% Cu
	in oxidized copper ores (sulphides)	$RCu_s = 100\% (1-0.08 Cu_s^{-0.8})$	Variable
	in oxidized copper ores (oxides)	$RCu_o = 100\% (1-0.40 Cu_o^{-0.3})$	Variable
	in copper-zinc ores	$R = 100\% (1-0.16 Cu^{-0.8})$	25.5% Cu
	in lead-zinc-copper ores	$R = 100\% (1-0.22 Cu^{-0.8})$	22.0% Cu
MoS_2	in straight molybdenite ores	$R = 100\% (1-0.04 MoS_2^{-0.8})$	88.0% MoS_2
	in copper-moly ores	$R = 100\% (1-0.06 MoS_2^{-0.8})$	Variable
Zn	in straight sphalerite ores	$R = 100\% (1-0.25 Zn^{-0.6})$	56% Zn
	in lead-zinc ores	$R = 100\% (1-0.32 Zn^{-0.6})$	53% Zn
	in copper-zinc ores	$R = 100\% (1-0.45 Zn^{-0.6})$	52% Zn
	in copper-lead-zinc ores	$R = 100\% (1-0.55 Zn^{-0.6})$	50% Zn
Pb	in straight galena ores	$R = 100\% (1-0.13 Pb^{-0.8})$	60% Pb
	in lead-zinc ores	$R = 100\% (1-0.18 Pb^{-0.8})$	53% Pb
	in copper-lead-zinc ores	$R = 100\% (1-0.28 Pb^{0.8})$	45% Pb

Miscellaneous Ores

Tungsten ores (gravity separation)	$R = 100\% (1-0.33 WO_3^{-0.5})$	75% WO_3
Nickel in nickel-copper ores	$R = 100\% (1-0.20 Ni^{-0.6})$	10% Ni
Uranium ores (flotation-leaching)	$R = 100\% (1-0.16 U_3O_8^{-0.8})$	77% U_3O_8
Iron ores (gravity-magnetic)	$R = 100\% (1-1.5 Fe^{-0.6})$	65% Fe

Precious Metal Ores and Precious Metals in Base Metal Ores: (head grades in oz per ton)

		Process
Gold in siliceous ores	$R = 100\% (1-0.013 Au^{-0.8})$	Cyanidation only
in pyrite ores	$R = 100\% (1-0.03 Au^{-0.8})$	Flot./Roast./Cyan.
in base metal ores	$R = 100\% (1-0.3 Au^{-0.8})$	Flotation
Silver in straight silver ores	$R = 100\% (1-0.22 Ag^{-0.6})$	Flot./Grav. Separation
in base metal ores (-1.0 oz/t)	$R = 100\% (1-0.40 Ag^{-0.6})$	Flotation

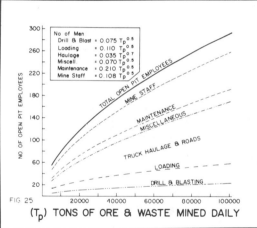

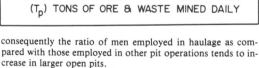

FIG 25

(T_p) TONS OF ORE & WASTE MINED DAILY

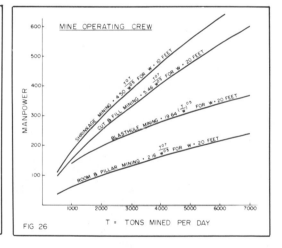

FIG 26   T = TONS MINED PER DAY

consequently the ratio of men employed in haulage as compared with those employed in other pit operations tends to increase in larger open pits.

Operating Personnel— Underground Mines

The number of men employed underground, as shown in Figure 26, varies for different stoping methods, different stope widths (W in feet) and different daily tonnage rates (T in tons per day) (Table 2).

Operating Costs—Open-Pit Mines

The operating costs per ton for open-pit mines are shown in Figure 27. Labour costs are based on basic pay rates as follows: operators - $9.50/hr, maintenance - $8.90/hr, and

TABLE 2. Underground mining personnel

	Percentage Distribution of Mine Crew			
	Blasthole	Cut & Fill	Shrinkage	Room & Pillar
Development	17%	12%	9%	14%
Stoping	23%	31%	31%	45%
Mine Service	22%	20%	29%	12%
Maintenance	23%	23%	18%	13%
Mine Staff	15%	14%	13%	16%
	100%	100%	100%	100%

staff - $10.40/hr, and an allowance of 35% of basic pay has been added to allow for fringe benefits. Cost of supplies per ton decreases less rapidly than labour costs as the daily mined tonnage increases.

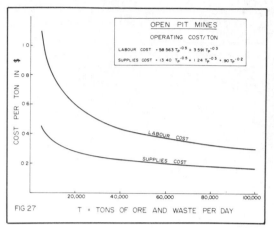

FIG 27

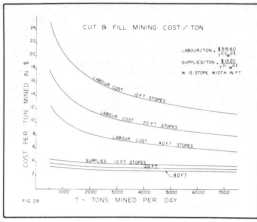

FIG 29

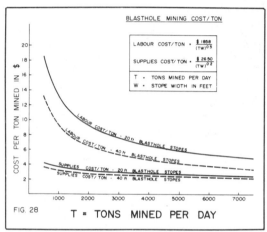

FIG. 28

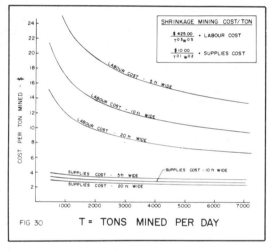

FIG 30

Operating Costs—Underground Mines

The operating costs for mines using blasthole, cut-and-fill, shrinkage or room-and-pillar stoping methods are shown on Figures 28 to 31. Labour costs are based on average basic pay rates of $8.50/hr for mine crew and $10.25/hr for mine staff, with an allowance of 35% to cover fringe benefits. The cost of supplies for development, stoping and general mine services typically varies with the number of men employed in these functions.

Employees Required—Process Plants

The number of mill employees and mill staff is shown in Figure 32 for differing process flowsheets. Typically, mills treating gold ores will require more employees in the cyanidation process than mills using flotation, and mills using selective flotation to produce two or more concentrates will require more employees in the flotation section than mills which produce only one flotation concentrate.

Mill Operating Cost per Ton

The operating costs of mills using differing processes are shown in Figure 33. Labour costs are based on an average pay of $7.40/hr for mill crew and $9.50/hr for mill staff, plus 35% to allow for fringe benefits. Grinding and crushing supplies normally constitute the largest item of mill supplies, and are also the most variable in terms of cost per ton because of differences in hardness and fineness of grind required for different ores.

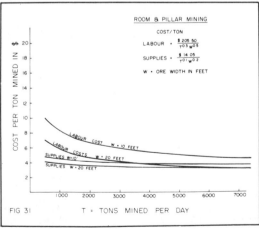

FIG 31

Administration and General Services—Employees

The employees required for administration and general plant services will tend to vary as a function of the total number of

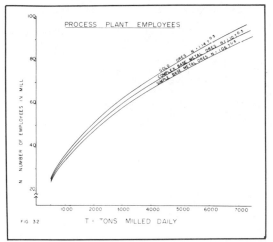

FIG 32   T = TONS MILLED DAILY

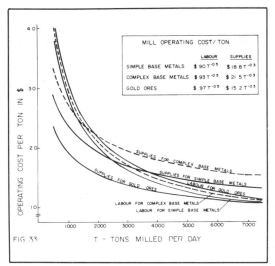

FIG 33   T = TONS MILLED PER DAY

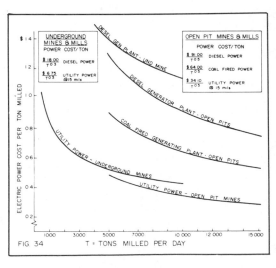

FIG 34   T = TONS MILLED PER DAY

staff, operating crew and maintenance crew in the mine and mill, and will be relatively independent of the tonnage throughput of the mine and mill (except insofar as the tons per day influences the size of the mine and mill crew).

Assuming total staff and crew of the mine and mill = Nm men

Employees required:
A. a) Electrical services = 0.03 Nm (power supplied by utility)
   b) Electrical services = 0.05 Nm (power diesel generated)
   c) Electrical services = 0.08 Nm (power generated by coal-fired plant)
B. Surface plant services & road maintenance = 0.04 Nm
C. (a) Townsite employees = none for existing town
   (b) Townsite employees = 0.03 Nm (bunkhouse-type mine camp)
   (c) Townsite employees = 0.05 Nm (family townsite—subsidized)
D. Concentrate transport (contracted out)
E. General administration = 0.07 Nm
Total employees N = Nm + (0.14 to 0.24)Nm, depending on type of townsite and electric power supply.

Administration & General Services— Operating Costs per Day

A. Electrical services—wages = $68 (0.03 to 0.08 Nm) (depending on type of electric power)
B. Surface plant services = $53 (0.04 Nm)
C. Townsite employees = $50 (up to 0.05 Nm) (depending on townsite type)
D. General administration = $85 (0.07 Nm)
E. Fringe benefits: 35% of above wages and salaries.
F. The electric power consumption and cost of electric power is shown in Figure 34 for underground mines and mills, and for open-pit mines and mills. Most of the electric power is usually consumed in the process plant, but the underground mine hoisting plant and compressor plant could be a significant consumer of electric power. Open-pit mines normally consume only a small fraction of the electric power required by the mill.
G. Supplies—general plant services = $6 T^{0.5} per day
H. Townsite operating cost
   (a) = $13 Nm (bunkhouse camp subsidy)
   (b) = $ 5 Nm (subsidized family townsite)
I. General administration expenses, including office and warehouse supplies, telephone, travel expenses, property taxes, insurance, and legal, auditing and consulting fees:
   Cost per day = $4 Nm

Notes on Computation of Formulae

Virtually all the formulae shown in this paper for estimating size, quantity and cost of the major components of a mine project have been developed by the author from detailed data on many Canadian and foreign mining projects over the last fifteen years. Size and quantity data from foreign projects were used in determining requirements for mine projects under differing physical conditions of climate, topography, orebody shape, etc., but cost data from foreign projects were utilized only when unit costs were judged to be comparable to Canadian unit costs. Cost data from completed Canadian mine projects and foreign projects with comparable unit costs have been escalated to 1978 by using appropriate indices, and expressed in terms of 1978 Canadian dollars, which had an average 1978 exchange rate of $1.00 Canadian = $0.877 U.S.

Most of the operating cost data utilized in computing operating cost formulae were from Canadian underground mines in the Canadian Precambrian Shield, or from open-pit mines in Western U.S.A. and Canada. Operating cost data from foreign mines, where the cost of labour and supplies was very different from that of North America, were not used in computing average operating costs, but foreign data were utilized in analyzing trends relating consumption of supplies and labour to increases in tonnages mined and milled.

The relationships between mine project requirements and costs relative to plant capacity, operating costs of supplies and labour relative to daily tonnage mined and milled, and operating performance and mill recovery relative to mill-head grade, were determined by computerized statistical analyses of the best fit of the data to an equation of the form $Q = KT^x$, where Q represents the actual data on quantities required or cost, and T represents the tonnage rate, milled head grade or other physical condition causing changes in quantities or costs.

The x values were determined to yield the lowest range of variation in K values across the widest range of T values for which reliable data were available. The x values tested were within a relatively narrow range which was judged to be consistent with technical considerations and operating experience. Whenever changes in the quantity Q were judged to result from simultaneous changes in two physical conditions T_1 and T_2, the K_1 values were first determined for the most influential condition T_1 to fit $Q = K_1T_1^x$, and then changes in the K_1 values were computed to fit the equation $K_1 = K_2T_2^y$, so that $Q = K_2T_2^yT_1^x$ where K_2 is a constant.

Because of the inevitable dispersion of data, the mathematical analyses did not rigorously consider each data point of equal value, but judgment factors were utilized in weighting data from individual mine projects according to the presence or absence of localized conditions that would result in unusually high or low values for costs or quantities.

Computers as Tools During Mineral Property Negotiations

R. E. Grant

ABSTRACT

The acquisition of a mineral property can take place at various stages of knowledge about the property. As a mineral property progresses from early exploration to exploitation, the requirements of additional capital may be met through negotiation with one or more sources of capital.

The parties to negotiations for needed capital can obtain information on the sharing of risk and profit through financial analysis. They can examine the profit indicators for the project and for each party to the negotiation. The uncertainty of relationships among tonnage, grade, production rate, recoveries, costs and other economic parameters can be examined to determine the sensitivity of profit indicators to changes in these variables and/or changes in the amount and method of financing.

Hypothetical examples are used in this paper to demonstrate the value of acquiring such information through the use of a computer. Computers can provide the information with time and cost requirements many times less than equivalent manual calculations.

The probability of finding a proposal which will meet the requirements of both parties to the negotiation is increased by the large number of proposals that can be examined when using a computer.

INTRODUCTION

THE ACQUISITION OF A MINERAL PROPERTY, which may or may not become an economic entity, is enacted at various stages in the state of knowledge of the mineral property (see *Figure 1*). The risks and unknowns involved in early stages of exploration are generally high, and costs of obtaining information or acquiring land are generally low.

As the mineral property is explored and developed, the state of knowledge increases, as does the cost of acquiring the information. Capital outlays increase and, depending

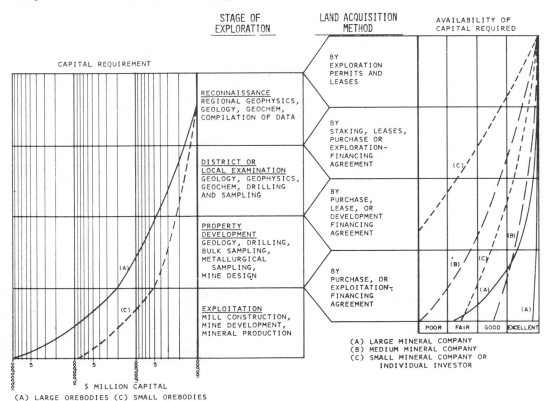

FIGURE 1 — Stages in the state of knowledge of a mineral property.

on the information obtained, the value of the property may increase as well. The prospector or owner of the property, faced with increasing costs, may or may not be able to provide the capital necessary to attain the next stage in taking the mineral property to production. Also, he may wish to share the risks and unknowns or acquire expertise in exploring, developing or exploiting the deposit.

The sources and costs of capital available to the owner vary with (1) the degree of risk and amount of capital involved in bringing the mineral property to production, and (2) the size of his assets. Common sources of desired capital are: (1) the sale of securities, often during early exploration and development of the property, (2) shared financing through royalty, equity on other types of agreements during development or (3) bank loans, bonds or debentures. This paper principally examines the latter two categories. Through acquiring capital for developing a deposit the owner agrees, in varying amounts, to share profit as well as risk with the source of capital. The sharing of risk and profit centralizes the objective of negotiations for capital. The basis for measuring profitability and degree of risk may vary widely between parties to the negotiation; certainly their propensity for risk aversion is generally different.

The concept of an economic mineral entity for the mineral property may vary widely for the parties to the negotiation, as their background and knowledge for evaluating the prospect differ. The tonnage and grade of ore depends, in addition to the physical distribution of metal, on the production rate, operating cost and capital costs, which in turn relate to the cut-off used in calculating the reserves.

The uncertainty of relationships among grades, recoveries and costs decreases with added knowledge and study of the prospect. It is in the late stages of examination and early stages of development, when capital needs sharply increase, that broader ranges of uncertainty prevail and that added information in terms of economic analyses provide valuable information in negotiations.

The purpose of this paper is to demonstrate practical aspects of using computers to obtain data which can provide insight during negotiations for financing mineral properties.

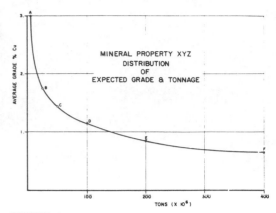

FIGURE 3 — Mineral Property XYZ — distribution of expected grade and tonnage.

Computer programs are used to model financial calculations, converting estimated values for tonnage of ore, value of ore, production rate, operating costs per ton of ore, capital costs, financing alternatives and tax structures into cash flows and profit indicators for the project and participating negotiators. The financial analysis does not have to be entirely performed on a computer. All or part of the calculations may be performed by hand. A large number of uncertainties, financing alternatives and profit indicators can be examined in a short time at low cost through the use of a computer; therefore, the information and savings in time and cost favour the use of computers for the bulk of the analysis. Negotiators can examine the potential of the orebody (*Figure 2*), select more likely targets, and examine the manner in which various financing agreements divide the risk and profit associated with the known or suspected orebodies (*Figure 4*).

FINANCIAL ANALYSIS OF THE MINERAL PROPERTY

There are several possible ways to analyze the mineral property. For each analysis, we can use exact numerical values for all economic variables, or we can use probability distributions for each variable. These have been characterized as "assumed certainty" and "assumed risk" (O'Brian, 1969).* Comparison of results for alternate targets, economic conditions and financing methods is generally desired. The "assumed certainty" technique is recommended for this aspect of the analysis.

Profit indicators are affected by a large number of variables. These variables enter into calculations of (1) mine cut-off(s) and resulting tonnage(s) and grade(s) of ore, (2) net smelter return value(s), (3) operating cost(s), (4) capital cost(s), (5) tax structure(s) and, (6) financing agreement(s). When examining the relative sensitivity of profit indicators to these variables, a percentage change in the variable tends to give a better insight into relative sensitivity than arbitrary unit change. If ranges or limitations of the variables(s) are relatively well known, these should be used to reflect the degree of certainty of that variable relative to the other unknowns.

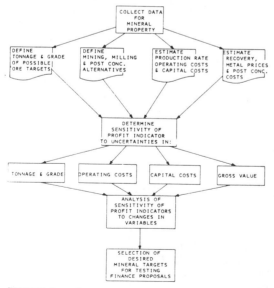

FIGURE 2 — Computer analysis of a mineral property.

*O'Brian, D. T., "Financial Analysis Applications in Mineral Exploration Development"; paper presented at 1969 Annual Meeting of AIME, Washington, D.C.

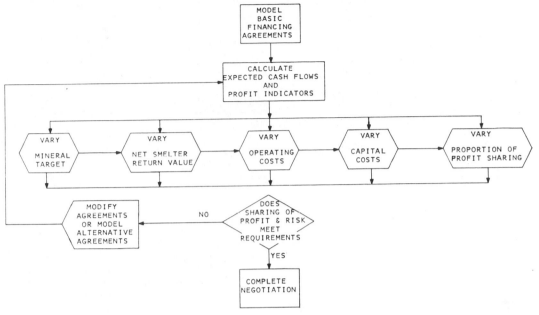

FIGURE 4 — Computer analysis of financing agreements.

TABLE 1 — Hypothetical Economic Parameters for Exploration of Property XYZ

NET SMELTER RETURN VALUE

Alternate	Tons (x 10⁶)	% Cu	Base $/ton	−20% $/ton	+20% $/ton
A	5	3.00	21.60	17.28	25.92
B	25	1.75	11.19	8.95	13.43
C	50	1.45	9.86	7.89	11.83
D	100	1.15	7.82	6.26	9.38
E	200	.85	5.78	4.62	6.94
F	400	.65	4.42	3.54	5.30

OPERATING COSTS

Alternate	Production Rate (tpy)	Life	Production Rate (tpd)	Base $/ton	−20% $/ton	+20% $/ton
A	333,333	15	1,111	8.30	6.64	9.96
B	1,250,000	20	4,167	5.20	4.16	6.24
C	2,500,000		8,333	3.80	3.04	4.56
D	5,000,000		16,667	3.50	2.80	4.20
E	10,000,000		33,333	3.35	2.68	4.02
F	20,000,000		66,667	3.20	2.56	3.84

CAPITAL COSTS

Alternate	Production Rate (tpd)	Exploration Cost (Constant) $ x 10⁶	Base $ x 10⁶	−20% $ x 10⁶	+20% $ x 10⁶
A	1,111	.500	7.555	6.044	9.066
B	4,167	.750	24.995	19.837	29.754
C	8,333	1.000	43.165	34.532	51.798
D	16,667	1.750	72.000	57.600	86.400
E	33,333	2.500	133.335	106.668	160.000
F	66,667	4.500	260.000	208.000	312.000

As a sensitivity demonstration, the mineral distribution and associated economic parameters in *Figure 1* and Table 1 were hypothesized. Let us assume that property XYZ (located in the U.S.) has been drilled by mining company B, who have approached mining company A concerning participation in further exploration, development and exploitation. Assume that company A and B share 50 per cent of the exploration costs listed in Table 1, and that the remaining capital will be shared as indicated by various financing methods.

In examining the sensitivity of profit indicators to change in variables, we shall say that company A is interested in a discounted cash flow rate of return of 12 per cent and company B is interested in the net present value discounted at 10 per cent. In reality, more than two indicators may be used and minimum desired values of the indicators may differ. Profit indicators may, or may not, be different for each of the parties to the negotiation. In this particular analysis, we determine changes in profit indicators resulting from changes in variables affected by a change in cut-off grade.

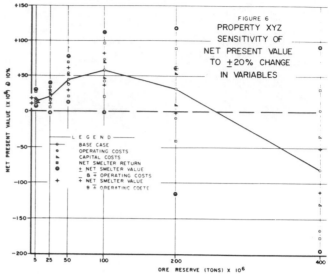

FIGURE 6 — Property XYZ — sensitivity of net present value to ±20% change in variables.

Sensitivity of Profit Indicators to Changes in Cut-off Grade

The total ore tonnage and average grade resulting from changes in the cut-off grade generally follow a curve similar to, but quite individually different from, the one shown in *Figure 3*. We imply an increased tonnage and decreased average grade at lower cut-offs. This is not actually true of all deposits, especially when they are vein or stratigraphically controlled. The lower cut-off may or may not be justified by decreasing operating costs and unit capital costs as we increase the production rate (assuming we fix the life at 20 years) for the increased tonnage. In the example, there is a substantial reduction in the average grade of ore when 25 million tons is considered rather than 5 million tons; however, the grade reduction in going from 25 million tons to 50 million tons is substantially less. The production rate for the 5-million-ton case has been

increased relative to other cases by assuming a shortened life (15 years as compared to 20 years used for other targets shown). Except for this adjustment, the example does not consider the problem of optimizing production life. In examining optimum life, there are a great many interactive restrictions such as company capital investment policies, personnel policies and administrative considerations which enter the analysis.

Base-case results for the example (see *Figure 5*) show that the discounted cash flow rate of return for the prospect exceeds the 12 per cent desired by company A for the 200-million-ton case and smaller-tonnage cases. *Figure 6* illustrates the results of interest to company B. The net present value discounted at 10 per cent increases with increasing tonnage to the 100-million-ton case, drops for the 200-million-ton case and is negative for the 400-million-ton case.

Sensitivity of Profit Indicators to Changes in Net Smelter Return Value

The variables that affect the net smelter return value include grade of metal(s) in ore, recovery of metal(s), concentrate grades(s), transportation and handling costs, smelting and refining costs, selling and related administrative costs, and price of metal(s). Each of these variables can be examined with other variables remaining fixed or, if appropriate, changing in a prescribed dependent manner. In the example (see *Figures 5* and *6*), we have varied the Net Smelter Return value by 20 per cent and we find that both the rate of return and the net present value discounted at 10 per cent are quite sensitive to the 20 per cent difference. The profit indicators are more sensitive to net smelter return than any of the other variables when tested alone. The example (*Figures 5* and *6*) shows that a 20 per cent decrease in value for the 200-million- and 400-million-ton cases would provide unacceptable profit levels. It may be desirable in the initial analysis to examine those variables which affect the Net Smelter Return value more closely to determine the likelihood of the net smelter return changing by as much as 20 per cent. Such a study is beyond the scope of the example.

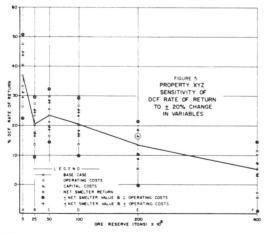

FIGURE 5 — Property XYZ — **sensitivity of DCF rate of return to ±20% change in variables.**

Sensitivity of Profit Indicators to Changes in Operating Costs

Operating costs can be affected by changes in the rate of production, method of mining, relationship of ore to waste, proximity of mine and concentrator to labour, power and supply centers, etc. The changes of profit indicators for respective changes in operating costs can be compared in *Figures 5* and *6*. Profit indicators are more sensitive to changes in operating costs as the tonnage increases and grade decreases. The effect of operating costs on profit indicators becomes almost as sensitive as the Net Smelter Return value for the low-grade deposits, but it is less than half as sensitive as the Net Smelter Return value for the high-grade deposit.

Sensitivity of Profit Indicators to Changes in Capital Costs

Capital costs, like operating costs, are affected by changes in production rate, mining methods, relationship of ore and waste, and the proximity of the mine and mill to labour, power and supply centers. To these dependent economic parameters might be added the degree of automation, availability of housing, transportation and/or smelting and refining facilities, and the type of climatic, topographic and environmental conditions. Changes in capital costs affect the profit indicators less than operating costs (see *Figures 5* and *6*). Capital costs also affect the DCF rate(s) of return differently than they affect net present value. Note that the change in net present value increases with increasing tonnage and decreasing grade, whereas the unit change of the DCF rate of return de-

creases as tonnage increases. The per-cent change in the DCF rate of return remains relatively constant to changes caused by other variables.

Sensitivity of Profit Indicators to Changes in Other Variables

There are several other economic variables that might be considered important to the analysis of the mineral property and deserve similar analysis. These could include possible changes in tax structure, or inflationary or deflationary trends of the economy which might affect costs and prices. Such variables were not examined in this example. Simultaneous 20 per cent changes for both operating costs and gross value were examined. The operating costs and Net Smelter Return value were also decreased or increased in opposite directions by 20 per cent to examine pessimistic and optimistic limits. Results are shown in *Figures 5* and *6*.

Post-Processing Analysis

A review of the effects of different variables on the profit indicators can give the parties to a negotiation an insight into the value of gathering more information for each of the variables during continued property development. It also will show those alternative target concepts which are more viable, provided that new risks are not generated, such as could be caused by raising the cut-off grade to a point where ore becomes discontinuous and impossibly distributed for mining. These risks can be assessed by additional processing if they have not been examined initially. One purpose of preliminary analyses is to define areas where profits are below or near the desired minimum level. Subsequent negotiations on sharing profit and risk for marginal targets would require a financing method that would minimize differences in risk and profit-sharing.

Selection of Desired Mineral Targets for Testing Finance Proposals

Obviously uneconomic target concepts may be dropped after analysis and more likely targets can be used for testing proposals for financing the mineral property. Targets may be selected by marginal analysis to see whether the rate of return from incremental investment is acceptable. Although one target could be selected for further appraisal at this point, especially at stages of development when unknowns have been reduced, the elimination of alternate targets prevents the determination of negotiated terms which may adversely affect such alternatives. For this reason, the added effort of analyzing multiple feasible targets under proposed financing schemes is generally enlightening and worthwhile. Let us again return to the hypothetical example and say that the target proposed by the geologist with company B is 25 million tons of 1.75 per cent Cu and the preconceived target for the geologist with company A is 200 million tons of .85 per cent Cu. After examining profit indicators for those targets, we might speculate that the ideas of each negotiating team with regard to sharing of profit and risk would differ.

ANALYSIS OF FINANCING AGREEMENTS

The mineral property has now been examined in our example. Time and information considerations may modify the analysis in actual negotiations. Independently of the previous example, each negotiator is normally concerned

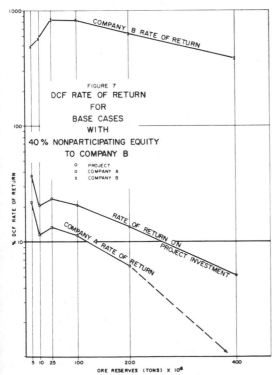

FIGURE 7 — DCF rate of return for base cases with 40% non-participating equity to Company B.

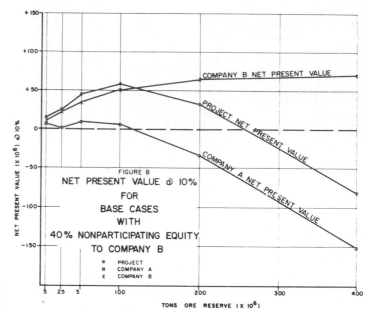

FIGURE 8 — Net present value at 10% for base cases with 40% non-participating equity for Company B.

by the thought that he may be "giving up too much" in return for what he is acquiring. Financial calculations can provide him with data as to how much he is giving up and whether he has gone beyond desirable limits for the risks involved. Possibly he could competitively give up more, which may make the difference as to whether he acquires capital to finance his property or, on the other hand, whether his money is put to work at a good rate of return.

At this point, alternative financing agreements can be modeled, and crucial variables can be tested to aid negotiation decisions.

Modelling of Financing Agreements

Selected mineral targets can now be analyzed with computer programs designed to handle alternative financing schemes. Basic financing agreements can be modeled to provide a basic system structure. These basic structures can be modified to meet the desired agreements. Tax effects for modified financing methods may require a sizable development of a new basic system(s).

Let us assume, for our hypothetical analysis, that company B controls the prospect and originally desires a 40 per cent carried equity in the property. Company A has run a preliminary one-pass calculation on the 200-million-ton case and has previously proposed a 5 per cent smelter return royalty. Our first analysis will be to examine the effects of these financing alternatives on the desired profit indicators. These finance agreements are basic to our system and no modification is necessary.

Calculating Expected Cash Flows and Profit Indicators for Varying Targets and Financing Agreements

Selected mineral targets can now be varied, changing critical variables such as net smelter return value, operating costs and/or capital costs to determine the effect of the financing agreements on the cash flows and profit in-

dicators of the participants in the financing agreement. The amount and source of capital and proportion of profit-sharing might also be changed.

We can examine the effect of the two basic proposed financing schemes on our basic models of mineral targets (see *Figures 7, 8, 9* and *10*). The operating costs and net smelter return value have been changed (±20 per cent) for the 100-million-ton case to determine the effect of these financing agreements on the profit indicators (see Table 2).

TABLE 2 — Effect of Changes in Variables on Profit Indicators for Basic Financing Agreements

(Target: 100 Million Tons of 1.15% Cu)

	DCF RATE OF RETURN (%)			
	Company A		Company B	
Variable Changed	40% Carried Equity	5% NSR Royalty	40% Carried Equity	5% NSR Royalty
Base Case	11.3%	19.1%	825.8%	149.0%
+20% NSR	15.5	25.3	1063.6	178.7
−20% NSR	6.5	12.5	588.0	119.2
+20% Operating Cost	9.5	16.6	733.2	149.0
−20% Operating Cost	13.0	21.7	918.4	149.0

	NET PRESENT VALUE AT 10% ($ x 10⁶)			
	Company A		Company B	
Variable Changed	40% Carried Equity	5% NSR Royalty	40% Carried Equity	5% NSR Royalty
Base Case	$ 6.3	$49.5	$ 51.9	$ 8.7
+20% NSR	29.1	85.6	67.2	10.6
−20% NSR	− 16.4	12.8	36.8	6.8
+20% Operating Cost	− 2.5	34.7	46.1	8.7
−20% Operating Cost	15.2	64.3	57.9	8.7

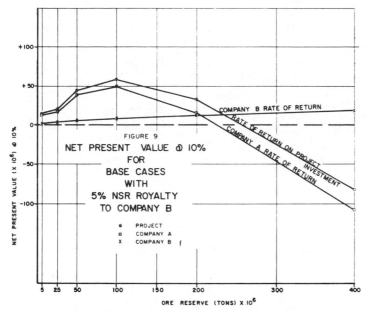

FIGURE 9 — Net present value at 10% for base cases with 5% royalty to Company B.

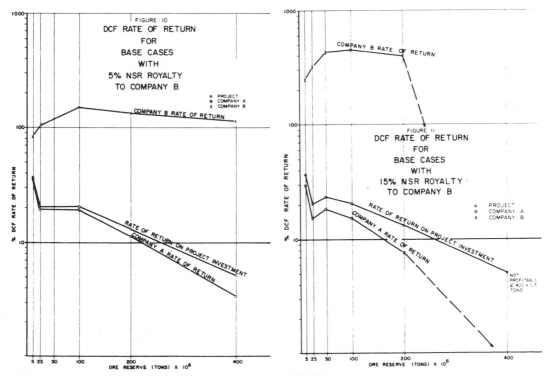

FIGURE 10 — DCF rate of return for base cases with 5% NSR royalty to Company B.

FIGURE 11 — DCF rate of return for base cases with 15% NSR royalty to Company B.

105

Determine if Risk and Profit-Sharing Meet the Negotiation Requirements

If we examine the alternatives for cases we have now run. we see that the 5 per cent NSR royalty reduced the DCF rate of return for company A in the 200-million-ton case below the desired level. As tonnage increases, the DCF rate of return for company B increased from about 1.5 times that of company A to approximately 30 times that of company A. The net present value (10 per cent) of company A with respect to that of company B increases to the 100-million-ton case and then decreases as company B continues to make a good return even if company A operates at a lower rate of return than the desired level. We can see (Table 2) that changes in net smelter return value affect profit indicators for both companies; however, changes in operating costs (or capital costs) do not affect company B as long as the property is operating.

The net smelter return value agreement at the 5 per cent level gives company B a guaranteed return as long as the property is operating, whereas company A takes the brunt of all risks in costs. Company B shares a small portion of the risk in changes in net smelter return value.

The 40 per cent non-participating equity shows extremely stiff restrictions on the rate of return for company A — only three targets out of six meet the desired level of profit. Company B now makes 50 to 100 times the rate of return of company A. The net present value at 10 per cent also shows an imbalance in the share of profits, as the present value always increases for company B with increasing tonnage, and company A obtains a very small present value on any of the targets and shows a negative present value on larger-tonnage targets. The spread shown between the project. company A, and company B profit indicators emphasizes the level of profit-sharing for the proposed financing scheme.

The 40 per cent non-participating equity gives company B fabulous return while increasing the risk for company A. An increase of 20 per cent for operating costs or a 20 per cent reduction in NSR value quickly reduces the rate of return for company A, even for the 100-million-ton case, to undesirable levels. Both companies would probably be looking for an alternative to these initial proposals.

Modify Financing Agreement and/or Economic Parameters

After the first round of analysis. there hopefully will be a closer understanding of the desired level of profit in respect to both risk assumed and the level of profitability of the venture. We can then modify the proportions of profit-sharing. the type of financing agreement and/or the source of capital. If a narrow view of the target or extractive processes has been taken at the outset. they also may be modified at this point. The cycle of calculations and analysis is then repeated until the desired information has been obtained to complete negotiations.

In our example, the NSR royalty was modified to 15 per cent. (*Figures 11* and *12*), raising the net present value returned to company B, but making the 200-million-ton orebody unattractive for company A, and resulting in a good deal more risk than the 5 per cent agreement for company A should operating and capital costs go up.

The 40 per cent non-participating equity was modified to a 10 per cent participating equity (*Figures 13* and *14*), which provided for nearly an equal rate of return for both company A and company B. (Company B is assumed to have shared 50 per cent of exploration costs, decreasing their capital-to-profit ratio slightly.) Company B obtains a very small net present value at 10 per cent in this case. however, and further negotiation is desired by both companies.

In an attempt to improve the rates of return and net present value. the source of capital is now changed. A loan of between 20 per cent and 60 per cent of the capital for plant construction and development has been investigated and is now deemed feasible. The loan is assumed to be paid back from initial project cash flows. Interest rate on

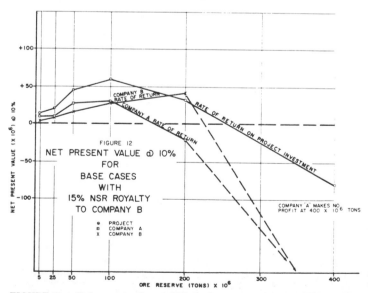

FIGURE 12 — Net present value at 10% for base cases with 15% NSR royalty to Company B.

the loan will be 8 per cent. Company A and company B will share the equity portion investment and cash flow on a 60-40 per cent basis respectively. The effect (in 10 per cent increments) of changing the proportion of plant capital financed by loan from 20 per cent to 60 per cent for the 200-million-ton case is shown in *Figures 15* and *16*. The effect of a ±20 per cent change in Net Smelter Return value is also shown.

Although the loan improves the rate of return for both companies (see *Figure 15*), the risk of loss is increased should adverse conditions be met, and company policy and financial conditions will most likely prevail in setting the level of debt assumed, although we can readily see from information available at what level the greater risks lie.

For the purpose of the hypothetical case, we shall leave the information-gathering process at this point. In real problems, the process would continue until the desired level of information was achieved.

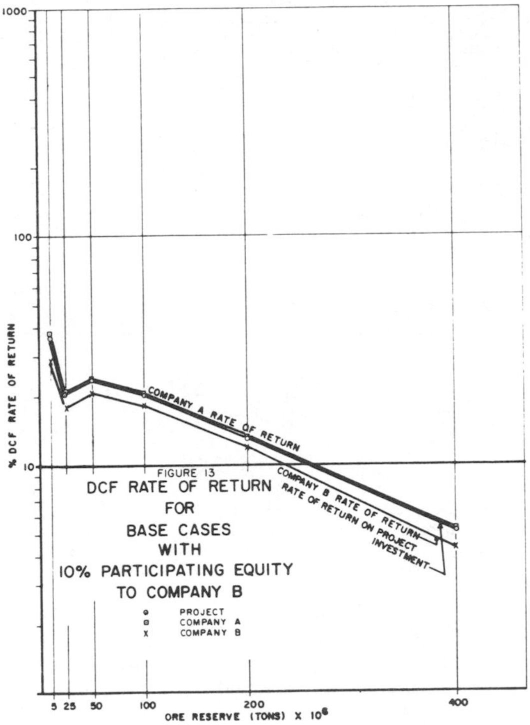

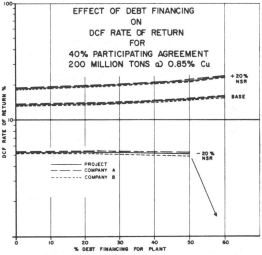

FIGURE 15 — Effect of debt financing on DCF rate of return for 40% participating agreement — 200 million tons at 0.85% Cu.

FIGURE 13 — DCF rate of return for base cases with 10% participating equity to Company B.

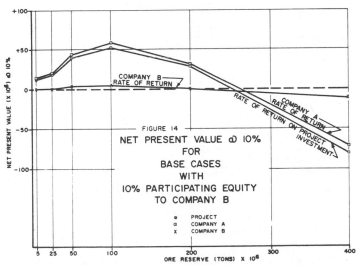

FIGURE 14 — Net present value at 10% for base cases with 10% participating equity to Company B.

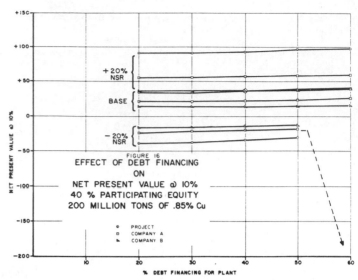

FIGURE 16 — Effect of debt financing on net present value at 10% — 40% participating equity — 200 million tons at 0.85% Cu.

CONCLUSION

The information derived from the case study has not been fully documented in this paper; only examples have been used from the data to demonstrate the value of such information during negotiations for mineral properties. Over 420 individual analyses were made, examining changes in variables and finance agreements. Computer output consists of listings of data entered, distribution of cash flow for the project, distribution of cash flow for financing agreements and profit indicators for the project (company A and company B). The data were obtained on a central processing unit at cost, time and manpower requirements many times reduced as compared to equivalent hand calculations.

The length of (or preparation for) negotiation may be expanded by the development of models to fit particular financing methods. Some items of negotiation, such as the value of expertise in mineral extraction, cannot be placed in the cash flow directly. Aversion to risk and availability of capital are also difficult to quantify in these calculations, although their effect might be implied by the method of analysis and information gathered rather than being directly observed in results.

The value of using the computer as a tool for analysis in negotiation lies in providing: (1) adequate economic information at low cost to eliminate particularly adverse financing agreements for either or both parties; (2) sufficient analyses to examine the sensitivity of mineral targets to a large number of unknowns; (3) quick and convenient analysis of alternative proposed agreements to adequately evaluate economic advantages or penalties prior to the completion of negotiations; and (4) the flexibility to examine a large number of proposals, one or more of which may come within the range of desirability of both parties to the negotiations, hence increasing the probability of satisfactorily completing the negotiations.

8

Reprinted from pages 283–288 of *Decision-Making in the Mineral Industry,* Canadian Institute of Mining and Metallurgy, Montreal, 1971

The Impact of Taxation Systems on Mine Economics

D. G. Krige

ABSTRACT

Comparisons are drawn of mining tax rates, bases of depreciation of capital assets and special allowances as applied in certain countries. It is shown how these can affect the flotation pay limit (cut-off grade) and hence the initial investment decision of a new proposition as well as the economics of producing mines. Suggestions are made for the ideal type of tax structure for mining.

GENERAL

AN IDEAL MINING TAXATION STRUCTURE cannot be designed without taking proper account of the risks, peculiar to mining, faced by the investor, of the extent to which these risks can be increased by mining taxation, and how this in turn can react to the detriment of the country's economy. The indirect and strategic value of mineral production should also be appreciated.

The need for concerted action by the international mining community is evident from recent and further proposed changes to the mining tax systems in various countries, and this symposium is an obvious platform for pooling information and suggestions in this regard.

MINING TAXATION IN PERSPECTIVE

Many arguments have been advanced from time to time in justification of the principle that mining should occupy a relatively favourable position in a country's taxation structure vis-a-vis other sectors of the economy, particularly secondary and tertiary industries. Rather than attempting to give a comprehensive review of these, only three aspects which are regarded as the most pertinent will be covered briefly.

Mining takes its undisputed place as a *primary industry,* the development of which creates the demand for a wide range of industrial products and thus normally leads to the establishment of industrial complexes, power, rail and road networks and the necessary infrastructure of a developed economy. The huge Witwatersrand industrial complex centered on Johannesburg is an excellent example of such a development which would not have taken place but for mining. The effective contribution of a new mine(s) to a country's economy cannot, therefore, be measured only by its direct contribution to the Gross National Product (G.N.P.); it must also take credit for at least part of the contributions made to the G.N.P. by secondary and tertiary industries which owe their creation and continued existence directly or indirectly to such a mine(s). Such 'continued existence' should be stressed where, as is frequently the case, the demand by the mines for industrial products forms the base load for these industries without which they would not be able to compete with imported products, or compete on the export markets. Mining is in fact a primary as well as a significant indirect contributor to capital formation and the G.N.P. These aspects are naturally far more important in younger developing countries such as South Africa, but are not insignificant even in highly developed industrial countries such as the U.S.A. and Canada.

A further important aspect is that in most countries *metals and minerals are of strategic importance.* It is obvious that every country must be prepared to face some financial burden in order to achieve a reduction in the extent of its dependence on imports of essential raw materials; where this is unavoidable, a reduction in vulnerability is frequently bought at the cost of maintaining strategic stock piles. If, therefore, there is a reasonable chance of developing sources of such raw materials internally, the cost of encouraging new mining propositions via tax concessions could, by comparison, be small. In countries where minerals contribute significantly to exports, such as South Africa, the strategic economic importance of mining as an earner of essential foreign currency also requires to be stressed.

In the case of the other main primary producer, i.e. agriculture, the above arguments also apply to varying degrees and have led in most countries to direct and/or indirect concessions and assistance measures. The total cost of such tax concessions, price subsidies, protective tariffs, etc. is usually very difficult to determine, but can be very substantial. Mining warrants similar encouragement, bearing in mind as well that most secondary industries are also given indirect assistance through tariffs, etc.

A third important aspect is that mining has always been and still is a *high-risk investment.* In fact, the risks are steadily increasing as higher-grade deposits become depleted and lower-grade ores have to be opened up on an ever-increasing scale; exploration efforts have to be stepped up and exploration and mining costs continue to escalate faster than the general rate of inflation. Metal and mineral prices have, of course, also risen, but generally not to levels which limit or reduce the over-all risks involved. This is in contrast to other industries where inflation and the rapid rate of real growth has resulted in an ever-expanding market and have therefore in fact eliminated a good deal of the risk element (Sadie, 1969).

In mining, as distinct from other industries, the bulk of the capital investment is at risk from the start for a longer non-productive period and is not eliminated when production starts. The capital assets become almost worthless if the mine turns out to be a failure as a result of any one or more of the initial uncertain factors (grade, price, life, etc.) being in fact unfavourable. Industrial maladies can generally be overcome by repositioning of factories, modifications of equipment, changes in product, development of new markets, etc., but mining suffers from a general inflexibility of position, equipment, product, price and labour.

Evidence of the high-risk element in mining investments is found in the wider range of market price movements of mining equities, particularly during the preproduction and initial production periods when the basic investment is being made.

IDEAL BASIC PRINCIPLES IN MINING TAXATION

The first basic principle is that *all expenditures* (capital and working costs) incurred directly or indirectly for mining purposes *should be allowed* for tax purposes. There is no valid reason why the cost of necessary surface rights and of mineral rights should not be deducted directly or indirectly through allowances.

Secondly, it should be recognized that in mining the distinction between capital expenditure (usually allowed via depreciation or redemption allowances) and current working costs (immediately deductible) is largely artificial. A mine, unlike other industries, does not have an unlimited life and, furthermore, its life always remains of uncertain duration and its profit expectations highly risky and largely outside its own control. Whereas the logical 'time' account for other industries can be accepted as a year, the only true 'time' account for a mine should cover its whole life, and in such an account the cost of all capital work and equipment would be items of working expenditure. There is justification, therefore, not only for *allowing all mine 'capital' expenditure immediately* as and when incurred, but also for the *carrying backward of losses* incurred in any year, so that past tax payments can be refunded when operations turn unsuccessful; any loss not used up in this way will then be carried forward as allowed at present.

Thirdly, in view of the direct and indirect benefits of mining to the State (economic and strategic), the State should ensure that, as far as possible, *taxation does not raise the current cost of production and hence the operating mine cut-off grade, nor the flotation cut-off grade;* if this is not done, the exploitation of ores or new mines will be rendered uneconomic. These requirements, together with those for ensuring the encouragement of new investments in a sector subject to high risk, can best be met by *special allowances on gross revenue* (e.g. the U.S.A. depletion allowances and the South African tax formula) *and on capital expenditure* (e.g. investment allowances on new capital expenditure, or a compound-interest allowance on unredeemed capital expenditure, as for new South African gold mines).

Finally, the tax rate for mines should not exceed that of other industries.

These principles will be dealt with by comparing the effects of the present mining tax systems of the U.S.A., Canada and South Africa on a specimen mine, together with some likely and suggested changes.

BASIS OF ANALYSES

Tax systems, and particularly those applicable to mining, are complex and cover many detailed provisions which render valid international comparisons difficult to effect. To be of any value, such comparisons must cover the main significant features only in order not to lose sight of the wood for the trees. For this reason, this analysis is confined to the following aspects only:

Capital Redemption — bases accepted for this comparison

(i) No redemption: items such as surface rights, and working capital — all three countries.

(ii) Immediate redemption: items such as shafts and development — all three countries.

(iii) Depletion allowances for mineral rights. Canada: accepted at one-third of taxable income, and, as an alternative, no allowance except actual cost redeemed on a reducing balance basis. U.S.A.: at from 0 per cent to 22 per cent of revenue, subject to a minimum covering actual cost redeemed over life, and a maximum equivalent to 50 per cent of taxable income before depletion. South Africa: the cost of mineral rights is not allowed; however, the new gold mines tax formula effectively allows a deduction of 8 per cent of revenue from profits before application of the flat tax rate, and this is equivalent in effect to the U.S.A. depletion allowances.

(iv) Redemption of remaining items (buildings, equipment, etc.): immediate in South Africa, straight line over life for U.S.A., and 25 per cent p.a. of reducing balance for Canada.

Special allowances

In South Africa, new gold mines are allowed an 8 per cent compound interest allowance on the unredeemed balance of capital expenditure from year to year. Investment allowances are accepted as no longer applicable in any of the three countries. For Canada, the position with and without the three-year tax-free allowance will be shown.

Tax rates

This has been accepted at 50 per cent for the U.S.A. and Canada. In South Africa, the rate for gold mines is 63 per cent, but the position for a 50 per cent rate will also be shown.

Specimen Mine

To illustrate the effects of the above factors, a specimen mine has been chosen with a relatively high level of capital investment required, and with variations in grade, metal and mineral price(s), and current production costs catered for through a range of annual revenue/cost ratios from 1.0 to 3.0. This is not necessarily a typical mine in any of the three countries concerned, but will serve to highlight differences between and desirable and undesirable features of the relevant tax systems.

The case examined, idealized for purposes of simplification, is as follows.

	Monetary Units
Annual costs, including royalties, local taxes and provincial (state in U.S.A.) taxes, but excluding all allowances for capital expenditure	1
Total capital investment, assumed expended at start of production	8
and comprising: —	
Items not allowed	½
Mineral rights	½
Items immediately allowed in all three countries	2
Other items	5
Annual gross revenue, assumed constant from inception in range from	1 to 3
Productive life of mine	20 years

The criterion accepted for comparing the effects of different tax systems on the capital investment in a new mine or in the expansion of an operating mine is that of the compound interest return indicated (or Discounted Cash Flow). For a new mine, the significance of the flotation pay limit (or cut-off grade) is also discussed, as well as the effect of taxation on the operating pay limit or cut-off grade of a producing mine.

IMPACT OF TAXATION
ON NEW INVESTMENTS

Figure 1 shows the compound interest return (D.C.F.) for the specimen mine under the tax systems in Canada, U.S.A. and South Africa (gold) for any annual revenue/cost ratio in the range from 1 to 3.

It is evident that for *highly profitable mines* (revenue/cost >2.0), the Canadian system (curve 2a) is by far the most favourable, mainly because of the three-year tax-free period, and the depletion of one-third taxable income. Comparing highly profitable mines in the U.S.A. and South Africa, it seems that the advantages of immediate capital redemption and the 8 per cent compound interest capital (or investment) allowance in South Africa (curve 3a) are balanced by the higher U.S.A. depletion allowances (14 to 22 per cent — South Africa, 8 per cent of revenue) and the lower U.S.A. tax rate (curves 4a and 4c). Furthermore, Canada, excluding only the three-year allowance, the U.S.A. with a 22 per cent revenue depletion and South Africa with a 50 per cent tax rate (curves 2b, 4a and 3b) are in close agreement. If the proposals to abolish the Canadian tax-free period and depletion allowance are both put into effect, the Canadian position will correspond to the second-lowest curve (no depletion) on the graph (curve 5b) and would change from the best to the worst system of the three countries concerned.

For *low- and medium-profit* mines (revenue/cost up to 2.0), it is interesting to note that the South African D.C.F. yield curve (No. 3a) merges with the no-tax curve (No. 1) at a D.C.F. of about 6¼ per cent (revenue/cost at 1.7), up to which level, therefore, no taxation is payable. In the U.S.A. and Canada, on the other hand, some tax is payable even if the mine shows a nil return (curves 2 and 4). In the South African case, the 8 per cent p.a. compound interest capital allowance does not result in a tax-free D.C.F. return of up to 8 per cent because certain

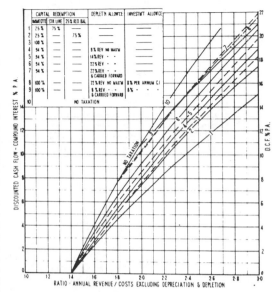

FIGURE 2 — D.C.F. returns for a specimen mine for various possible taxation systems.

items of capital expenditure (surface and mineral rights) are not allowed for tax purposes.

Figure 2 shows, on the same basis, the relative D.C.F. returns for a series of theoretical cases, not related to any country, where additional desirable features are added progressively until the 'ideal' cases are reached. The 'basic' case (curve 1) allows all capital expenditure (mineral and surface rights at cost) to be redeemed, shafts and development immediately, and the balance on a straight-line basis over the life of the mine. The 'ideal' cases (curves 7/9) correspond to the immediate redemption of all capital expenditures (or excluding only cost of mineral rights), and

- (i) a 'depletion' or special allowance of 22 per cent of revenue, with no maximum, carried forward from year to year (curve 7);
- (ii) same as for (i), but the allowance is not carried forward, and an 'investment' or capital allowance of 8 per cent compound interest p.a. on unredeemed capital expenditure is added (curve 8); and
- (iii) a depletion allowance of 8 per cent of revenue, with no maximum and carried forward, and an 8 per cent 'investment' allowance, same as for (ii) above (curve 9).

For all the cases shown on *Figure 2*, the tax rate has been accepted at 50 per cent. For the 'ideal' cases, no tax is payable until the D.C.F. return approaches or exceeds 8 per cent p.a.

Still referring to *Figure 2*, let us accept an 8 per cent p.a. D.C.F. return either as the minimum objective (in which case the flotation pay limit or cut-off grade will be defined by the corresponding ratio of revenue/cost) or as the level below which the investor will show an effective loss relative to alternative safer investments. The effects of various tax systems are evident from the revenue/cost figures on the horizontal axis corresponding to an 8 per cent D.C.F. return. For the 'ideal' cases, tax has no effect and the flotation limit is R/C = 1.8, whereas the 'basic' case shows a substantial tax effect and an R/C ratio of

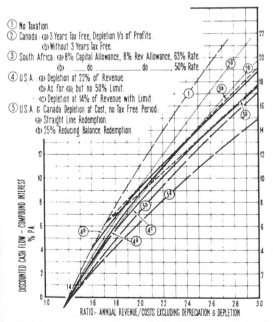

FIGURE 1 — D.C.F. returns for a specimen mine under U.S.A., Canadian and South African taxation systems.

almost 2.2; i.e. about 20 per cent higher. Tax systems other than the 'ideal' could, therefore, render the investment of capital in propositions with an R/C ratio in the range between 1.8 and 2.2 economically unjustified. The results of a survey of all the mineral potential within this range in countries such as the U.S.A., Canada and South Africa would be˙most revealing.

Even propositions with indicated R/C ratios above 2.2 would also be affected, as the criterion for investment must take account not only of the indicated return, but also of the chances, i.e. the risks, of the R/C ratio being in fact such that the corresponding return is below some specified minimum (8 per cent p.a. accepted above). From a standard statistical viewpoint, the criterion normally accepted would be the D.C.F. return corresponding to the *expected* R/C ratio, i.e. the mean of the probability distribution of R/C ratios, and, if the return is above 8 per cent p.a., this would be acceptable provided the investor had an adequate spread of simultaneous investments of this type to make; the values above the expected would make up for those below the expected. For most mining concerns, however, the decision to invest or not relates to a single proposition to be considered in isolation, and these events do not occur at such frequent intervals for the same company that the 'law of averages' would operate effectively. The chance, say 1 in 5, of the proposition being a bonanza will therefore, in practice, not necessarily outweigh a similar chance of the investment being in fact uneconomic; in the latter event, the company might have to go into liquidation.

The State, on the other hand, effectively has the maximum spread of interests, as it becomes a shareholder in *all* mining propositions (without risking any capital!) and can afford to encourage investment in propositions indicated as marginal, through a system such as the 'ideal' on *Figure 2*, so as to ensure that no tax is levied on mines which in fact turn out to be marginal. The State can only gain from such a policy in the long run, because if, without this encouragement these marginal propositions are not opened up, the State will not obtain any tax revenue, whereas if the propositions are turned to account, some will in fact turn out well above marginal and become tax payers, whereas those below marginal will pay no tax, and the assessed losses for tax purposes, if any, will in most cases not be claimable against other income. The State's 'expected' return from propositions indicated as marginal can, therefore, be positive even after taking account of its capital investments by way of the required infrastructures for such propositions.

The logical conclusion drawn is that, from the viewpoint of both the State and the mining investor, it must be advantageous to have a tax structure which ensures that no tax is payable unless the return on the investment exceeds some reasonable minimum level.

The curves on *Figure 2* (and those on *Figure 1*) clearly show that this can be achieved partly by a percentage revenue allowance, such as the present U.S.A. depletion allowances, provided that the maximum limit of 50 per cent of taxable income is abolished and the allowance can be carried forward (curve 7), or by a special 'investment' or 'capital' allowance calculated as compound interest on all unredeemed capital expenditure at a rate (p.a.) corresponding to the reasonable minimum level of return on the investment, combined with a revenue allowance on alternative bases (curves 8 and 9). These schemes will stimulate new investments without significantly affecting the tax accruals from existing producers. Representations for positive and against negative changes in mining tax systems should, therefore, be aimed at the range of 'ideal' cases shown.

THE TRUE NATURE OF THE VARIOUS TAX SYSTEMS

It is revealing to examine and compare the effective mining tax rates in the U.S.A., Canada and. South Africa after taking account of the relevant depletion allowances in the former two countries and the tax formula in the latter, but ignoring the differences in capital depreciation allowances and the special allowances in Canada and South Africa (tax-free period and 8 per cent capital allowance respectively). For this purpose, therefore, '*profit*' *is accepted in each case as taxable income after depreciation but before depletion,* and the effective tax rates applicable to these profits are compared.

In the U.S.A., a flat 50 per cent rate is taken to apply after a depletion allowance of b per cent of revenue, i.e. tax paid = 50 per cent (profit − b per cent revenue).

The effective tax rate on 'profit' as defined is determined as follows:

$$\text{Effective tax rate \%} = (\text{tax paid} / \text{profit}) \times 100$$
$$= 50 \left(\frac{\text{profit}}{\text{profit}} - \frac{b\% \text{ revenue}}{100\% \text{ profit}} \right)$$
$$= 50 - \frac{50\,b}{P/R\,\%}$$

i.e. for a 22% depletion rate,

$$\text{effective tax rate \%} = 50 - \frac{1100}{X}$$

$$\text{where } X = \text{profit/revenue \% = P/R \%}$$
$$= \text{measure of relative profitability}$$

The maximum depletion rate of 50 per cent of profit means that this effective tax formula is subject to a minimum tax rate of 25 per cent of profit.

This tax formula can be further developed as follows:

$$\text{tax} = 50\% \, (\text{profit} - 22\% \text{ revenue})$$
$$= 50\% \, (78\% \text{ revenue} - \text{costs})$$
$$= 39\% \text{ revenue} - 50\% \text{ costs}.$$

Provided that the minimum rate does not apply, the company is left, after tax, with

$$61\% \text{ revenue} - 50\% \text{ costs};$$

i.e. it retains 61 per cent of any revenue improvement and 50 per cent of any cost saving, or the State effectively bears 39 per cent of any drop in revenue and 50 per cent of all cost increases.

This effective tax formula is of exactly the form applicable directly as a sliding-scale formula for gold mines in South Africa, where for new mines,

$$\text{the tax rate \%} = 63 - \frac{504}{X}$$

i.e., the 'b' factor above = 8. and
the tax paid becomes = 63% (profit − 8% revenue).

The South African system is therefore effectively the same as the U.S.A. system, with an 8 per cent revenue allowance equivalent to depletion, but with no maximum limit of 50 per cent of profit for the latter. As a mine cannot qualify for negative taxation, i.e. the tax rate cannot be negative, the effective limit for the 8 per cent allowance will be 100 per cent of profit. In both the U.S.A. and South African cases, the concessions to small (low income) mines and other refinements are ignored.

In Canada, the depletion allowance of one-third of profit, as defined above, and a 50 per cent flat tax rate is effectively equivalent to a net flat rate equal to 33⅓ per cent of profit.

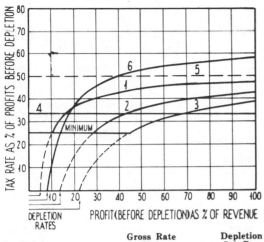

FIGURE 3 — True graduated nature of the U.S.A. and South African mining tax rates.

	Gross Rate	Depletion
1. U.S.A.	50%	5% Rev.
2. U.S.A.	50%	14% Rev.
3. U.S.A.	50%	22% Rev.
4. Canada	50%	1/3 Profit
5. U.S.A. & Canada	50%	Nil
6. South Africa	63%	8% Rev.

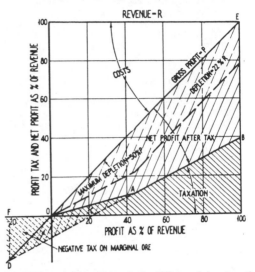

F = EFFECTIVE CUT—OFF GRADE AT P = -22% R
O = CUT-OFF GRADE WITHOUT DEPLETION I.E. 28% HIGHER

FIGURE 4 — The effect of the U.S.A. mining taxation system on the operative cut-off grade.

Figure 3 shows the true nature of the effective tax rates. The U.S.A. and South African rates rise from 25 per cent and 0 per cent, respectively, along curves which approach their theoretical maxima as a mine's profitability reaches a maximum of 100 per cent P/R ratio, i.e. when costs are nil; the mines are taxed on a graduated scale according to their relative abilities to pay in a way similar to the income tax for individuals in most countries. This differential will, of course, disappear if the 'depletion' allowance in the U.S.A. is reduced to nil per cent of revenue, i.e. the rate will then be a flat 50 per cent, as shown in *Figure 3*. The effective Canadian tax rate is shown as a flat 33⅓ per cent.

Apart from the usual argument based on the differential 'ability to pay', a further pertinent argument in favour of a differentiated tax, as in the U.S.A. and South Africa, is raised by Sadie (1969); i.e., that an undifferentiated tax which absorbs an equal portion of the profits of small and big companies (or individuals), thus reducing the funds available for re-investment, must have a greater adverse effect (in absolute, if not relative terms) on the former than on the latter. Their ability to develop and offer competition to the larger firms is impeded. The tax system should provide a positive incentive 'to reduce the tempo of domination of the economy by a decreasing number of giants'. The depletion allowance as applied in the U.S.A. serves this objective and should be preserved.

Further arguments will be raised in the next two sections.

ADDITIONAL CAPITAL INVESTMENTS IN ESTABLISHED PRODUCERS

In the case of additional capital investments being made, for example to expand production, the general position of a producing mine will be similar to that for a new proposition as analysed above, except that where the company is already paying taxation the incremental capital expenditure could be reduced through negative taxation effectively ac-cruing on this expenditure, i.e. through the State paying for part of the investment via reduced taxation on existing operations. This effect can substantially improve the D.C.F. return on such an incremental investment compared to that in a new venture where no negative tax effect will operate. A simple example will clarify this effect.

Mine A is in production with an annual gross revenue of $20 million, and costs for tax (before depletion) of $8 million:

Profit	= $12 m	
Depletion	= $2.8 m	(14% of revenue)
Taxable income	= $9.2 m	
Tax @ 50%	= $4.6 m	

For expansion, a capital expenditure of $6 m is incurred, $3 m on shafts, etc. (redeemed immediately) and $3 m depreciated at, say, $0.2 m p.a. Thus, the position in the year during which this occurs will change as follows:

Normal profit	= $12 m
Additional redemption	= $3.2 m
Depletion	= $2.8 m
Taxable income	= $6.0 m
Tax @ 50%	= $3.0 m

Effectively, therefore, there is an immediate negative taxation of $1.6 m on the investment of $6 m, which must assist in raising the incremental D.C.F. return.

The question arises as to why, if negative taxation does in fact operate under certain conditions as shown, should this principle not be extended by, for example,

(i) allowing assessed losses to be carried backward (as well as forward) so that a mine is in fact assessed on its over-all achievements *over periods longer than a year* — mining, being subject to a variety of risks, can show substantial variations from year to year; or preferably

(ii) accepting the principle, which appears rather radical at first sight, that if a negative tax is indicated, the company would in fact receive that amount from the State — this principle has already been incorporated in the South African gold mines assistance formula, to which further reference will be made later.

The logical application of (ii) above will ensure that for a tax rate of, say, 50 per cent, the State will effectively be contributing and therefore risking its half share of the capital invested before and during production instead of, as at present, contributing it indirectly at a later stage by way of reductions in the tax payable as and when the capital is redeemed for tax purposes.

TAXATION AND THE CUT-OFF GRADE

The fact that the South African gold mines tax formula effectively lowers the operating cut-off grade (pay limit) has been appreciated for many years (McKechnie, 1948) and formed the basis of the formula devised recently for the State-assisted gold mines (Krige, 1968). Because, as shown above, the U.S.A. depletion allowances in fact result in a similar form of tax formula, it is interesting to analyse the effect of this form of tax on a producing U.S.A. mine subject to a 22 per cent of revenue depletion allowance (maximum = 50 per cent of profits before this allowance) and a 50 per cent tax rate.

The results of this analysis are shown on *Figure 4*. All figures are shown as percentages of revenue. The horizontal scale covers the range of profitabilities from a loss of 20 per cent to a profit of 100 per cent of revenue and the vertical scale reflects the corresponding tax payments and net profits (after tax). Above a profit level of 44 per cent of revenue, the depletion operates at the 22 per cent revenue level and taxation (along line AB) equals 50 per cent of profits after deducting 22 per cent revenue. Below a profit level of 44 per cent revenue (at which level the 22 per cent depletion allowance = 50 per cent of profits), taxation follows the line AO at an effective rate of 25 per cent of profits.

However, if a mine operates *as a whole* at a profit level above 44 per cent of revenue, the tax paid on profits or losses on *individual ore blocks* or units mined will extend along the straight line BD; i.e. it will be nil when profit equals 22 per cent of revenue and in fact negative for profits (or losses) below this level. Similarly, the net profit after tax for individual ore units will be determined by the vertical interval between lines DE and DB and will only reduce to nil at the cross-over point D when

Profit = −22% Revenue
Tax   = 50% (Profit − 22% Revenue)
      = 50% (−44% Revenue)
      = −22% Revenue
and hence profit − tax = nil.

The effective cut-off grade is, therefore, determined at the level when

Profit = Revenue − Costs = − 22% Revenue
i.e. 122% Revenue = Costs
  or Revenue   = 82% Costs

as compared with the normal cut-off grade when revenue = costs.

A simple example will illustrate this effect.

	Existing Position	Additional Unpay. Ore Mined	Combined Position
Revenue	$1000	$100	$1100
Costs	$ 400	$122	$ 522
Profit	$ 600	−$ 22	$ 578
Depletion	$ 220	$ 22	$ 242
Taxable income	$ 380	−$ 44	$ 336
Tax @ 50%	$ 190	−$ 22	$ 168
Net cash flow	$ 410	Nil	$ 410

i.e., the company's net cash flow is not affected by the mining of the unpayable ore.

The cut-off grade is therefore effectively lowered (as a result of a 22 per cent revenue depletion allowance) by 18 per cent and the abolition of the depletion allowance will raise this effective cut-off grade by 28 per cent. The same position will, of course, apply to mines operating at a profit level below 44 per cent of revenue if the maximum limit (of 50 per cent of profit) for the depletion allowance is lifted.

This provides an additional argument against further reduction in the U.S.A. depletion allowances, as such action must progressively reduce the total reserves of economically mineable ores in established mines; the country's reserves of potential ores will also be affected, as discussed earlier, because of the effect on the flotation cut-off grade for new mines.

The question can and should, of course, be raised as to whether mining to the lower effective cut-off grade, although yielding an increased total tonnage, mineral output and profit, will not result in a lower D.C.F. because the lower annual profits could have a greater effect on present values than an extension of life. This argument is invalid where the mine and treatment plant both have spare capacity and the concurrent mining of additional ore, even at a marginal net profit, is fully justified. Where this is not the case, comparable realistic assessments are required of revenue and cost figures, and hence cash flows, with and without the mining of the marginal ore and allowing fully for likely escalation of costs, and also of mineral prices. Assessments made for South African gold mines, for example, show clearly that if costs continue to escalate and no increase in the gold price occurs, the highest D.C.F. is obtained by mining now to the cut-off grade approaching that based on the cost level expected at the end of the mine's life. However, if a substantial increase in price is accepted as likely, even if deferred for years, the D.C.F. will be lower for such a policy, and the most economic scheme will be to concentrate on marginal, or even some sub-marginal, ore at this stage.

CONCLUSION

It is hoped that this paper will be of some assistance in ensuring that in the public domain decisions on mining taxation are taken which will encourage the exploitation of minerals to their full potential and to the maximum advantage not only of the mining sector as such, but of the economy as a whole.

The permission of Anglo-Transvaal Consolidated Investment Company Limit for the publication of this paper is appreciated, as well as the valuable assistance and advice obtained from my two colleagues, Messrs. G. M. Hitchcock and A. G. Carson.

REFERENCES

Bucovetsky, M. W., "The Taxation of Mineral Extraction", Royal Commission on Taxation, Ottawa, 1966, Study No. 8.
Krige, D. G., "Some Implications of the New Assistance Formula for South African Gold Mines", *Jour, S. A. Inst. Min. Metall.*, Vol. 68, No. 10, pp. 408-434, May 1968; Vol. 69, pp. 340-353, February 1969.
Lentz, O. H., "Mineral Economics and the Problem of Equitable Taxation", *Quarterly, Colorado School of Mines*, April 1960.
McKechnie, C. A., "Some Aspects of Taxation and Lease Consideration Applicable to Gold Mines in South Africa", unpublished Master's thesis, University of the Witwatersrand, 1948.
Sadie, J. L., "Company Taxation", *S. Afr. J. Economics*, Vol. 37, No. 4, pp. 345-371, December 1969.

9

Reprinted from pages 279-284 of *Economic Geology in Massachusetts,*
O. C. Farquhar, ed., University of Massachusetts, 1967

EVALUATION OF MINERAL-BEARING LANDS ACQUIRED BY EMINENT DOMAIN

Lawrence A. Wing

Exercise of the power of eminent domain has increased in geometric proportion to recent population growth. Modern highways and high-voltage transmission may take as much as 50 to 100 acres per line-mile, and added to this is the acreage taken for pipelines, reservoirs, and recreational facilities. Some of the lands taken may serve more than one function, but rarely will mining or quarrying operations be compatible with the contemplated public use. The owner of mineral lands acquired by eminent domain is entitled to just and fair compensation in the same way that the public should not be charged unreasonable sums for the land-taking itself. A larger land acquisition, such as for a reservoir or park, may include an entire mineral deposit or several deposits, while the taking of narrow, linear parcels for highway or transmission purposes is more likely to involve part of a deposit or the severing of a deposit.

When mineral-bearing lands are taken by eminent domain, geologists may find that their services are sought by one or other party, in an attempt to arrive at a valuation of the lands taken and any damages or benefits to the remainder. A geologist who accepts such an assignment will most likely find that many of the yardsticks with which he is familiar for purposes of dollar valuation are not acceptable to the court, will not reach the jury, and are, therefore, of little use to the client or anyone else. The experienced expert witness will find little in this paper that is new to him, but it is hoped that some of these comments will be helpful to the unsuspecting geologist facing his first court appearance as an expert witness or appraiser (evaluator of special-purpose property).

The expert witness should always keep in mind that hundreds of questions may be posed to him during a trial, but his responses should be an answer, or partial answer, to only four basic questions: (1) What are your qualifications? (2) What method of valuation did you use? (3) What was the basis for your consideration of value? (4) What is the value? The order of the last two questions may be reversed in some courts. If the expert has these questions well in mind while carrying out his investigations and while writing his report, the courtroom examination should present no serious problems. Each of these basic questions will be further explored, but before doing this it is necessary to consider just what is an expert witness along with his relationship to client, attorney, court, and jury.

THE EXPERT WITNESS

Any geologist who contemplates appearing as an expert witness would be well advised to review papers by Free (1957) and Horgan (1959). The sole function of an expert witness is to testify; the attorneys will argue; the judge will rule; and the jury will award. It is alarmingly clear from reviews of court records that many experts are attempting to advocate rather than giving testimony.   Judges and jurors who allow their functions to be pre-empted by a witness are rare indeed.   If a jury is to give much weight to an expert's opinion as to value, that opinion must be supported by well-documented facts and presented in a straightforward and sincere manner.   The use of the phrase "in my opinion" may be the only way to answer some questions, but such an opinion becomes much more effective if it can be strengthened by reference to fact and measured values.   Obvious bias and evasive answers will greatly weaken and may entirely negate the usefulness of an expert witness.   For a judge to say "It is generally safer to take the judgments of unskilled jurors than the opinions of hired and generally biased experts" is a serious reflection upon expert testimony that can only be corrected by improved conduct on the part of the experts who are called upon to bear witness.

WITNESS-CLIENT-ATTORNEY

When a geologist is first approached by a client or, more likely, the client's attorney, a logical sequence for finding the value of the condemned property should be worked out and clearly understood by all concerned before any actual work is undertaken.   Appraisal of a mineral property differs substantially from other types of real estate valuation in that most of the desired data requires excavation or borings to establish the presence, quantity, and quality of the mineralization.   It is often necessary to spend substantial sums before the geologist has anything upon which to base his opinion as to value.   A reasonable sequence of work might be (1) preliminary examination of the property and available data (at this time the geologist may decide his services may be of little value and so advise his client), and (2) detailed examination, utilizing all needed surveys, borings, tests, etc., and the preparation of a report stating all the facts and the expert's opinion of value and/or damages.   At this point the client and his attorney must review the report and decide whether or not they wish to continue with the services of the geologist in preparation for trial and the trial itself.   If the services are to be continued, (3) exhibits must be prepared in conjunction with pre-trial conferences with the attorney and then, (4), the trial itself takes place.   Cost estimates and the basis for reimbursement should be clarified in advance for each of these four steps.

It is natural and quite common for the client or his attorney to suggest values to the geologist which the latter may feel are either too high or too low. If the expert accepts these suggestions and arrives at a value which is not truly his own, his expertise must be questioned, as it probably will be during cross-examination.   In this sense a sound rule to follow is that his appraisal should be essentially the same if it were done for either party to litigation.   There is obviously some leeway in determining value, but there can be no reasonable justification for one expert arriving at two widely differing values.

WITNESS AND JURY

The jury may pay as much attention to the qualifications of an expert as to his subsequent testimony, and both the witness and his attorney should be thoroughly prepared before the trial starts.   Nothing can be as frustrating to a witness as having his own attorney ask a question for which he is poorly prepared or omit questions which they had previously agreed were important. An unexpected answer from the witness must be equally annoying to the attorney. Long hesitations, evasive answers, and rewording of questions can only lead the jury to suspect poor preparation or, far worse, inadequate knowledge.

Responses to questions should be brief and avoid highly technical terms when more common words can be used. Exhibits such as maps and photographs should be large enough to be readily legible to the seated jurors. Visual aids that clearly show many of the things the witness is trying to say are invaluable. They not only remain before the eyes of the jury during the trial but also are available for the deliberations of the jury long after the precise verbal descriptions have been forgotten. Enlarged aerial photographs with data presented on transparent overlays make excellent exhibits.

Returning now to the four basic questions, they are considered in the order in which they will probably arise during the trial.

QUALIFICATIONS

Many witnesses make the mistake of assuming that qualifications are a routine matter to be disposed of modestly and quickly. This is not true because, when a witness finally gets to the all-important question of value, opinion is involved. The weight a jury gives to an opinion must be strongly influenced by the expert's previous experience and qualifications.

Questioning about qualifications generally begins with education, moves on to general experience, and concludes with specific experience in the geographic area and with the specific commodities involved. The witness should be well prepared with dates and durations of time associated with similar work. Dates and courts in which he has previously qualified are important, and these appear to count heavily with some judges.

METHOD OF VALUATION

Geologists and mining engineers have used a number of methods to arrive at the value of a mine or mineral deposit. All require knowledge of tonnage and grade in order to anticipate life expectancy, size of operation, plant requirements, and return on the invested dollar. All methods utilize a capitalization technique; discussions by Hoskold (1905), Hoover (1909), Morkill (1918), Parks (1957), and others are well known to geologists involved in mine evaluation. Many of these discussions refer to some form of an annuity or estimate of "present worth." The simpler way to look at all of these methods is to use straightforward mid-year discount of future earnings. Stated in another way, what would one accept today for a dollar promised 10 or 20 years from now with that dollar being earned by a mining venture. The present worth of that dollar involves a factor of interest normal to any annuity and another factor of interest depending on risk inherent to the particular property or commodity. It is only the rare investor who would consider a single mining property as akin to a government bond or a blue chip stock, and for this reason present worth on a single mining property commonly will involve discount rates (present-worth factors) in the range of 10 to 25 percent, or higher on undeveloped properties. As a rough example, a mineral property that could return $1,000,000 per year for a period of ten years would show a total return of $10,000,000 but the present worth of that property would be only slightly more than $3,500,000, discounted at a rate of 20 percent.

Any calculation of future earnings must involve assumptions for the future regarding costs, and must assign interest rates or discount factors (risk factors). All of this is common business practice in mining and elsewhere, but is very rarely accepted in eminent domain taking of a mineral deposit because of the element of "speculation." The staggering difference in value noted in many court cases is purely the result of the method of appraisal or the optimism of one party versus the pessimism of the other in arriving at discount rates.

Most expert witnesses who have testified about value are aware that capitalization methods are generally unacceptable in courts. This leaves the expert witness in the awkward position of not being able to use those methods with which he is most familiar. The three methods of appraisal commonly in use for property in general are: replacement cost, income, and market value. Replacement cost rarely applies to a mineral deposit; income is inadmissible; and market value (comparable sales and offerings) is hard to determine. The latter method, however, is by far the most acceptable in court and deserves careful consideration as to just how a witness can use this approach to establish what he believes to be true value. Some further comments also are warranted on the income approach and these are discussed in the conclusions.

If sales of comparable (not identical) mineral-bearing land can be found within reasonable distance, dated within a few years of the land taking, the expert has a good foundation upon which to determine value. This may be quite common with such commodities as granite, limestone, gravel, peat, diatomite, clay, and the like; i. e., deposits that often occur in groups and where numerous property transfers may be a matter of record. Most of these commodities have in common relatively low royalty or market values per ton, and land prices in the area may not be significantly different as between mineralized and non-mineralized ground.

The other extreme is represented by a high-grade base metal or precious metal deposit with an astronomically large "present-worth" value per acre. This large value may be further complicated by a total lack of comparable sales over the area of an entire state or more. One further complication is that, in much of this country and in most other countries, the owner may have acquired title to mineral claims for very modest sums, and it does not seem likely that these can be considered in arriving at "market value."

It may now be apparent that the expert witness must consider the market approach (comparable sales) to whatever extent he feels is right. The key word of the previous sentence is <u>consider,</u> which leads to the third question.

CONSIDERATION OF VALUE

A common answer by expert witnesses to the question "What did you consider in arriving at this value?" is "Everything." Even if true, this answer will hardly be of much help to the jury. The reasoning of the witness must be guided, in most situations, by the willing seller-willing buyer rule; i. e., both are fully informed of the facts and act intelligently. The witness may then <u>consider</u> recent transfers, nature of the surface, improvements, operating feasibility, overburden, tonnage and grade, prices, market trends, and anything else that is relevant. He may then indicate that of the many elements considered some were rejected and some were weighed heavily. It is in the answer to this question about consideration of value that an expert witness finds he must use the qualifying phrase "in my opinion," and the weight that opinion will be accorded by the jury will actually depend upon how adequately previous questions have been answered.

THE VALUE

The introduction of the expert's suggested appraisal is the climax of his appearance, but, if all of the preliminary questions have been answered without the witness discrediting himself, the answer may be short and simple.

CONCLUSIONS

The expert witness will usually find that he must consider market value in his estimate. If he places more reliance on other methods, his opinion will be challenged. Geologists should be aware that several recent cases throughout New England, involving such commodities as limestone, granite, and sand and gravel, have attempted to introduce and use capitalization methods of appraisal but, when both sides have been properly represented by qualified expert witnesses, the methods have met with little success. Capitalization methods will almost always lead to much higher present-worth sums than can be obtained by the market approach with such relatively common rock formations as have thus far been condemned in New England. The promise of high awards insures that this approach will continue to be used. While it is very difficult to remove the element of speculation from any present-worth calculation, the situations which offer most promise are long-term royalty arrangements with minimum guaranteed annual returns regardless of pro- duction. The royalty receiver (owner) can be under a long-term arrangement with a reputable producer that essentially removes such uncertainties as labor costs, market prices, etc. However, the span of time cannot be very long for such terms, nor will capitalization of this type of income result in the very high figures obtain- able by assuming full production for long spans of time. Contracts of this type (several are now in existence and all apply to higher value base metal deposits) encourage methods of appraisal for condemnation that are more in line with business practice.

Finally, after the trial is over, the expert witness should obtain a transcript of the court record. This should serve to indicate any weaknesses in his replies to interrogation and should help to improve the nature of his testimony in future cases.

REFERENCES

Free, R. L. , 1957, Preparing a condemnation appraisal: School of Business Administration, Law School of the University of Pittsburgh.

Hoover, H. C. , 1909, Principles of mining, valuation, organization and administration: Hill Publishing Co. , New York.

Horgan, J. P. , 1959, Ten courtroom commandments for appraisers: Right of Way, October.

Hoskolk, H. D. , 1905, The engineer's valuing assistant: Longmans, Green and Co. , London.

Morkill, D. B. , 1918, Formulas for mine valuation: Mining and Scientific Press, v. 117.

Parks, R. D. , 1957, Examination and valuation of mineral property: 4th ed. , Addison-Wesley, Reading, Massachusetts.

Part II

SAMPLING, EXAMINING, AND ESTIMATING

Editor's Comments
on Papers 10 Through 19

The variogram and the variance approaches to the determination of spatial variation in geochemical sampling and statistical estimation of ore deposits were compared and contrasted by Miesch (Paper 10). Rendu noted that statistical decision theory could be used to quantify the merits of any exploration decision; emphasis was on the various applications of objective and subjective probabilities and classical and Bayesian interpretations of probability assessment (Paper 11). Such papers are fundamental in risk analysis and have been well received by those involved in mining-risk analysis and decision making. A fine example of sampling-prospecting case studies is Paper 12 in which Godwin and Sinclair used multiple regression analysis and models that related value measures to geological, geochemical, and geophysical variables in estimating grade potential.

Paper 13 is an epitome translated from Matheron's monumental *Traité de geóstatistique,* which is highly significant in the quest for accuracy in comprehensive grade and ore estimation. Although brief, Mickle's article is representative of the many good general works stressing the failings of traditional methods of ore-reserve estimation (Paper 14). He suggests such improved procedures as estimating of probable accuracy, providing a means for determining the number of samples required to evaluate a given deposit within specified accuracy ranges, and using a more reasonable assumption about continuity of grades between samples. Parker's well-presented paper with superb figures, tables, and formulas explains how the volume-variance relationship can be used in conjunction with kriging to determine grade and tonnage, planning, and suitability of mining method (Paper 15).

Papers 16 and 17 are good examples of the abundance of works considering specific metallic deposits. Krige's modern and classic paper applied the lognormal frequency curve to the observed distribution of borehole values in the Main Sector of the Orange Free State Goldfield. Trend-surface analysis was employed by Harris and Zodrow as an aid to the grading of magnetite by expressing the interrelationship of the statistical and geological aspects of a complex orebody in Newfoundland. Good papers dealing with reserve estimations in nonmetallic deposits (other than fuels) are not so abundant as those dealing with metallic deposits. Nemec's article (Paper 18) is a very innovative contribution and, because of inadequacy of traditional methods, gives a new approach by using space models based on moving averages, contouring, and regression analysis for quality control and estimation of cement raw materials; microblock, macroblock, and megablock systems concomitantly are employed for calculating reserves. There are many concise papers explaining the numerous methods of oil- and gas-reserve estimation, but one of the most resourceful and interesting is Paper 19 by Fertl and Vercellino, which is based on computer interpretation of log data and is concerned with variables that cannot easily be evaluated by charts.

123

10

Reprinted courtesy of the Geological Society of America from pages 333–340 of *Quantitative Studies in the Geological Sciences,* E. H. Whitten, ed., Geol. Soc. America Mem. 142, 1975

Variograms and Variance Components in Geochemistry and Ore Evaluation

A. T. MIESCH

U.S. Geological Survey
Denver, Colorado 80225

ABSTRACT

Investigations of spatial variability in ore deposits and other rock bodies have been approached through classical analysis of variance methods and through the theory of regionalized variables, as developed by the French school of geostatistics. The analysis of variance approach leads to estimates of variance components that can be used to form the variogram employed in geostatistics. The equivalence of the two approaches allows a possibly useful interchange of methods for sampling and statistical estimation.

INTRODUCTION

One of the most fundamental properties of a rock body is the type of chemical and mineralogic variability within it. Many volcanic units, for example, are notably uniform in chemical and mineralogic composition in both vertical and lateral directions. Granitic plutons tend to vary laterally in composition toward their contacts with older country rock, but vertical variations over the limited intervals that we commonly see are typically small. In sedimentary units, the stratigraphic variability is indicative of changes in the sedimentary environment or source area with time, whereas lateral variability reflects the configuration of the sedimentary basin.

Variability within a rock body is also important in sampling, and efficient sampling plans depend on a knowledge of the degree and type of variability present. A compositionally uniform volcanic unit or ore deposit, for example, can be adequately represented for most purposes by relatively few samples, whereas adequate representation of a more variable deposit can be much more difficult. It is well known that the efficiency of sampling can be significantly improved through a knowledge of the type of variability; the most familiar example is the common practice of sampling sedimentary units across the bedding, but significant improvement in sampling efficiency can also be achieved where directional variability, or other types of nonuniform variability, is far more subtle.

Although the degree and type of compositional variability in a rock body are always observed by geologists, they are only rarely measured in any rigorous way. Most computed standard deviations in the geochemical literature suffer from biases brought about by selective sampling, and only a few investigators have bothered to assess the directional and spatial properties of the variability in designing sampling programs. Some notable exceptions, which have involved estimation of variance components, are investigations by Youden and Mehlich (1937), Krumbein and Slack (1956), Baird and others (1967), and Connor and others (1972).

The practical advantages of employing a knowledge of the type of variability in a rock unit in sampling are well recognized in the mining industry, and interest in the subject has prompted intensive research during the past decade into mathematical methods of ore evaluation. The outstanding research appears to have been centered in France and led by Professor Georges Matheron of the École Nationale Supérieure des Mines de Paris. Matheron's work has led to a theory of regionalized variables. The theory is based on the concept of a variogram; an estimated variogram summarizes the information available on the type of spatial variation present in an ore deposit. My own knowledge of the theory comes principally from papers by Matheron (1963), Blais and Carlier (1968), David (1969, 1970), and Olea (1972). A comparison of the two approaches to the problem of assessing spatial variation, one using variograms and the other using variance components, has been a subject of interest, however, and I find that the approaches are fundamentally the same. The purpose of this paper is to support this contention in the hope of bringing proponents of the two approaches closer together.

VARIANCE COMPONENTS

The first application of classical analysis of variance methods to a problem in field geochemistry that I have been able to find was by Youden and Mehlich (1937), who investigated the variability in soil pH over two areas in New York and New Jersey. Their principal interest was in determining the type of variation present in order to design an efficient sampling program. This was done by estimating components of the total variance associated with various sampling intervals. Krumbein and Slack (1956) used a similar approach in the study of radioactivity in a thin shale bed that overlies Coal No. 6 throughout much of the Illinois Basin. They explained the general philosophy and model, the sampling scheme, and the computational procedures in sufficient detail to provide a basic reference for future work. According to the model they used, a radioactivity measurement (actually the square root of the measurement after transformation) is viewed as being determined by the grand mean for the entire area studied, plus a deviation (α) characteristic of the supertownship (a group of nine townships) from which the sample came, plus a deviation (β) characteristic of the township within the supertownship, plus a deviation (γ) characteristic of the mine within the township, plus a deviation (δ) characteristic of the sample from within the mine. Thus, in the regional phase of their study, the total variance was viewed as containing four components:

$$\sigma_x^2 = \sigma_\alpha^2 + \sigma_\beta^2 + \sigma_\gamma^2 + \sigma_\delta^2. \tag{1}$$

The first variance component (σ_α^2) is a function of the differences among supertownships; the second (σ_β^2), a function of differences among townships within supertownships; the third (σ_γ^2), a function of differences among mines within townships;

and the last (σ_δ^2) a function of differences among measurements on individual samples from within the same mine.

Krumbein and Slack (1956) also estimated the components of variance of radioactivity in the shale bed within a single mine and, as in the regional study, viewed the total variance (σ_m^2) as consisting of four components associated with varying sampling intervals. A fifth component consisted of variance arising from measurement errors. Thus,

$$\sigma_m^2 = \sigma_\epsilon^2 + \sigma_\theta^2 + \sigma_\lambda^2 + \sigma_P^2 + \sigma_\phi^2, \qquad (2)$$

where σ_ϵ^2 is the variance arising from differences among sections across the shale bed (called "grids") that are about 1.02 m thick and 3.05 m wide, σ_θ^2 is the variance among major units of the grids that are about 51 cm by 1.53 m, σ_λ^2 is the variance between minor units that are about 25 cm by 76 cm within the major units, and σ_P^2 is the variance among samples from within minor units. The final variance component (σ_ϕ^2) was estimated by making duplicate measurements of the radioactivity on all samples.

The estimated variance components are given in Table 1 along with the average spacings between the various sampling units. The estimated total variance, which includes variance on all local and regional scales as well as variance due to measurement error, is 0.1906. The estimated components indicate that if sampling were confined to any single supertownship, the total observed variance would tend to be less than this amount by only 1.8 percent. Similarly, if sampling were confined to any single township, only an additional 7.2 percent of the variance would be lost, and only an additional 6.2 percent would be lost if the sampling were confined to a single mine. Thus, most of the variance in radioactivity tends to be associated with small scales, and the variance within a single mine tends to be almost 85 percent of that found for the entire area, which is about 320

TABLE 1. AVERAGE SPACINGS BETWEEN SAMPLING LOCATIONS AND ESTIMATES OF VARIANCE COMPONENTS FOR RADIOACTIVITY IN A SHALE BED IN SOUTHERN ILLINOIS*

Source of variation	Average spacing	Estimated variance component	Estimated variance component (% of total)
Regional study			
Between supertownships	..	0.0035	1.8
Between townships within super-townships	15.0 km (9.3 mi)	0.0137	7.2
Between mines within townships	5.0 km (3.1 mi)	0.0119	6.2
Between samples within mines	0.8 km (0.5 mi)	0.1615	84.7
Total		0.1906	99.9
Local study			
Between grids within a mine	1.13 km (0.7 mi)	0.0000[†]	0.0
Between major units within grids	1.40 m (4.6 ft)	0.0106	8.9
Between minor units within major units	0.67 m (2.2 ft)	0.0561	47.3
Between samples within minor units	0.15 m (0.5 ft)	0.0482	40.6
Between replicates within samples	..	0.0038	3.2
Total		0.1187	100.0

*From Krumbein and Slack (1956).
[†]Computed estimate is negative, but is set equal to zero as is conventional.

km long. Within the single mine, about 47 percent of the total variance tends to be associated with sampling intervals of 15 to 67 cm, and another 40 percent with intervals of less than 15 cm. Only about 3 percent of the total variance in radioactivity arises from measurement error.

The purpose in reviewing the experimental results of Krumbein and Slack here is to point out, prior to the following discussion of variograms, that application of the classical analysis of variance approach leads to estimation of the variance and the percent of the total variance that is associated with various sampling intervals. The estimates are used, as shown by Krumbein and Slack (1956, p. 749-753), both to improve efficiency in sampling and, therefore, statistical estimation, and to assess the relative importance of geologic processes that act over local and regional scales.

VARIOGRAMS

Although the theory of regionalized variables is only about ten years old, its application to problems of ore evaluation, at least, appears to be further developed than the classical techniques of analysis of variance. Analysis of variance, on the other hand, has received more application in geochemical investigations not directly related to economic matters. Application of the theory of regionalized variables begins with sampling at regular intervals, if possible, down a drill hole, along a traverse or a grid, or within some three-dimensional network. The sample data are then used to construct the estimate of a variogram, which is a curve representing the degree of continuity of mineralization within an ore body (Matheron, 1963, p. 1250). The variogram is defined by

$$\gamma(h) = \frac{1}{2V} \int \int \int_V [f(M + h) - f(M)]^2 \, dV, \tag{3}$$

where $\gamma(h)$ is the value of the variogram (the ordinate) for sampling interval h (the abscissa). The triple integration is over the volume of the deposit, V, and $f(M + h)$ and $f(M)$ are values of the compositional variable at two points separated by the interval h. Variograms are similarly defined for areas and for traverses, with corresponding decreases in the number of integrals. In practice, most variograms are initially estimated for drill holes or traverses across the deposit in various directions in a search for any anisotropy that may be present. The variograms are then averaged as found to be appropriate.

The variogram for a traverse is estimated by

$$\gamma(h) = \frac{1}{2n} \sum_i (X_{i+h} - X_i)^2, \tag{4}$$

where n is the number of pairs of measurements (X_{i+h} and X_i) made at points separated by the interval h. Equation (4) is identical to one commonly used by chemists and others to estimate the analytical variance from duplicate analytical measurements on n specimens (Youden, 1951, p. 17) and is obviously a measure of variance, as Matheron (1963) pointed out.

Ore deposits that display completely random or chaotic compositional variation would be represented by variograms that are generally parallel to the abscissa over the entire range of h; that is, the value of $\gamma(h)$ would be the same for each possible sampling interval. Thus, a given number of samples from any small region are just as representative of the deposit as the same number of samples from widely separated points throughout the deposit. More commonly, the variogram increases with increasing h, thereby indicating that the differences are greater for samples taken far apart than for those collected close together. However, most variograms show a plateau or flattening beyond some value of h, indicating that a maximum variance has been attained. Large values of $\gamma(h)$ at small sampling intervals (for example, $h = a$) are said to reflect a nugget effect; that is, samples collected only "a" units apart tend to differ greatly, reflecting an erratic distribution of the ore on a small scale.

Typically, $\gamma(h)$ is based on a large number of sample pairs where the interval h is small and on many fewer pairs where h is large. Because of this, the form of the variogram is generally erratic at large values of h and may contain some negative slopes between points where it has been estimated. The negative slope is caused by the same properties of the data that lead to negative estimates of variance components in classical analysis of variance.

Following the estimation of the variogram, the theory of regionalized variables calls for the fitting of one of several types of models that have been developed (Blais and Carlier, 1968). Selection of the model is not entirely objective, and considerable experience in the theory and its application appears to be required. After the estimate of the variogram has been obtained and an appropriate model has been fitted, the parameters of the model can be used to determine weighting factors for computing moving averages, with confidence intervals, for the various ore blocks throughout the deposit. The method is referred to as *kriging*, and the confidence intervals are determined from the *kriging* variance (David, 1970). The models are also used to estimate the volume of a deposit, as well as its overall value and the associated confidence intervals.

COMPARISON OF METHODS

Proponents of the classical analysis of variance approach and of the variogram approach to the determination of spatial variation in rock bodies view the total variance to exist at a range of scales. That is, any rock body may exhibit variation on a regional scale, a very local scale, or any scale in between. The difference between the two approaches is merely one of variance measurement. If variance components are estimated for the sampling intervals h and $2h$, the component σ_h^2 is a measure of all the variance at scales less than h, and σ_{2h}^2 is a measure of all the variance on scales between h and $2h$. If a variogram is estimated at the points h and $2h$, the value of $\gamma(h)$ is a measure of all the variance at scales less than h ($\gamma(h) = \sigma_h^2$), but the value of $\gamma(2h)$ is a measure of all variance at scales less than $2h$ ($\gamma(2h) = \sigma_h^2 + \sigma_{2h}^2$).

The equivalence of the two approaches as described here is more or less intuitive; the notation becomes cumbersome when an attempt is made to demonstrate the equivalence mathematically. The equivalence might be verified if we were able to compute both variance components and a variogram for the same set of real data, but the methods are normally based on different sampling plans, and no single set of data is satisfactory for both. However, it is possible to generate some hypothetical data for a sampling plan that is suitable for both approaches. The data are given and the sampling plan is illustrated in Figure 1. Computed values of the variogram, using equation (4), and of the variance components, using the procedure given in detail by Krumbein and Slack (1956, p. 754), are given

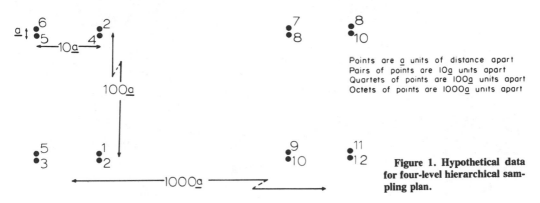

Figure 1. Hypothetical data for four-level hierarchical sampling plan.

in Table 2. The final column of Table 2 shows the cumulative sums of the variance components, which are exactly equal to the corresponding values of the variogram.

It is possible now to construct variograms for radioactivity in the shale bed investigated by Krumbein and Slack from the variance components given in Table 1. The variograms are shown in Figure 2, but inexperience in the theory of regionalized variables prohibits any attempt to fit a model or to carry the procedure further. It may suffice for the present to point out that the flatness of the regional variogram is in accord with Krumbein and Slack's finding that most of the variation is local.

CONCLUDING REMARKS

The apparent equivalence of variograms and cumulative sums of variance components estimated by use of the classical analysis of variance methods should be of interest and benefit to the proponents of both approaches. At least one limitation of the variogram approach is the requirement that observations be made at intervals that are at least approximately equal over the ore deposit or rock body. This is possible in much exploration drilling if the requirement is known beforehand, but can present a difficulty where the variogram is to be constructed from previous data. Hierarchical sampling plans of the type used for the analysis-of-variance approach, on the other hand, may be easy to simulate using previous data and might be used in situations where regular spacing of drill holes or sampling localities is not possible. The hierarchical sampling designs used by Youden and

TABLE 2. VARIOGRAM VALUES AND VARIANCE COMPONENTS*

Sampling interval (h)	Variogram $\gamma(h)$[†]	Variance component (σ_h^2)[§]	Cumulative sum of variance components
a	1.06250	1.06250	1.06250
10a	2.87500	1.81250	2.87500
100a	3.53125	0.65625	3.53125
1,000a	19.87500	16.34375	19.87500

*Computed from data of Figure 1.
[†] Values of $\gamma(h)$ based on comparison of 8, 16, 32, and 64 pairs of values, respectively.
[§] Variance components represent, respectively, differences between points within pairs (8 d.f.), between pairs within quartets (4 d.f.), between quartets within octets (2 d.f.), and between octets (1 d.f.).

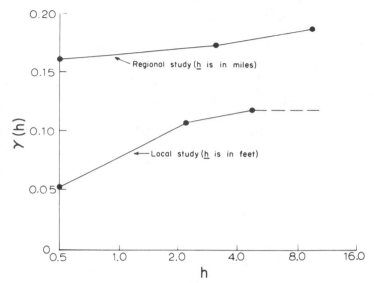

Figure 2. Variograms for radioactivity in shale bed overlying Coal No. 6 in Illinois constructed from variance component estimates of Krumbein and Slack (1956) given in Table 1.

Mehlich (1937) are particularly adaptable to some situations where regular-interval sampling cannot be used, such as in areas where outcrops are sparse and irregularly distributed. If the variograms constructed on the basis of hierarchical sampling plans are satisfactory, the possibility of using variograms can be extended to many investigations where they are currently impractical.

The theory of regionalized variables appears to offer a great deal to those who, heretofore, have approached investigations of spatial variability solely through conventional analysis-of-variance methods. The variance components may be more easily interpreted if cast into a variogram or some similar graphical device. Where variograms are used, it may be possible to employ all or at least some of the theory that has been developed for interpreting them and to derive soundly based confidence intervals for geochemical abundance estimates and for estimates of the mean values that are commonly contoured or otherwise displayed on geochemical maps. Application of the theory of regionalized variables to contouring procedures was shown by Olea (1972).

REFERENCES CITED

Baird, A. K., McIntyre, D. B., and Welday, E. E., 1967, Geochemical and structural studies in batholithic rocks of southern California: Pt. II, Sampling of the Rattlesnake Mountain pluton for chemical composition, variability, and trend analysis: Geol. Soc. America Bull., v. 78, p. 191-222.

Blais, R. A., and Carlier, P. A., 1968, Applications of geostatistics in ore evaluation: Canadian Inst. Mining and Metallurgy, Spec. Vol. 9, p. 41-68.

Connor, J. J., Feder, G. L., Erdman, J. A., and Tidball, R. R., 1972, Environmental geochemistry in Missouri—A multidisciplinary study: Internat. Geol. Cong., 24th, Montreal 1972, Symp. 1, p. 7-14.

David, M., 1969, The notion of "extension variance" and its application to the grade estimation of stratiform deposits, *in* Weiss, Alfred, ed., A decade of digital computing in the mineral industry: New York, Am. Inst. Mining, Metallurgy, and Petroleum Engineers, p. 63-81.

——1970, Geostatistical ore estimation—A step-by-step case study: Canadian Inst. Mining and Metallurgy, Spec. Vol. 12, p. 185-191.

Krumbein, W. C., and Slack, H. A., 1956, Statistical analysis of low-level radioactivity of Pennsylvanian black fissile shale in Illinois: Geol. Soc. America Bull., v. 67, p. 739–762.

Matheron, G., 1963, Principles of geostatistics: Econ. Geology, v. 58, p. 1246–1266.

Olea, R. A., 1972, Application of regionalized variable theory to automatic contouring: Univ. Kansas Center for Research, Inc., Spec. Rept. to Am. Petroleum Inst., Research Proj. 131, 191 p.

Youden, W. J., 1951, Statistical methods for chemists: New York, John Wiley & Sons, Inc., 126 p.

Youden, W. J., and Mehlich, A., 1937, Selection of efficient methods for soil sampling: Yonkers, N.Y., Contr. Boyce Thompson Inst., v. 9, p. 59–70.

11

Reprinted from pages 435–445 of *Advanced Geostatistics in the Mining Industry,*
M. Guarascio, M. David, and C. Huijbregts, eds., D. Reidel, Dordrecht, Holland, 1976

BAYESIAN DECISION THEORY APPLIED TO MINERAL EXPLORATION
AND MINE VALUATION

J.M. Rendu

Head Operations Research Section
Anglo-Transvaal Investment Company, Johannesburg
South Africa

ABSTRACT. The decision process in mineral exploration and mine
valuation is analysed. Statistical decision theory is a powerful
tool which can be used to quantify the value of any exploration
decision. A short review of this theory is given emphasizing the
various applications of subjective and objective probabilities,
and the difference between Bayesian and classical approaches to
probability assessment. Indications are given on how the concepts
in this theory can be progressively introduced in an exploration
company to improve any existing decision process.

1. INTRODUCTION

The aim of exploration is to prove the presence and estimate the
value of economic mineral deposits. Exploration is a multistage
process the stages of which can be defined as follows:

Stage 1: Exploration feasibility planning
Stage 2: Regional exploration, or reconnaissance
Stage 3: Follow-up exploration
Stage 4: Detailed investigation
Stage 5: Ore body valuation

At the beginning of each stage, a decision must be made,
whether or not to continue exploration, and in the affirmative,
which exploration technique should be used. During each stage
of exploration, information is obtained concerning the presence
or economic value of mineralizations. Once a stage of exploration
has been completed, this information must be taken into consideration to make a decision concerning the following stage. The last

decision to be made is whether or not to open a mine. The factors
which will influence the decisions made are:
- The probability that there are mineral deposits in the
  geographical area explored, the probability that these
  deposits will be discovered, and the probability that they
  will be of economic value.
- The economic value of the deposits to be discovered assuming
  that they exist.
- The cost of discovering these deposits, ie. the cost of
  exploration.

We are faced with a typical multistage problem of decision
under uncertainty, where the "State of the World", ie. the geology
in the area considered and the economic conditions at the time of
exploration and mining, is unknown, but more information about it
can be obtained at a cost. The uncertainty about the State of the
World is due to the lack of knowledge of polytico-economic and
geological variables. These two sets of variables are not
independent; a mineralization will be of economic value provided
the polytico-economic conditions are right but conversely a
profitable operation might affect the polytico-economic climate.
In this analysis, the only uncertainty considered is the geological
uncertainty, or uncertainty about the "State of Nature." The
polytico-economic uncertainty could however, be included and treated using the same techniques.

2.   THE BAYESIAN STATISTICAL DECISION THEORY

For obvious reasons, all the models which have been proposed in
the literature, to find a mathematical solution to the problem of
decision making in mineral exploration, have made use of the
statistical decision theory. Both Bayesian and non-Bayesian
("classical") solutions to the problem can be found in the
literature. A brief review of the theory is given here. An
excellent introduction to statistical decision theory has been
published by Raiffa (1968). More mathematically oriented
decision makers can refer to Raiffa and Schlaiffer (1961). A
general description of how this method can be used in exploration
is given by Davis, Kisiel and Duckstein (1973).

Statistical decision theory is used for the making and
analysis of decisions where the factors affecting the decisions
are uncertain. Let Θ be a possible state of nature eg. presence
of an orebody of economic value. Before exploration, we do not
know whether Θ exists, but we have some idea of whether or not it
exists. We can define the a - priori or prior probability that
Θ exists: $P'(\Theta)$. Let's consider the feasibility of an exploration
project e, for example a geochemical survey of the area. If e is
completed, an observation z will be made, e.g. geochemical

anomalies will be observed. Given z, we could reestimate the probability distribution of θ and obtain an a - posteriori, or posterior distribution of θ given z, $P''(\theta/z)$. But x has not yet been observed. To make the decision whether or not to choose e we must also estimate the chances that z will be observed, ie. the probability $P(z)$.

To use the statistical decision theory, it is necessary to be able to estimate the three types of probabilities mentioned. These probabilities are not independent but are related as follows:

$$P'(\theta) = \sum_z P''(\theta/z) \; P(z) \tag{1}$$

If $P''(\theta/z)$ and $P(z)$ are estimated directly and $P'(\theta)$ is deduced using the above formula, the approach to decision making is classical, or non-Bayesian. Let's define $P(z/\theta)$ as the probability that z will be observed given that the state of nature is θ. For example $P(z/\theta)$ is the probability that a geochemical anomaly z will be observed, assuming the presence of an economic orebody θ. The following relations between probabilities exist:

$$P(z) \quad = \sum_\theta P(z/\theta) \; P'(\theta) \tag{2}$$

$$P''(\theta/z) = \frac{P(z/\theta) \; P'(\theta)}{P(z)} \tag{3}$$

Equation (3) is known as Bayes Theorem. If $P'(\theta)$ and $P(z/\theta)$ are estimated directly and $P(z)$ and $P''(\theta/z)$ are deduced using equations (2) and (3) the approach to decision making is known as Bayesian.

3. THE DECISION TREE

All exploration decision problems are composed of at least two stages. In the first stage, or exploration stage, an exploration decision e must be made. In the last stage, or mining stage, a mining decision m must be chosen. The exploration decision e can be chosen among a set of possible decisions. For example, one might have to decide between not doing any exploration ($e=e_0$), or choosing any of two exploration methods, $e=e_1$ or $e=e_2$. The result of exploration will be an observation z function of the decision e chosen and of the unknown state of nature θ. The mining decision can also be chosen among different options. One may decide to give-up the venture ($m=m_0$) or to open a large low-grade mine ($m=m_1$) or a selective operation ($m=m_2$). The validity of the choice made will be a function of the state of nature θ.

The logical relation between the different possible decisions and observations can be represented on a decision tree (Figure 1).

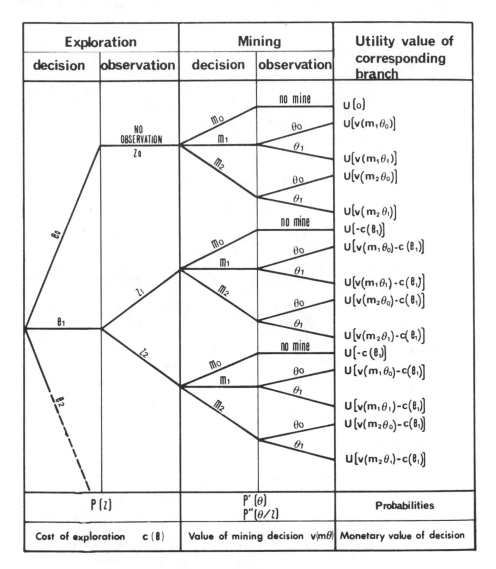

Figure 1;   Simplified decision tree

4.   UTILITY DEFINITION

To choose between two exploration decisions, say e_1 and e_2, we must compare the cost of completing exploration according to the decision chosen, with the value of the information we might obtain from exploration. The cost of exploration can be estimated with reasonable precision, but the value of the information obtained

135

is much more difficult to quantify. A geochemical anomaly has
no value per se : its value comes from the fact that its presence
might be due to an underlying economic mineralization. The value
of an information can be measured by the price at which the
company who owns this information would be willing to sell it.
This price should have nothing to do with the amount of money
spent to get the information but should be a function of the value
of any mine which might be discovered using the information, and
of the chances of discovering such a mine. Calculating the
monetary value of a potential mine assuming a known state of
nature is a classical mine valuation problem. Detailed calculation
must be made during the later stages of exploration, but simplifying assumptions can be made when only very little information
is available about the state of nature. This point is illustrated
by Cooper, Davidson and Reim (1973).

However, a monetary unit is not necessarily a valid unit to
use for comparison of options in the presence of uncertainty. To
illustrate this point, consider the following example. Two
options a_0 and a_1 are available to you. You can choose a_0, do
nothing and gain or lose nothing, or choose a_1 which gives you a
chance to gain or lose $1 000 000. If your assets are valued at
$100 000 000 and there is at least a 50/50 chance that action a_1
will result in a success, you might choose a_1. In other words
you might consider that a 50/50 chance to lose or win $1 000 000
is equivalent to the certainty of a zero gain. But if your assets
are only $1 000 000 or less, you would never choose a_1 unless
there was no chance whatsoever that you might lose. The same
opportunity to play the game a_1 does not have the same value,
depending on the assets of the gambler. Also two gamblers
having the same assets will give a different value to the same
opportunity because of their different attitude towards risk.

When dealing with mining exploration problems, it is therefore necessary to define a "utility function" which represents
the policy of the exploration company towards risk. How to
estimate a utility function has been described by Raiffa (1968),
and examples of utility functions of petroleum exploration
companies have been published by De Geoffroy and Wignall (1970)
and Harbaugh (1973). An example follows which illustrates how
a utility function can be calculated.

Let's assume that a company has assets of $1 000 000 and we
wish to calculate the utility curve of the company. We know that
the utility of $ - 1 000 000 must be infinitely negative. Let's
set arbitrarily the utility of $0 equal to zero and the utility
of $500 000 equal to 1. To calculate the utility of any other
dollar amount, three types of questions must be asked of the
management.

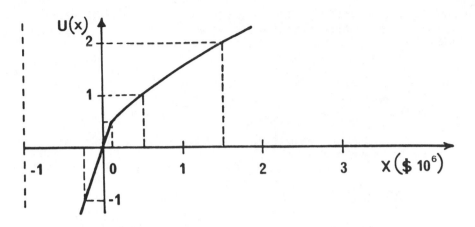

Figure 2;   **Example of utility curve**

Question 1. The management must choose between two options. Either to toss a coin and gain \$500 000 or \$0, depending on whether head or tail is observed, or to take a fixed sum Vo. For which value of Vo would the management consider the two alternatives as equivalent? Let's assume that the answer is Vo = \$150 000. We deduce:

U (150 000) = ½ U (500 000) + ½ U (o) = 0,5

Question 2. The choice is now between a 50/50 chance to gain \$0 or \$$V_1$ and the certainty to gain \$500 000. Let's assume that the first alternative is preferred if V_1 is at least equal to \$1 500 000. We deduce:

U (500 000) = ½ U (o) + ½ U (1 500 000)

hence U (1 500 000) = 2

Question 3. The management must choose between a 50/50 chance to gain \$500 000 or \$$V_2$, and a certainty to gain \$0. Let's assume that the second alternative is preferred as soon as V_2 is smaller than \$- 250 000:

U (o) = ½ U (500 000) + ½ U (-250 000)

hence U (-250 000) = -1

The relationship between utility and monetary values is plotted on figure 2. Repeating the same questions as above, but for values other than \$0 and \$500 000, we can obtain any number

of points on the utility curve. Utilities should be used in the
early stages of exploration, but monetary values give acceptable
results in the later stages. The utility must also be used
whenever the gains or losses cannot be measured in monetary terms
only.

5. PROBABILITY ASSESSMENT

We saw earlier that to decide the value of a project, it is
necessary to be able to estimate:
- $P'(\Theta)$, the prior probability that a given state of nature
  Θ is present in the area considered.
- $P(z)$, the probability that the exploration project e
  will result in an observation z.
- $P''(\Theta/z)$, the posterior probability that Θ prevails,
  given that z has been observed. Note that $P''(\Theta/z)$ must
  be estimated before exploration has taken place.

To estimate these probabilities, we can use the classical
approach, estimate $P(z)$ and $P''(\Theta/z)$ directly, and calculate
$P'(\Theta)$ using equation 1 above. Or we can use the Bayesian
approach, estimate $P'(\Theta)$ and $P(z/\Theta)$ directly and calculate
$P''(\Theta/z)$ and $P(z)$ using equations 2 and 3 above. Both approaches
have been used with success. Also the probabilities can be
estimated subjectively by asking the geologist to quantify his
opinion about the chances of success of a project, or objectively
by measurements made in similar areas already explored.

As soon as enough samples have been taken from an orebody
so that a geostatistical model can be built, a non-Bayesian
objective approach to probability assessment can be used, as
illustrated by Matheron and Formery (1963) and Rendu (1971-a).
In some circumstances, subjective probabilities and/or Bayesian
approach could be considered (Krige, 1961; Rendu 1971-b).

In the early stages of exploration, it is also possible to
use both subjective and objective probabilities. The methods
of probability assessment can be either classical or Bayesian. A
non-Bayesian approach to the estimation of objective probabilities
for reconnaissance in large areas has been proposed in the
original paper by Alais (1957). The method used consists in
defining control areas which have been well explored and whose
metal endowment is well known. The necessary probabilities can
be calculated in these areas. The assumption is made that the
same probabilities apply to the unexplored area considered. This
methodology is also followed by Harris (1965) and Agterberg (1971).
Let Θ be the state of nature and z some information which is
available in the control area and could be or has been made
available in the study area using exploration. Using a non-

Bayesian objective approach the probabilities $P'(\Theta)$, $P''(\Theta/z)$ and $P(z)$ are calculated in the control area.

A Bayesian approach to assessment of objective probabilities has been preferred by De Geoffroy and Wignall (1970). In this analysis $P'(\Theta)$ and $P''(\Theta/z)$ are calculated directly from the control area.

Subjective posterior probabilities have been estimated by Harris, Freyman and Barry (1971). The probabilities $P''(\Theta/z)$ are calculated directly by asking geologists to quantify their opinion about the state of nature Θ taking into consideration their experience and knowlege z of the area considered. If subjective probabilities are used the notion of control area may disappear. A similar, but extremely simplified approach to the same problem is given by Cooper, Davidson and Reim (1973) to help individual geologists in their daily decisions.

A thorough analysis of how subjective probabilities of the type $P(z/\Theta)$ and $P'(\Theta)$ can be calculated, and how the posterior probabilities $P''(\Theta/z)$ can be deduced using a Bayesian approach has been published by Harris and Brock (1973). The Bayesian approach for utilization of subjective probabilities is also illustrated by Fuda (1971) and King (1974).

6. OPTIMIZATION

The value of any exploration decision is a function of the cost of exploration, the value of any mine which might be discovered and the probability that such a mine will be discovered. We briefly mentioned how these quantities can be calculated separately. We must now see how they can be combined to give a final answer.

Let's consider the decision tree on figure 1. Each branch is studied successively. For example, let's assume that we choose e_1, observe z_1, then choose m_2 and observe Θ_o. The cost of exploration being $c(e_1)$ and the value of the mining decision being $v(m_2\Theta_o)$ we can give to this path a utility:

$$u(\ e_1\ z_1\ m_2\ \Theta_o) = U(v\ (m_2\ \Theta_o) - c\ (e_1)\ )$$

where the function U is given by figure 2. For each branch we can thus calculate a "terminal utility", as shown in the last column of figure 1. Moving towards the trunk of the decision tree we can calculate the expected utility of choosing m_2:

$$u(e_1\ z_1\ m_2) = \underset{\Theta}{E}\ u(e_1\ z_1\ m_2\ \Theta) = \underset{\Theta}{\sum}\ u(e_1\ z_1\ m_2\ \Theta)\ P''(\Theta/z_1)$$

If the decision is made to stop exploration ($e = e_o$) then the prior probability $P'(\Theta)$ must be substituted for the posterior probability $P''(\Theta/z)$.

Let's now define the utility of choosing e_1 and observing z_1. Once z_1 has been observed it is possible to maximize the utility by choosing the decision m which has the highest utility $u(e_1\ z_1\ m)$. Therefore the utility of observing z_1 is:

$$u(e_1\ z_1) = \max_m\ (u\ (e_1\ z_1\ m)\ )$$

The utility of choosing e_1 is given by the expected utility of $u(e_1\ z\ )$:

$$u(e_1) = \underset{z}{E}\ u\ (e_1\ z) = \sum_z u(e_1\ z)\ P\ (z)$$

We can now decide which exploration decision should be made. We should choose the exploration decision which has the highest utility $u(e)$. The utility value of the information we now have is given by:

$$u^* = \max_e\ u(e)$$

Reading u^* on the vertical axis of figure 2 we deduce its monetary equivalent. This is the price at which we should consider to sell the information we presently have to an eventual buyer.

7. CONCLUSION

Statistical analysis appears to be a logical tool for optimization of decisions in mineral exploration. However, geologists might find it difficult to quantify their opinion, decision makers might consider that a utility function gives an over-simplified representation of their preferences. Some aspects of the statistical decision theory, such as the use of a decision tree and the rigorous structurization of the decision process will be more easily accepted and can be used as a starting point for introduction of the theory in an exploration company.

Permission by Anglo-Transvaal Cons. Inv. Co. Limited for the publication of this paper is acknowledged.

REFERENCES

AGTERBERG, F.P. (1971) A probability index for detecting favourable geological environments, Decision Making in the Mineral Industry, Canadian Inst. Min. and Metall., Special Vol. 12, 82 - 91.

ALAIS, M.(1957) Methods of appraising economic prospects of mining exploration over large territories, Management Science, 3, 4, July 1957, 285 - 367.

COOPER, D.O., DAVIDSON, L.B., and REIM, K.M. (1973) Simplified financial and risk analysis for minerals exploration, Eleventh International Symposium on Computer Applications in the Minerals Industry, University of Arizona, J.R. Sturgul ed., B1 - B14.

DAVIS, D.R., KISIEL, C.C. and DUCKSTEIN, L. (1973) Bayesian methods for decision - making in mineral exploration and exploitation, Eleventh International Symposium on Computer Applications in the Minerals Industry, Unversity of Arizona, J.R. Sturgul ed., B55 - B67.

DE GEOFFROY, J. and WIGNALL, T.K. (1970) Application of statistical decision techniques to the selection of prospecting areas and drilling targets in regional exploration, Canadian Min. and Metall. Bulletin, August 1970, 893 - 899.

FUDA, G.F. (1971) The role of decision-making techniques in oil and gas exploitation and evaluation, Decision Making in the Mineral Industry, Canadian Inst. Min. and Metall., Special Vol. 12, 130 - 138.

HARBAUGH, J.W. (1973) The Kansas oil exploration (KOX) decision system, Eleventh International Symposium on Computer Applications in the Minerals Industry, University of Arizona, J.R. Sturgul ed., B15 - B54.

HARRIS, D.P. (1965) Multivariate statistical analysis - a decision tool for mineral exploration, Short Course and Symposium on Computers and Computer Applications in Mining and Exploration, University of Arizona, J.C. Dotson and W.C. Peters ed., C1 - C35.

HARRIS, D.P. and BROCK, T.N. (1973) A conceptual Bayesian geostatistical model for metal endowment: a model that accepts varying levels of geologic information with a case study, Eleventh International Symposium on Computer Applications in the Minerals Industry, University of Arizona, J.R. Sturgul ed., B113 - B180.

HARRIS, D.P., FREYMAN, A.J. and BARRY, G.S. (1971) A mineral resource appraisal of the Canadian northwest using subjective probabilities and geological opinion, Decision Making in the Mineral Industry, Canadian Inst. Min. and Metall., Spec. Vol. 12, 100 - 116.

KING, K.R. (1974) Petroleum exploration and Bayesian decision
  theory, Twelfth International Symposium on the Applications
  of Computers and Mathematics in the Minerals Industry,
  Colorado School of Mines, T.B. Johnson and D.W. Gentry ed.,
  D77 - D117.

KRIGE, D.G. (1961) Developments in the valuation of gold mining
  properties from borehole results, Seventh Commonwealth Min.
  and Metall. Congress, Johannesburg, April 1961, 20.

MATHERON, G. and FORMERY, P. (1963) Recherche d' optimum dans
  la reconnaissance et la mise en exploitation des gisements
  miniers, Annale des Mines, Mai 1963, 23 - 42, June 1963, 2-30.

RAIFFA, H. (1968) Decision analysis - introductory lectures on
  choices under uncertainty, Addison - Wesley, U.S.A., 309.

RAIFFA, H. and SCHLAIFFER, R. (1961) Applied statistical
  decision theory, Harvard University, U.S.A., 356.

RENDU, J.M. (1971 -a) Some applications of geostatistics to
  decision-making in exploration, Decision Making in the Mineral
  Industry, Canadian Inst. of Min. and Metall., Spec. Vol. 12,
  175 - 184.

RENDU, J.M. (1971 -b) Some applications of statistics to
  decision making in mineral exploration, unpublished Eng. Sc. D.
  thesis, Columbia University, New York, 273.

12

Copyright © 1979 by The Institution of Mining and Metallurgy

Reprinted from *Inst. Mining and Metallurgy Trans.* **88B**:B93–B106 (1979)

APPLICATION OF MULTIPLE REGRESSION ANALYSIS TO DRILL-TARGET SELECTION, CASINO PORPHYRY COPPER-MOLYBDENUM DEPOSIT, YUKON TERRITORY, CANADA

C. I. Godwin and A. J. Sinclair

Multiple regression analysis was examined as a quantitative method for the evaluation of large volumes of costly mineral exploration data with a view to selecting optimal drill targets for additional exploration. This technique was tested on the Casino porphyry copper—molybdenum deposit,[1] which is owned by Casino Silver Mines, Ltd.; it is in the Dawson Range (62°43′ N, 138°49′ W) about 300 km northwest of Whitehorse, Yukon Territory, Canada (Fig. 1). Casino was chosen for this study because intensive exploration on this property from 1968 to 1973 by Brameda Resources, Ltd., produced a large amount of quantitative or quantifiable data, including the following information: geological and alteration information from surface mapping and drill core logging,[1–2] geochemical analyses of soils and rocks;[4,5] geophysical surveys from ground surface and airborne;[4] and diamond and rotary drill-hole assay data (Fig. 2).[4]

Dependent and independent variables selected from these data are listed in Table 1. Drill-target selection at Casino to date has been based on subjective evaluation of exploration data, including visual projection of grade trends determined from completed drill-holes. Drilling to 1970[6] outlined an ore reserve, based on preliminary open-pit design with a waste : ore ratio of 1.67 : 1.0, of 162 000 000 t grading 0.37% copper and 0.039% molybdenite. Because Casino is a porphyry-type deposit, changes in grade can be expected to follow variations in geological, geophysical and geochemical parameters systematically.[7–11] Other significant features of the Casino deposit include an association of copper minerals with the latest Cretaceous sub-volcanic Casino complex that intruded the mid-Cretaceous Klotassin batholith,[12] supergene depleted and enriched zones that were preserved because the deposit area was unglaciated in the Pleistocene[13–16] and a primary zoning of ore and alteration minerals in a manner similar to that of many other porphyry deposits.[9]

Analysis

The method of analysis was similar to that of Sinclair and Woodsworth[17] and Kelly and Sheriff,[18] but was applied at a more detailed scale. Sinclair and Woodsworth[17] studied value estimates of 128 cells 4 miles square and developed multiple-regression models based on sets of two different control areas (for example, total dollar values of all deposits in a cell were equated to a number of geological parameters). Cells of this size are, however, too large for defining drill targets.

The work was done to evaluate the usefulness of multiple-regression analysis in forecasting, from commonly available exploration data, the grade of

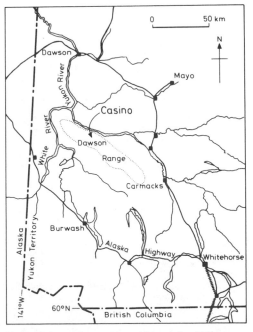

Fig. 1   Casino porphyry copper—molybdenum deposit, Yukon Territory, Canada

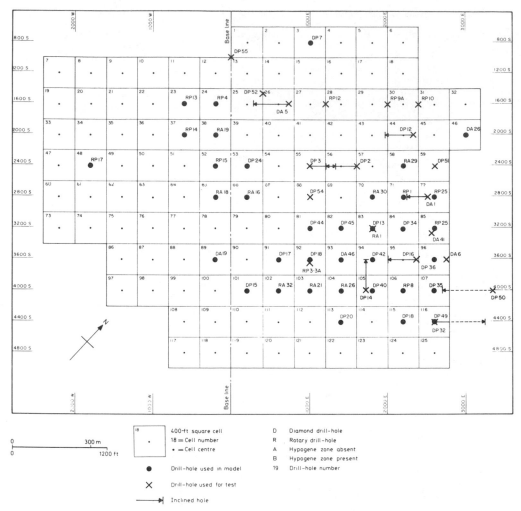

Fig. 2   Location (by property coordinates) of drill-holes

mineralization that might be intersected in a drill-hole centred on any cell within the grid that covered the Casino property. This grid covered the best exposed and most comprehensively studied part of Patton Hill, the geographical centre of exploration activity on the Casino deposit (Fig. 2). The 400-ft cell size was chosen to conform with the existing drill-hole pattern and property grid. The grid was positioned so that as many vertical drill-holes as possible coincided with the centre points of the cells (Fig. 2); these cells could then be used as a control on which the statistical models could be based.

Backward stepwise multiple-regression analysis[19] was used to build statistical models to predict the change in the mean of a particular dependent variable as independent variables changed. Dependent and independent variables are shown in Table 1. This approach results (Table 2) in equations of the form

$$Y = A_0 + A_1 X_1 + A_2 X_2 + \ldots e$$

and specifies statistics for each model derived; the most significant one can then be selected. Backwards stepwise regression has been performed with part of a 'triangular regression package' available as library software at the Computing Centre, University of British Columbia.[19] A normal multiple-regression equation is obtained from all potential independent variables. Standard errors of all regression coefficients are estimated and are used to evaluate the significance of the coefficients. An F-ratio is calculated as follows.

$$F_1 (1, M-m-1) = \frac{A_i}{S_i}$$

where A_i is the estimated ith coefficient, s_i is the standard error of the ith coefficient, M is the overall degree of freedom of the data set and m is the number of independent variables. This calculated F-ratio can be compared with the appropriate F distribution to determine the probability of a value A_i resulting if the true value

144

Table 1 Cell variables and transformations for regression analysis

Dependent variables			Independent variables		
Symbol	Name	Transform	Symbol	Name	Transform
EZCu	Enrichment zone, Cu%	Log 10	**Geochemical variables**		
			Cu	Copper, cell average, ppm	Log 10
EZMo	Enrichment zone, Mo%	Log 10	Mo	Molybdenum, cell average, ppm	Log 10
			Pb	Lead, cell average, ppm	Log 10
HZCu	Hypogene zone, Cu%	Log 10	Zn	Zinc from cell centre, ppm	Log 10
HZMo	Hypogene zone, Mo%	Log 10	**Geophysical variables**		
			GMA	Ground magnetic average cell value, gammas	Log 10
			GMR	Ground magnetic range in cell, gammas	Log 10
			RA	Resistivity average cell value, Ω m	Log 10
			RR	Resistivity range in cell, Ω m	Log 10
			DGM	Distance from ground magnetic high to cell centre, ft	None
			DAM	Distance from main airborne magnetic low to cell centre, ft	Log 10
			DSM	Distance from a secondary airborne magnetic low to cell centre, ft	None
			AM	Airborne magnetic value at cell centre, gammas	Log 10
			AMR	Airborne magnetic range in cell, gammas	Log 10
			Lithology		
			AQ	Area of cell that consists of inequigranular quartz monzonite, %	Arcsin
			AD	Area of cell that consists of Patton dacite, %	Arcsin
			AB	Area of cell that consists of breccia, %	Arcsin
			Hydrothermal alteration		
			AAP	Area of cell that consists of phyllic alteration, %	Arcsin
			AAK	Area of cell that consists of potassic alteration, %	Arcsin
			AHM	Area of cell that contains visible hematite and/or magnetite, %	Arcsin
			AHMT	Area of cell that contains both visible hematite and/or magnetite and visible tourmaline, %	Arcsin

is zero. If the probability is high, A_i is not significantly different from zero, and *vice versa*. The variable whose coefficient provides the lowest probability of being different from zero is dropped from the calculation and a new multiple-regression equation is calculated. This procedure is repeated until only those variables remain that have coefficients different from zero at some predetermined level of significance. $\alpha = 0.05$ was used in this work. Regression models are estimated from control data obtained for those cells with the best drill-hole assay data (Table 1 and Fig. 2).

Quantitative variables
Several dependent variables (value measures of the cells) were chosen for investigation; they are expressed as grade %. Because different geological factors were responsible for copper and molybdenite concentrations in the supergene enrichment and the hypogene zones, separate

Table 2 Regression models for enrichment zone, log 10 Cu% and Mo%*

Dependent variable (Y), log 10†	EZCu Step 10‡		EZCu Step 12‡		EZMo Step 8‡	
Independent variable	Symbol	Coefficient	Symbol	Coefficient	Symbol	Coefficient
Constant	Constant	−21.7616	Constant	−24.8834	Constant	−15.9918
X_1	Cu	0.4726	Cu	0.4919	Cu	0.4237
X_2	Pb	−0.1280	GMA	−0.6516	Pb	−0.5037
X_3	GMA	−1.0501	RA	0.6299	GMA	−1.5063
X_4	GMR	0.2534	RR	−0.1999	GMR	0.1999
X_5	RA	0.5798	DGM	0.4760D-03	RA	0.4623
X_6	RR	−0.1498	AM	9.0239	DGM	0.4817D-03
X_7	DGM	0.5566D-03	DAM	−0.8186	AM	8.3368
X_8	AM	8.3155	AD	0.1596	AMR	−1.3412
X_9	DAM	−0.9639	AHMT	−0.2161	DAM	−1.2735
X_{10}	AD	0.1573	−	−	DSM	− .9653D-03
X_{11}	AHMT	−0.1929	−	−	AQ	−0.4473
X_{12}	−	−	−	−	AD	−0.1010
X_{13}	−	−	−	−	AHM	−0.6316
R squared	0.9440		0.9143		0.9882	
Standard error of Y	0.0983		0.1132		0.0865	

*Control data for model from cells in Fig. 2 with drill-holes identified by large dots. Abbreviations explained in Table 1.
†Dependent variable (Log 10) = Y = Constant + $A_1 X_1$ + $A_2 X_2$ + . . . c
‡Step number selected from backwards stepwise regression analysis.
R, multiple correlation coefficient.

Table 3   Comparison of observed values from control data* with calculated values from regression models†

Cell	Hole	EZCu			EZMo		HZCu			HZMo	
		Observed	Calculated Step 10	Calculated Step 12	Observed	Calculated Step 8	Observed	Calculated Step 17	Calculated Step 20	Observed	Calculated Step 10
3	DP7	0.227	0.267	0.283	0.011	0.012	0.12	0.153	0.118	0.021	0.025
23	RP13	0.140	0.193	0.189	0.007	0.009	0.08	0.075	0.070	0.011	0.009
24	RP4	0.329	0.255	0.278	0.042	0.032	0.06	0.130	0.085	0.005	0.006
37	RP14	0.170	0.178	0.189	0.006	0.005	0.10	0.104	0.063	0.010	0.010
38	RA19	0.110	0.108	0.125	0.005	0.006	–	–	–	–	–
46	DA26	0.390	0.147	0.121	0.012	0.012	–	–	–	–	–
48	RP17	0.020	0.020	0.022	0.001	0.001	0.02	0.025	0.023	0.001	0.001
52	RP15	0.205	0.139	0.135	0.013	0.018	0.13	0.156	0.106	0.013	0.013
53	DP24	0.220	0.266	0.255	0.037	0.037	0.10	0.102	0.132	0.019	0.017
58	RA29	0.230	0.404	0.450	0.026	0.033	–	–	–	–	–
65	RA18	0.242	0.193	0.153	0.021	0.020	–	–	–	–	–
00	RA10	0.100	0.102	0.104	0.010	0.004	–	–	–	–	–
70	RA30	0.344	0.141	0.155	0.019	0.010	–	–	–	–	–
71	RP1	0.545	0.401	0.341	0.069	0.074	0.19	0.123	0.141	0.048	0.054
72	RP25	0.367	0.379	0.434	0.076	0.082	0.17	0.108	0.107	0.027	0.025
81	DP44	0.277	0.333	0.309	0.020	0.021	0.22	0.187	0.206	0.024	0.016
82	DP45	0.255	0.275	0.266	0.019	0.017	0.16	0.212	0.227	0.014	0.019
83	DP13	0.310	0.269	0.347	0.018	0.017	0.18	0.193	0.244	0.032	0.035
84	DP34	0.315	0.351	0.329	0.058	0.055	0.13	0.203	0.238	0.102	0.101
85	RP27	0.456	0.514	0.481	0.041	0.032	0.16	0.126	0.137	0.034	0.033
89	DA19	0.264	0.111	0.087	0.024	0.013	–	–	–	–	–
91	DP17	0.225	0.228	0.217	0.016	0.017	0.24	0.267	0.173	0.033	0.036
92	DP10	0.453	0.416	0.358	0.035	0.033	0.28	0.299	0.264	0.028	0.029
93	DA46	0.112	0.459	0.656	0.015	0.015	–	–	–	–	–
94	DP42	0.333	0.329	0.413	0.039	0.040	0.37	0.247	0.274	0.051	0.047
96	DP36	0.240	0.253	0.192	0.016	0.018	0.26	0.173	0.105	0.026	0.017
101	DP15	0.197	0.203	0.190	0.020	0.019	0.13	0.165	0.186	0.018	0.016
102	RA32	0.140	0.168	0.154	0.006	0.021	–	–	–	–	–
103	RA21	0.400	0.077	0.066	0.012	0.004	–	–	–	–	–
104	RA26	0.390	0.118	0.125	0.012	0.006	–	–	–	–	–
105	DP40	0.062	0.075	0.086	0.007	0.006	0.08	0.120	0.102	0.008	0.010
106	RP8	0.350	0.330	0.382	0.010	0.012	0.18	0.069	0.151	0.013	0.017
107	DP35	0.233	0.190	0.177	0.006	0.007	0.11	0.091	0.094	0.006	0.005
113	DP20	0.090	0.083	0.086	0.010	0.001	0.12	0.090	0.107	0.002	0.002
115	DP18	0.150	0.142	0.139	0.001	0.001	0.07	0.065	0.123	0.001	0.001
116	DP32	0.140	0.133	0.126	0.006	0.005	0.02	0.036	0.039	0.003	0.003
N	36	36	36	36	36	36	25	25	25	25	25
r	–	0.679	0.595	–	0.852		–	0.842	0.842	–	0.988

*Control data are from those cells in Fig. 2 with drill-holes identified by large dots.
‡Cells and drill-holes are shown on Fig. 2.
N, number of data points; r, simple correlation coefficient (based on 1n values) between observed and calculated values.

multivariate analyses were done for each category to avoid unnecessary complications because different causative sources of value measure were combined. The usefulness of combining molybdenite and copper to a single variable — the 'copper equivalent' — was investigated, but results were much poorer than those for individual elements and were not considered further. The dependent variables that were investigated were the percentages of Cu and Mo in the supergene enrichment and the hypogene zones (Tables 3 and 4). The actual values that were used to derive statistical models (Table 3 and Fig. 2) were arithmetic averages for each of 36 drill-holes in the case of supergene data and arithmetic averages for each of 25 drill-holes in the case of hypogene data. The use of arithmetic averages is justified by the symmetry of grade distributions that is apparent in histograms (not reproduced here). The thicknesses over which grades were averaged obviously varied from one hole to another, but commonly exceeded 200 ft. A few holes that penetrated only the more highly enriched upper portion of the enrichment zone could not be used because of a biased high average value.

A large number of independent geochemical, geophysical, lithological and hydrothermal alteration variables were determined initially for each of the 125 cells. Those of most use (Table 1) were selected by the avoidance of obviously redundant variables (for example, geochemical analyses of copper for both rock and soil samples), the omission of variables that were applicable to only a very few cells (for example, propylitic alteration occurred in only three cells and was, therefore, omitted), and the rejection of variables that showed no significant linear correlation with the dependent variables in preliminary correlation matrices (for example, gold geochemistry and induced polarization chargeability values).

In this way the 20 independent variables (a number acceptably less than the minimum of 25 control points) were selected (Table 1). From a geological viewpoint, each of these variables can be related directly, in a subjective manner, to characteristics that might influence metal values or

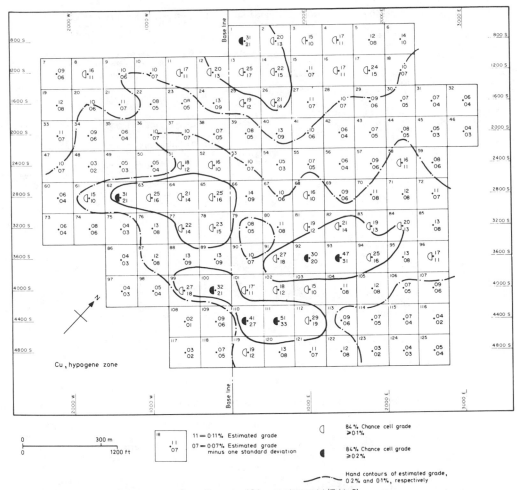

Fig. 3   Cu% in supergene enrichment zone estimated from step 10 in regression model (Table 2)

grade. Where these variables are inter-related by a regression equation their direct meaning is less obvious.

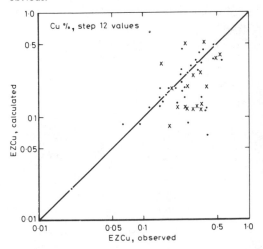

Fig. 4   Scatter diagram: calculated step 10 values (Table 2) *vs* observed values for Cu% in supergene enrichment zone. ●Values calculated by comparison of model with control data used in construction of model (Tables 3 and 6); x values calculated by comparison of model with values from test data (Tables 4 and 6). Straight line represents perfect correspondence of observed and calculated values

The geochemical variables (Table 1) were all analyses of rock chips that were obtained from rare outcrops or from residual overburden. Copper, molybdenum and lead values were averaged from two or more sites per cell. Zinc samples were taken only at cell centres.

The geophysical data (Table 1) included resistivity results from an induced-polarization survey and magnetic intensity measurements from both ground and airborne magnetometer surveys. The range of ground magnetometer readings in each cell was also recorded as a measure of local variability in magnetic response. The airborne magnetic value at the cell centre was recorded because it represented a general background value. The range of airborne magnetic values in a cell was calculated as a means of approximating, without directional information, one aspect of the first

Table 4   Comparison of observed values from test data* with calculated values from regression models†

Cell	Hole	Inclination, °	EZCu Observed	EZCu Calculated Step 10†	EZCu Calculated Step 12†	EZMo Observed	EZMo Calculated Step 8†	HZCu Observed	HZCu Calculated Step 17†	HZCu Calculated Step 20†	HZMo Observed	HZMo Calculated Step 10†
1,12,13	DP55	−90	0.245	0.416	0.508	0.060	0.046	0.125	0.253	0.143	0.064	0.072
25,26	DP52	−90	0.343	0.108	0.118	0.049	0.014	0.105	0.195	0.113	0.021	0.054
25,26	DA5	−50	0.292	0.108	0.118	0.023	0.014	−	−	−	−	−
27,28	RP12	−90	0.174	0.088	0.082	0.007	0.006	0.114	0.106	0.101	0.005	0.017
29,30	RP9A	−90	0.251	0.101	0.121	0.016	0.005	0.062	0 078	0.112	0.023	0.006
30,31	RP10	−90	0.255	0.169	0.197	0.010	0.009	0.115	0.068	0.116	0.019	0.004
43,44	DP12	−60	−	−	−	−	−	0.122	0.075	0.092	0.017	0.011
44	DP12	−60	0.348	0.123	0.131	0.025	0.013	−	−	−	−	−
56	DP3	−50	−	−	−	−	−	0.171	0.055	0.111	0.026	0.035
56	DP2	−50	0.245	0.095	0.123	0.023	0.009	0.160	0.055	0.111	0.023	0.035
59	DP51	−90	0.180	0.132	0 189	0.062	0.043	0.095	0.083	0.110	0.047	0.024
68	DP54	−90	0.147	0.379	0.321	0.000	0.083	0.138	0.168	0.180	0 015	0 028
71,72	DA1	−50	0.529	0.390	0.387	0.032	0.078	−	−	−	−	−
83	RA5	−90	0.337	0.269	0.347	0.012	0.017	−	−	−	−	−
85	DA41	−90	0.373	0.514	0.481	0.025	0.032	−	−	−	−	−
92	RP3-3A	−90	0.478	0.416	0.358	0.056	0.033	0.289	0.299	0.264	0.027	0.029
94	DP14	−65	−	−	−	−	−	0.260	0.247	0.274	0.120	0.047
94,105	DP14	−65	0.289	0.202	0.250	0.046	0.023	−	−	−	−	−
95	DP16	−75	0.315	0.158	0.126	0.047	0.010	0.197	0.127	0.117	0.131	0.017
96	DA6	−90	0.392	0.253	0.192	0.017	0.018	−	−	−	−	−
107	DP50	−50	−	−	−	−	−	0.134	0.091	0.094	0.063	0.005
116	DP49	−50	0.207	0.133	0.126	0.006	0.005	−	−	−	−	−
N		18	18	18	18	18	18	14	14	14	14	14
r			−	0.381	0.328	−	0.375	−	0.443	0.673	−	0.175

*Test data are from those cells in Fig. 2 with drill-holes identified by a cross.

†Tables 4 and 5.

N, number of data points; *r*, simple correlation coefficient (based on 1*n* values) between observed and calculated values.

derivative map. The distance from an airborne magnetic high and low to each cell centre was also measured because of the prior recognition of the possibility of a zonal distribution that involved at least one magnetite-rich zone.

Lithological variables were recorded as the percentage of the area in each cell that was underlain by a specific rock type (Table 1). Such a variable commonly leads to peculiar density distributions. For example, many relatively abundant rock types underlie the whole of some cells, are absent from many others and are present in variable percentages in others. A more or less U-shaped distribution results.

Hydrothermal alteration variables of the type used here (Table 1)[20] have not been considered in regression analysis by other authors, as far as is known. The close genetic relationship between alteration and copper—molybdenum sulphides emphasized their importance. Each alteration type was defined in terms of percentage area of cell occupied, in a manner analogous to lithological variables, which have comparable density distributions. The precision of such estimates cannot be great, but they were used as a generalized characterization of the distribution of alteration types; their importance was emphasized by the regularity of alteration zones in the ideal models

Table 5   Regression models for hypogene zone for log 10, Cu% and Mo%*

Dependent variable (Y), log 10 Independent variable	HZCu Step 17 Symbol	Coefficient	HZCu Step 20 Symbol	Coefficient	HZMo Step 10 Symbol	Coefficient
Constant	Constant	−3.4467	Constant	−13.7713	Constant	−10.3395
X_1	Cu	0.4518	RA	0.4842	Mo	0.2712
X_2	Zn	−0.1895	DGM	0.2075D-03	RA	0.3205
X_3	RA	0.5252	AM	3.9515	RR	0.2805
X_4	DGM	0.2163D-03	−	−	AM	4.3494
X_5	—	−	−	−	AMR	−0.9126
X_6	−	−	−	−	DAM	−0.7946
X_7	−	−	−	−	DSM	−0.6737D-03
X_8	−	−	−	−	AQ	−0.8568
X_9	−	−	−	−	AB	−0.3510
X_{10}	−	−	−	−	AAP	−0.5882
X_{11}	−	−	−	−	AHM	−0.7219
R squared	0.7082		0.7092		0.9764	
Standard error of *Y*	0.1807		0.1760		0.1071	

*Control data for model from cells in Fig. 2 with drill-holes identified by large dots.

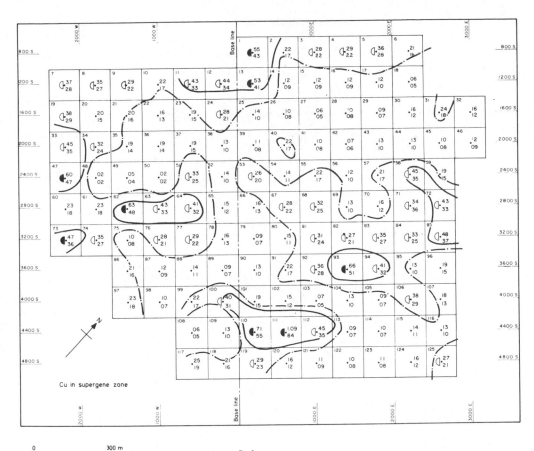

Fig. 5   Cu% in supergene enrichment zone estimated from step 12 in regression model (Table 2)

that have been presented by various authors.[7,8,9]

The transformation of data prior to statistical analysis presents problems. The main purpose of transformation is the production of density distributions that are moderately close to Gaussian

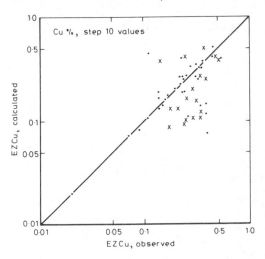

form, so that quantitative statistical models can be subjected to fairly rigorous statistical evaluation; otherwise, error estimates, in particular, have little meaning. Furthermore, it is useful if all variables are of approximately the same order of magnitude. Variables were transformed, therefore, where it was felt to be warranted by density distributions, but only two functions were used (Table 1). Positively 'skewed' distributions were modified by a logarithmic transformation; percentage data for lithological and hydrothermal alteration information were divided by 100 to give proportions and arcsin transformed.[21]

Control and test cells

Dependent-variable drill-hole data and independent-variable data in coincident cells were needed to establish the regression equations. The best cells for this purpose were those cells with nearly centred vertical drill-holes that either had a

Fig. 6   Scatter diagram: calculated step 12 values (Table 2) *vs* observed values for Cu% in supergene enrichment zone. ●Values calculated by comparison of model with control data used in construction of model (Tables 3 and 6); x values calculated by comparison of model with values from test data (Tables 4 and 6). Straight line represents perfect correspondence of observed and calculated values

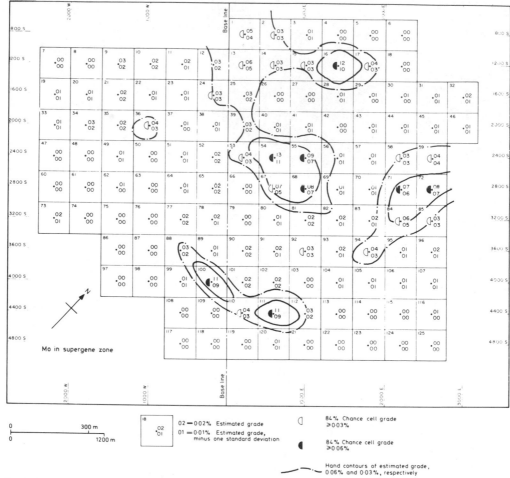

Fig. 7   Mo% in supergene enrichment zone estimated from step 8 in the regression model (Table 2)

nearly complete intersection through the supergene enrichment zone or, preferably, also penetrated the hypogene zone. Thus, 36 cells were used for the derivation of equations for evaluation of the supergene enrichment zone and 25 cells were used in models for the hypogene zone (Table 3 and Fig. 2). All drill data used for the control cells were based on work during or prior to 1970.

Data for an independent check on the values predicted by the regression models were not easy to obtain because most drill data that were centered on cells were used for control data to make the statistical models as reliable as possible, but inclined holes, holes located between two cells rather than at cell centres and a few holes drilled in 1973 provided comparative or test data for 18 points in the supergene zone and for 14 points in the hypogene zone (Table 4 and Fig. 2). The test data do not relate as specifically to a given cell as do the control data; consequently, predicted values for cells cannot be expected to match very closely.

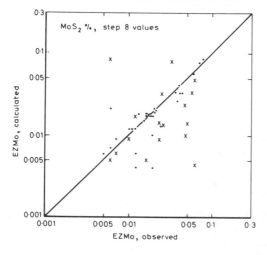

Fig. 8   Scatter diagram: calculated step 8 values (Table 2) *vs* observed values for MoS₂% in supergene enrichment zone. ●Values calculated by comparison of model with control data used in construction of model (Tables 3 and 6); x values calculated by comparison of model with test data values (Tables 4 and 6). Straight line represents perfect correspondence of observed and calculated values.

Table 6   Simple correlation coefficients to compare model predictions with control data used in construction of model and test data from assay data not used in development of model

Dependent variable*	EZCu		EZMo	HZCu		HZMo
Step	10	12	8	17	20	10
Figure reference.	3,4	5,6	7,8	9,10	11,12	13,14
Model prediction compared to control data						
N	36	36	36	25	25	25
r	0.679	0.595	0.851	0.842	0.842	0.985
Level, %†	<0.1%	<0.1%	<0.1%	<0.1%	<0.1%	<0.1%
Model prediction compared to test data						
N	18	18	18	14	14	14
r	0.381	0.328	0.375	0.443	0.673	0.175
Level	<15%	<20%	<15%	<15%	<1%	>40%

*Statistical tests done on $1n$ of dependent variable data.
†Level at which hypothesis — that two variables are independent — is rejected.
N. number of data pairs.
r, simple correlation coefficient.

Evaluation of regression models

Backward stepwise regression analysis provided a large number of possible solutions for each dependent variable (Y). The equation for step 1 included all independent variables (X_i, $i = 1, n$). As it is doubtful whether all X variables are significantly related to Y, the equation would be more useful if those that were not significantly related were eliminated. Variables were dropped one at a time as described previously.

Final equations were chosen (Table 2) on a statistical basis; dependent variables had already been screened for geological relevance. In the evaluation of the quality of the final equations two statistical measures were particularly significant. One measure, the square of the multiple correlation coefficient (Table 2), expresses the proportion of the total variance of the transformed dependent variable that is accounted for by the equation. The R^2 values in Table 4 were all greater than 0.9 and suggested that the intercorrelations indicated in the equations were real. A second measure specifies the error attached to the value estimate standard error of Y (Table 4). The standard error of the dependent variable (Y) must be small enough for some probabilistic significance to be attached to the mean estimate of Y. For example, Figs. 3, 5, 7, 9, 11 and 13 show expected values forecasted by the regression model for each of the 125 cells as well as this value minus one standard error. There is an 84% chance that the true value of a cell would exceed a value that is one standard error less than the expected value, as calculated by the statistical model. For example, cell 110 (Fig. 3) has a calculated mean value ($\overline{X}_c$) of Cu 0.81 and a 'mean minus one standard error' ($\overline{X}_c - Se$) value of Cu 0.64%.

Equations from early steps in the evaluation of variables are clearly not useful; they generally have abnormally high standard errors for Y values (e.g. for HZCu, step 2, $Y = 0.228$; on Table 5 in steps 17 and 20 it is 0.181 and 0.176, respectively). Many steps are so closely related, however, that there is little distinction between them. Thus, enrichment copper values probably are estimated as well from step 10 (Table 4 and Figs. 3 and 4) as from step 12 (Table 2 and Figs. 5 and 6). The map patterns in Figs. 3 and 5 are closely related; similarly, patterns for hypogene copper in step 17 (Fig. 9) and step 20 (Fig. 11) are alike.

An additional method to evaluate results from the regression models is to examine plots and simple linear correlation coefficients of predicted values *vs* observed assays. Observed assays can be either the control data that are used to construct the models or the test data. Table 6 summarizes, for all control and test cases, the correlation coefficients (also Tables 3 and 4) and the levels at which they define that the two populations are significantly different. Figs. 4, 6, 8, 10, 12 and 14 are scatter diagrams

Table 7   Forecast values (%) within ± 2 standard deviations of control* and test† values

Dependent variable (Y)	EZCu Step 10	Step 12	EZMo Step 8	HZCu Step 17	Step 20	HZMo Step 10
Standard error of Y	0.0983	0.1132	0.0865	0.1807	0.1760	0.1071
Forecast values within ± 2S of control data, %	81	81	81	96	96	100
Forecast values within ± 2S of test data, %	50	50	44	86	100	43

*Table 1.
†Table 2.

151

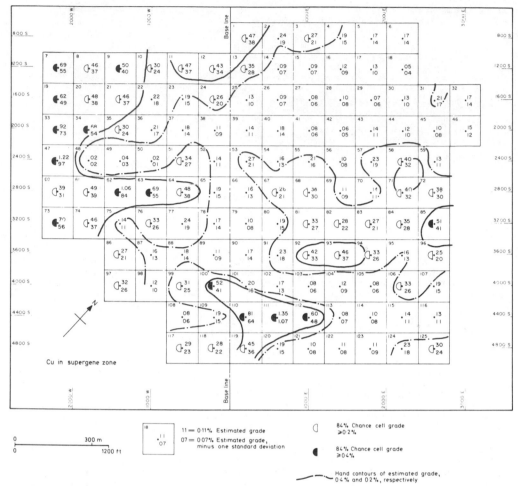

Fig. 9   Cu% in hypogene zone estimated from step 17 in the regression model (Table 5)

that relate predicted values to the model control data and to test data.

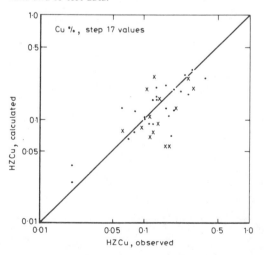

Fig. 10   Scatter diagram: calculated step 17 values (Table 5) *vs* observed values for Cu% in hypogene zone. ●Values calculated by comparison of model with control data used in construction of model (Tables 3 and 6); x values calculated by comparison of model with test data values (Tables 2 and 6). Straight line represents perfect correspondence of observed and calculated values

Extremely high correlations between forecast values and calculated values are apparent from the consistently high correlation coefficients in Table 6, and in the visual clustering of points along the diagonal line in the scatter diagrams. Correlations between forecasted values and test case values are less pronounced, but it must be recalled that most test data are poorly located relative to grid centres and thus depart somewhat from the model that is being tested.

Table 7 also shows that values forecast by the model are closer to those values used in constructing the model than to those obtained from the test case. Hypogene copper grade seems to be predicted particularly well in both the model and the test case. Observed and calculated hypogene molybdenite grades agree well for the data control used (Table 6, $r = 0.985$) but poorly for the

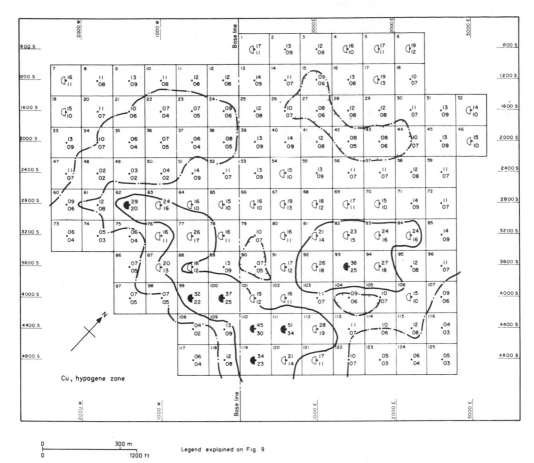

Fig. 11   Cu% in hypogene zone estimated from step 20 in regression model (Table 5)

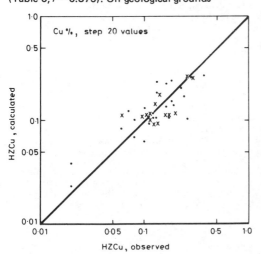

Fig. 12   Scatter diagram: calculated step 20 values (Table 5) *vs* observed values for Cu% in hypogene zone. ● Values calculated by comparison of model with control data used in construction of model (Tables 3 and 6); x values calculated by comparison of model with test data values (Tables 4 and 6)

observed test data (Table 6, $r = 0.175$). Whereas this is not particularly encouraging, the statistical data for the test case comparison of molybdenite grade in the enrichment zone are somewhat better (Table 6, $r = 0.375$). On geological grounds molybdenite grade in the enrichment zone should correspond closely to grades in the hypogene zone. Indeed, the equations for molybdenite grade in the enrichment (Table 2) and hypogene (Table 5) zones specify similar independent variables Fig. 7 has, predictably, a pattern that compares closely with Fig. 13. Consistency of these patterns and the clustering of high values on corresponding scatter diagrams provides an intuitive 'feel' for the quality of the model and, in this case, suggests that it might have more validity than can be demonstrated statistically. In other words, grouped variables might give overall patterns that are more reliable than the precise values given for cell centres.

Forecast values for the 125-cell grid

Forecast values for all cells in the grid are plotted at each cell centre, together with predicted value minus one standard deviation, in Figs. 3, 5, 9, 11 and 13. Forecast predicted values are hand contoured; those cells with an 84% chance of exceeding the values chosen for the contours are

Table 8   Proportion of cells in arbitrary low-, intermediate- and high-value groups*

	Cell value at cell centre					
	Low		Intermediate		High	
EZCu	<0.2% Cu		0.2 to <0.4% Cu		≥0.4% Cu	
HZCu	<0.1% Cu		0.1 to <0.2% Cu		≥0.2% Cu	
EZMo, HZMo	<0.03% MoS_2		0.3 to <0.06% MoS_2		≥0.06% MoS_2	
Evaluation of cell	Drilled	Undrilled	Drilled	Undrilled	Drilled	Undrilled
EZCu†						
Cells in value¶ group, %	40	60	39	61	25	75
All cells in grid, %	20	30	12	19	5	14
EZMo‡						
Cells in value group, %	37	63	37	63	50	50
All cells in grid, %	28	48	6	10	4	4
EZCu and EZMo						
Cells in value group, %	38	62	38	62	32	68
All cells in grid, %	24	40	9	14	4	9
HZCu						
Cells in value group, %	26	74	30	70	33	67
All cells in grid, %	8	22	16	37	5	11
HZMo¶						
Cells in value group, %	24	76	46	54	27	73
All cells in grid, %	17	54	10	11	2	6
HZCu and HZMo						
Cells in value group, %	25	75	35	65	31	69
All cells in grid, %	12	38	13	24	4	9
EZCu and EZMo and HZCu and HZMo						
Cells in value group, %	32	68	36	64	32	68
All cells in grid, %	18	39	11	19	4	9

*Within each value category, cells are further classed as drilled or not drilled to compare results of statistical models for cells that have been tested (drilled) and those that have not been tested (not drilled).
†Average of steps 10 and 12.
‡Step 8.
§Average of steps 17 and 20.
¶Step 10.

emphasized. Relatively low contour values have been chosen to aid grouping of the data on the figures. Predicted values represent the average for the 250-ft thick supergene zone[15] and the top 400 ft of the hypogene zone as drilled in the control cells. Significant zones of higher-grade mineralization are not precluded in this analysis.

Regression equations for supergene and hypogene copper grade look useful statistically (Tables 2, 5 and 6). This is supported by the coherent patterns developed when these equations are used to forecast values in the entire grid (Figs. 3, 5, 9 and 11). The general similarity between the different regression steps (Figs. 3, 4, 5, 9, 10 and 11) has been noted previously. In the hypogene zone, step 20 (Fig. 11) is slightly better by statistical tests (Tables 5 and 6), but step 17 (Fig. 9) might be preferred by geologists because a positive independent variable related to geochemical copper is retained in the regression equation (Table 5).

Copper-forecast maps (Figs. 3, 5, 9 and 11) consistently show four anomalous areas clustered about cells 1, 62, 93 and 100—111. The area that surrounds cell 93 is the one that has been tested extensively by drilling (Fig. 2). One drill-hole (DP55, Table 4) at the intersection of cells 1, 12 and 13 is the only exploration of this anomaly. No drilling was done in the anomalous area around cells 100—111. This area might be a zone as much as 2000 ft long (Fig. 9). The values for hypogene

and supergene copper are strong. The anomalous area that surrounds cell 62 has been tested at its margin by hole RA18 in cell 65 (Fig. 2); the grade predicted for cell 65 is approximately correct (Table 3). High copper grades are predicted for the area west and north of cell 62, which is most apparent in Fig. 3, but might not be an attractive area for exploration. The supergene enrichment zone might be largely removed by erosion because only a very shallow intersection was encountered in hole RP17 in cell 48. Furthermore, forecast values are not high enough in the hypogene zone (Figs. 9 and 11) to constitute a target or to suggest an original source of supergene copper.

Molybdenite-grade prediction does not compare well with test data, as discussed previously, but there are some general coincidences with the copper anomalies that are outlined above. This is to be expected because copper and molybdenum generally correlate well on the property.[3] A combined molybdenum—copper zone coincides with cells 100—111 and enhances this anomaly.

Table 8 is an evaluation of drilling that has been done on the basis of forecast values by regression analysis. 24—40% of the cells forecast as 'low' by the regression models have been drilled. Similarly, only 25—50% of the holes have been drilled in cells forecast as 'high' in assay values. Thus, 50—75% of the high-priority areas defined by the regression models have not been drilled. The situation for

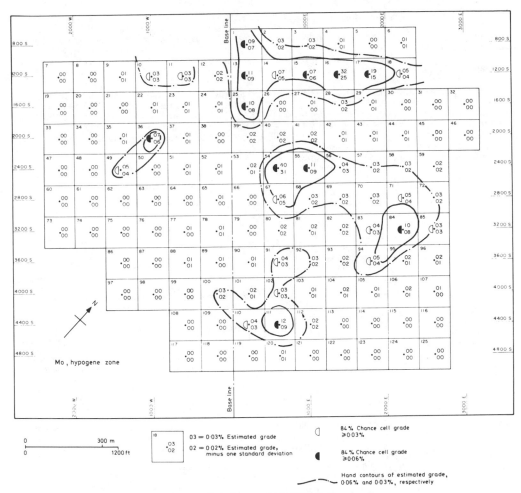

Fig. 13   Mo% in hypogene zone estimated from step 10 in the regression model (Table 5)

copper and molybdenite in both the enrichment and hypogene zones is summarized in Table 8. 43% (54 cells) of the 125-cell grid are forecast as 'intermediate' or high, but less than half of them have been tested by drilling. 13% (16 cells) of the grid are forecast as high, but 68% (10 cells) of these cells remain to be drilled.

Conclusions

Multiple-regression analysis can be an effective way of defining favourable drill targets where exploration data that have been collected systematically are available. Regression equations selected for forecasting copper and molybdenum grades in enrichment and hypogene zones of copper and molybdenite at Casino are statistically meaningful for the reasons listed below.

(1)   They explain a large portion of the total variance.

(2)   They have standard errors in Y that are small enough for the mean value forecast to have a high probability of being a close estimate of true values.

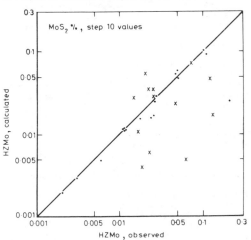

Fig. 14   Scatter diagram: step 10 calculated values *vs* observed values for MoS_2% in hypogene zone. ●Values calculated by comparison of model with control data used in construction of model (Tables 3 and 6); x values calculated by comparison of model with test data values (Tables 4 and 6)

(3)   They predict closely the values on which the models were based.
(4)   They appear to predict the values in the test case in a satisfactory manner.
(5)   Predicted values cluster into zones of high, anomalous values — forecast highs are not scattered erratically but are grouped.

All variables used in this evaluation were thought, on geological grounds, to relate to potential mineralization. Regression techniques have been used to select a subset of these important variables that show a statistically significant inter-relationship among themselves and with metal grade. This methodology is an empirical one, superimposed on already screened geological data. It is difficult, therefore, to read more than general significance into individual variables retained in the regression equations when it is their interdependences that are important. In a qualitative way the regression equations can be interpreted geologically. For example, in the first equation in Table 5 for hypogene copper (HZCu, step 17) all independent variables can be related to systematic variations that are characteristic of porphyry-type systems: rock copper (Cu) is obviously related to grade; soil zinc (Zn) represents a characteristic base-metal halo that surrounds a pyrite halo and distant from copper mineralization; average resistivity (RA) distinguishes sulphide-rich (pyrite halo) and sulphide-poor (core) parts of the porphyry system; and distance to ground magnetic high (DGM) is a function of the magnetite-rich potassic core of the system. The signs for variables are consistent with the relationships given above.

The grid area, according to the values of the cells predicted from the multiple regression equations, has not been adequately drilled. Clustering of high values in Figs. 3, 5, 7, 9, 11 and 13 has defined four anomalous areas, only one of which has been drilled extensively. Forecasted grades in these untested zones are comparable to the zone that has been drilled extensively. Analysis of drilling to date (Table 8) indicates that more than 50% of the high cells have not been drilled. A test of our analysis would be to drill the anomalous cells noted above and to compare the grades encountered to those forecast by our calculations. This, of course, is the ultimate test — one which might be made if exploration in the area progresses.

A number of case histories of this type might lead to recognition of those exploration parameters (independent variables) that are consistently part of equations for grade (dependent variables) prediction. In particular, the technique probably would be most applicable to porphyry-type deposits in which variations in grade are reflected by systematic changes in measurable, indirect exploration parameters. If only a few variables turn out to be consistently useful, exploration expenditure on indirect exploration methods could be reduced.

Application of this type of analysis would be facilitated if all field data were collected systematically on a grid that would be amenable to division into cells. It is important that data on dependent variables, especially, should be collected consistently near cell centres. Early availability of data on surface exploration and early drilling would make regression analysis more effective as a guide for subsequent drilling.

Acknowledgement
The authors thank Casino Silver Mines, Ltd., and Brameda Resources, Ltd., for their assistance and for their permission to publish data. R. E. Hindson and the late M. J. Carr, both of Teck Corporation, Ltd., were particularly helpful. The support of M. Phillips of Archer Cathro and Associates, Ltd., is appreciated. Early guidance in this research by the late J. A. Gower is gratefully acknowledged. S. S. Wong helped in the calculations and drafting. Funds from the University of British Columbia and the National Research Council of Canada assisted this project.

References
1.   Godwin C. I. Casino. In *Porphyry deposits of the Canadian Cordillera* Sutherland Brown A. ed. (Montreal: Canadian Institute of Mining and Metallurgy, 1976), 344—54. (*CIM spec. vol.* 15)
2.   Phillips M. P. and Godwin C. I. Geology and rotary drilling at the Casino Silver Mines property. *West. Miner*, **43**, Nov. 1970, 43—9.
3.   Godwin C. I. Geology of Casino porphyry copper—molybdenum deposit, Dawson Range, Y.T. Ph.D. thesis, University of British Columbia, 1975.
4.   Brameda Resources, Ltd. Company reports and maps.
5.   Archer A. R. and Main C. A. Casino, Yukon — a geochemical discovery of an unglaciated Arizona-type porphyry. In *Geochemical exploration* Boyle R. W. and McGerrigle J. I. eds (Montreal: Canadian Institute of Mining and Metallurgy, 1971), 67—77. (*CIM spec. vol.* 11)
6.   Menzies M. M. Report on the Canadian Creek Project. Casino Silver Mines, Ltd. (N. P. L.) fourth annual report for the year ended May 31, 1970.
7.   Jerome S. E. Some features pertinent in exploration of porphyry copper deposits. In *Geology of the porphyry copper deposits southwestern North America* Titley S. R. and Hicks C. L. eds (Tucson: University of Arizona Press, 1966), 75—85.
8.   Lowell J. D. and Guilbert J. M. Lateral and vertical alteration-mineralization zoning in porphyry ore deposits. *Econ. Geol.*, **65**, 1970, 373—408.
9.   Drummond A. D. and Godwin C. I. Hypogene mineralization — an empirical evaluation of alteration zoning. Reference 1, 52—63.
10.   Pilcher S. H. and McDougal J. J. Characteristics of some Canadian Cordilleran porphyry prospects. Reference 1, 79—82.
11.   Sutherland Brown A. Morphology and classification. Reference 1, 44—51.
12.   Godwin C. I. Alternative interpretations for the Casino Complex and Klotassin Batholith in the Yukon crystalline terrain. *Can. J. Earth Sci.*, **12**, 1975, 1910—6.
13.   Hughes O. L. *et al.* Glacial limits and flow patterns, Yukon Territory, south of 65 degrees north latitude. *Pap. geol. Surv. Can.* 68-34, 1969, 9 p.
14.   Prest V. K. Grant D. R. and Rampton V. N. Glacial map of Canada. *Map geol. Surv. Can.* 1253A, 1968.
15.   Godwin C. I. Supergene enrichment, Casino porphyry deposit, Y.T. (Abstract). In *Geomorphology of the Canadian Cordillera and its bearing on mineral deposits: Program and Abstracts geol. Ass. Can. Cordilleran Sect.*, 1976.
16.   Ney C. S. *et al.* Supergene copper mineralization. Reference 1, 72—8.
17.   Sinclair A. J. and Woodsworth G. J. Multiple regression as a method of estimating exploration potential in an area near Terrace, B.C. *Econ. Geol.*, **65**, 1970, 998—1003.
18.   Kelly A. M. and Sheriff W. J. A statistical examination of the metallic mineral resources of British Columbia. In *Proceedings of a symposium on decision-making in mineral exploration II, Vancouver, February 1969* Kelly A. M. and Sheriff W. J. eds (Vancouver: University of British Columbia, 1969), 221—43.
19.   Le C. and Tenisci T. *UBC TRP triangular regression package* (Vancouver: University of British Columbia, Computing Centre, 1978), 197 p.
20.   Blanchet P. H. and Godwin C. I. 'Geolog System' for computer and manual analysis of geologic data from porphyry and other deposits. *Econ. Geol.*, **67**, 1972, 796—813.
21.   Dixon W. J. and Massey F. J. Jr. *Introduction to statistical analysis* (New York: McGraw-Hill, 1957), 370 p.

156

13

PRINCIPLES OF GEOSTATISTICS

G. MATHERON

ABSTRACT

Knowledge of ore grades and ore reserves as well as error estima-tion of these values, is fundamental for mining engineers and mining geologists. Until now no appropriate scientific approach to those esti-mation problems has existed: geostatistics, the principles of which are summarized in this paper, constitutes a new science leading to such an approach. The author criticizes classical statistical methods still in use, and shows some of the main results given by geostatistics. Any ore deposit evaluation as well as proper decision of starting mining operations should be preceded by a geostatistical investigation which may avoid economic failures.

RESUME

Pour tout mineur et géologue minier, la connaissance des teneurs et du tonnage et l'appréciation des erreurs sur ces grandeurs est fonda-mentale. Or, jusqu'à présent, il n'existait pas d'approche scientifique correcte de ces problèmes.

La *géostatistique*, dont les principes sont résumés dans cet article, est la nouvelle science qui permet cette approche. L'auteur indique les méthodes statistiques antérieures et encore courantes et donne quel-quesuns des résultats principaux de la géostatistique.

Toute évaluation de gisement et toute décision de mise en ex-ploitation devrait être précédée d'une étude géostatistique permettant de limiter le risque d'une déconvenue ultérieure.

INTRODUCTION AND SHORT HISTORICAL STATEMENT

GEOSTATISTICS, in their most general acceptation, are concerned with the study of the distribution in space of useful values for mining engineers and geologists, such as grade, thickness, or accumulation, including a most important practical application to the problems arising in ore-deposit evaluation.

Historically geostatistics are as old as mining itself. As soon as mining men concerned themselves with foreseeing results of future works and, in particular as soon as they started to pick and to analyze samples, and com-pute mean grade values, weighted by corresponding thicknesses and in-fluence-zones, one may consider that geostatistics were born. In so far as they take into account the space characteristics of mineralization, these traditional methods still keep all their merit. Far from disproving them, modern developments of the theory have adopted them as their starting point and have brought them up to a higher level of scientific expression.

However, assuming they could provide a correct evaluation of mean values, the traditional methods failed to express in any way an important character of mineralizations, which is their variability or their dispersion. Some scores of years ago, classical probability calculus techniques began to be used in order to take into account this characteristic. If an unskillful application of those techniques has sometimes led to absurdities, it remains certain that, on the whole, results have been profitable. In a way this is a paradox, for classical statistical methods, in so far as they are not concerned with the spatial aspect of the studied distributions, actually cannot be applied. As a matter of fact, the South-African school, which has recorded the most remarkable results with Krige, Sichel, used to say, and believed that they were applying classical statistics. But the methods they were developing differed more and more from classical statistics, and adjusted themselves spontaneously to their object.

The second decisive change appeared when the insufficiency of classical probability calculus was clearly understood as well as the necessity of re-introducing the spatial characters of the distributions. It consisted in realizing on a higher level the synthesis between traditional and statistical methods. Hence, geostatistics started elaborating their own methods and their own mathematical formalism, which is nothing else than an abstract formulation and a systematization of secular mining experience. This formalism has inherited from its statistical origin a language in which one still speaks of variance and covariance, including however in those notions a new content. This similarity in vocabulary must not deceive. At the end of a protracted evolution, the geostatistical theory had to admit that it was facing, instead of random occurrences, natural phenomena distributed in space. And, therefore, its methods are approximately these of mathematical physics and more specially those of harmonic analysis.

INSUFFICIENCY OF CLASSICAL STATISTICAL CONCEPTS

To be brief, we shall limit ourselves, in what follows, to the distribution of ore-grades in a deposit. The results that will be obtained will however have a general range and will be applicable to any character owned by a spatial distribution. In an usual statistical approach, the grades of samples picked in a deposit are classified on a histogram. Such a procedure does not take into account the location of samples in the deposit. But it is not enough to know the frequency of a given ore-grade in a deposit. It is also necessary to know in what way the different grades follow each other on the field, and specially what is the size and the position of economic orebodies. At the starting point of the theory we have to face one fact: the inability of common statistics to take into account the spatial aspect of the phenomenon, which is precisely its most important feature.

More precisely, the aim of the classical probability calculus is the study of aleatory variables. The mere example of the heads or tails game shows clearly what is going on. Let us record $+1$ each time the coin falls on tails and -1 in the opposite case. Before throwing the coin, there is no way of forecasting whether $+1$ or -1 will be recorded; we only know that there is one chance out of two for one or the other of these two opportunities. An aleatory variable has classically two essential properties: 1) The possibility, theorically at least, of repeating indefinitely the test that assigns to the variable a numerical value; we can for example, throw the coin as often as we want.   2) The independence of each test from the previous and the next ones; if all the 100 first attempts have given tails, there remains however one chance out of two for the 101st attempt to give heads.

It appears clearly that a given ore-grade within a deposit cannot have those two properties.   The content of a block of ore is first of all unique. This block is mined only once and there is no possibility of repeating the test indefinitely.   When the grade of a sample is concerned, which may be a groove sample of a given size for example, the result is exactly the same, because the grade of a groove located in a point with coordinates (x, y) is unique and well determined.   However it is possible to pick a second sample close to the first, then a third one, etc. . . . which shows an apparent possibility of repeating the test.   Actually, it is not exactly the same test but a slightly different one.   But even assuming this possibility of repetition, the second property will surely not be respected.   Two neighboring samples are certainly not independent.   They tend, in average, to be both high-grade if they originate from a high-grade block of ore, and vice-versa.   This tendency, more or less stressed, expresses the degree of more or less strong continuity in the variation of grades within the mineralized space.

The misunderstanding of this fact and the rough transposition of classical statistics has sometimes led to surprising misjudgments.   Around the fifties, in mining exploration, it was advised to draw lots to locate each drilling (i.e., to locate them exactly anywhere).   Miners of course went on still using traditional regular grid pattern sampling, and geostatistics could later prove they were right.   Or else again, it was urged that the accuracy of ore evaluation of a deposit depended only on the number of samples (and not on their location) and varied as the square root of this number.   This unskillful transposition of the theory of errors led to absurdities.   For example, if a given deposit is explored by drilling, it would suffice to cut the cores in 5 mm pieces instead of 50 cm pieces to obtain 100 times more samples, and therefore 10 times higher accuracy.   This, of course, is wrong.   The multiplicity of samples thus obtained is a fallacy, and does nothing more than repeat indefinitely the same information, without yielding anything else.   Geostatistics actually show that accuracy is the same with pieces of 5 mm and 50 cms, as every miner understands instinctively.

NOTION OF REGIONALIZED VARIABLE

Thus a grade cannot in any way be assimilated to an aleatory variable. We speak of regionalized variables precisely in order to stress the spatial aspect of the phenomena.   A regionalized variable is, *sensu stricto*, an actual function, taking a definite value in each point of space.

In general such a function has properties too complex to be studied easily through common methods of mathematical analysis.   From the point of view of physics or geology, a given number of qualitative characteristics are linked to the notion of regionalized variable.

a) In the first place, a regionalized variable is *localized*. Its variations occur in the mineralized space (volume of the deposit or of the strata), which is called *geometrical field* of the regionalization. Moreover such a variable is in general defined on a *geometrical support* (*holder*). In the case of an ore-grade, this support is nothing but the volume of the sample, with its geometrical shape, its size and orientation. If, in the same deposit, the geometrical support is changed, a new regionalized variable is obtained, which shows analogies with the first one, but does not coincide with it.

For instance, samples of 10 Kg corresponding to drill cores are not distributed in the same way as samples of 10 tons corresponding to blasts. Often the case of a punctual support will be considered. A punctual grade, for example, will take value 0 or value $+1$ according to whether its support will fall into a barren or mineralized grain.

b) Secondly, the variable may show a more or less steady continuity in its spatial variation, which may be expressed through a more or less important deviation between the grades of the two neighboring samples mentioned above. Some variables with a geometrical character (thickness or dip of a geological formation) are endowed with the strict continuity of mathematicians. Fairly often (for grades or accumulations) only a more lax continuity will exist or, in other words, a continuity "in average." In some circumstances, even this "in average" continuity will not be confirmed, and then we shall speak of a *nugget effect*.

c) Lastly the variable may show different kinds of *anisotropies*. There may exist a preferential direction along which grades do not vary significantly, while they vary rapidly along a cross-direction. Those phenomena are well known under the names of runs, or zonalities.

To those general characters, common to any regionalized variable, specific features can be superimposed. For example, in the case of a sedimentary deposit, a *stratification effect*, will be noted. Large-scale stratification provides individualizable and separately minable strata. Inside each strata it may appear by the existence of beds following one another vertically, and separated by discontinuity surfaces. The grade, almost constant or barely varying inside a given bed, will vary abruptly from one bed to another; however common and familiar this phenomenon appears to be, it is still fundamental, and a theoretical formulation of the problem that would not take it into account would miss the point. It will happen as well that to those vertical discontinuities, stressed by jointing, will be added lateral discontinuities, owing to the lenticular endings of beds. This *bed-relaying phenomenon*, when it does exist, shows up at each stratigraphic level a partitioning of the sedimentation area into micro-basins with almost autonomous evolution, and may appear during operation through *grade-limit effect*.

In the same way in stockwerk types of deposits, high-grade veinlets or granules individualized in a more or less impregnated mass will be observed. This *stockwerk effect*, just as the stratification and bed relaying effects, expresses the appearance of a discontinuity net-work within a homogeneous geometrical field. On a very different scale, that of granularity, the nugget effect appears as a phenomenon of the same nature, the net-work of discontinuities being here that one separating barren from mineralized grains.

Those different specific aspects of spatial distribution of regionalized variables—far apart from classical probability calculus—must compulsorily be taken into account by geostatistics. This is made possible owing to a simple mathematical tool: the *variogram*.

THE VARIOGRAM

The variogram is a curve representing the degree of continuity of mineralization. Experimentally, one plots a distance d in abcissa and, in ordinate, the mean value of the square of the difference between the grades of samples picked at a distance d one from the other. Theoretically, let $f(M)$ be the value taken in a point M of the geometrical field V by a regionalized variable defined on a given geometrical support v (in general support v will be small and the limit may be considered as punctual). The *semi-variogram* $\gamma(h)$, or law of dispersion, is defined, for a vectorial argument h, by the expression:

$$\gamma(h) = \frac{1}{2V} \int \int \int_V [f(M+h) - f(M)]^2 dV. \qquad (1)$$

In general, the variogram is an increasing function of distance h, since, in average, the farther both samples are one from the other, the more their grades are different. It gives a precise content to the traditional concept of the *influence zone* of a sample. The more or less rapid increase of the variogram represents, indeed, the more or less rapid deterioration of the influence of a given sample over more and more remote zones of the deposit. The qualitative characteristics of regionalization are very well expressed through the variogram:

a) The greater or lesser regularity of mineralization is represented by the more or less regular behavior of $\gamma(h)$, near the origin. It is possible to distinguish roughly four types (Fig. 1). In the first type the variogram has

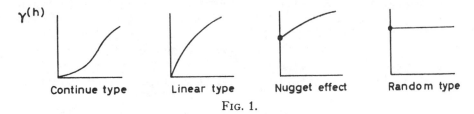

$\gamma(h)$

Continue type   Linear type   Nugget effect   Random type

FIG. 1.

a parabolic trend at the origin, and represents a regionalized variable with high continuity, such as a bed-thickness.

The second type, or linear type, is characterized by an oblique tangent at the origin, and represents a variable which has an "in average" continuity. This type is the most common for grades in metalliferous deposits.

The third type reveals a discontinuity at the origin and corresponds to a variable presenting not even an "in average" continuity, but a nugget effect.

The fourth type is a limit case corresponding to the classical notion of random variable. Between type I (continuous functional) and type 4 (purely random) appears a range of intermediates, the study of which is the proper object of geostatistics.

b) The variogram is not the same along different directions of the space. Function $\gamma(h)$ defined in (1) does not only depend upon the length, but also upon the direction of vector h. Preferential trends, runs, and shoots are revealed through the study of the distortion of variogram when this direction is altered. Geological interpretation of such anisotropies is often instructive.

c) Structural characters are also reflected in the variogram. For instance, the bed-relaying phenomenon appears in the experimental curve as a level stretch of the variogram beyond a distance, i e., a range equal to the mean diameter of the autonomous micro-basins of sedimentation. And the

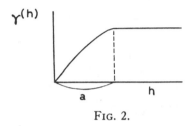

FIG. 2.

fact that these ranges are not the same along different directions makes it possible to determine the directions of elongation, and the average shape of the micro-basins.

This tool, the variogram, does not represent the totality nor the local details of the mineralizing phenomenon, but it expresses in a synthetic form their essential characters. The harmonic analysis of a vibratory phenomenon assigns for each harmonic a phase and an amplitude. The local outline of the phenomenon depends mostly upon phases, but energy depends only upon the square of amplitudes. The spectral curve giving the squares of the amplitudes does not describe the whole phenomenon but gives an account of the essential, i.e., the energetic characteristics. The variogram (or more precisely its Fourier's transformed curve) plays exactly the part of such a spectral curve.

In the following paragraphs, a few of the possible applications of variograms will be run over. It is obviously out of the question to give here a systematic study. I will merely mention some examples and several characteristic formulae. For more details I kindly ask the reader to refer to my "Treatise of Applied Geostatistics." [1]

[1] Editions *Technip*, Paris, Tome I (1962)—Tome II (Le Krigeage) (in press). Tome III (l'effet de pépite et les phénomènes de transition) to be published.

ABSOLUTE DISPERSION (OR INTRINSIC) LAW

The semi-variogram defined in (1) is bound to the geometrical field V of the regionalized variable.   If, instead of the total field V, only a portion V' of it would have been considered, a function $\gamma'(h)$ possibly different from $\gamma(h)$ would have been obtained.   However, we have the intuitive notion that in a geologically homogeneous geometrical field there might be something intrinsic, independent from location, in the characteristics representing the variabilities of regionalized variable.

Formulated in an accurate way, that intuition leads to the hypothesis of an absolute dispersion or intrinsic law expressed through the equation:

$$\gamma'(h) = \gamma(h)$$

which means that the variogram is independent from the portion V' of the deposit V selected for its calculation.   It may be said at once, that this hypothesis is not really essential to the development of the theory and it is possible to eliminate it because of some mathematical complications.[2]

Nevertheless it makes the statement of the theory much easier, and for that reason it will be followed here.   A slow deviation of the variogram in space is generally ascertained through experience and if this drift does not take too much importance, the results yielded by the hypothesis of an absolute dispersion law provide an excellent approximation of reality (on the condition that $\gamma(h)$ actually employed has been calculated from the actual portions of the considered deposit).

When this hypothesis is verified, the semivariogram $\gamma(h)$ itself acquires an intrinsic significance.   It is often designated under the name of intrinsic (or absolute) *dispersion law* or, more shortly, intrinsic function of regionalized variable.

VARIANCES AND COVARIANCES

Let us consider in the first place, a regionalized variable (which will be called grade in order to simplify) defined in a field V, on a punctual support and submitted to an intrinsic dispersion law $\gamma(h)$.   Let $f(M)$ be the value taken by the grade in a point M of the field V.   Instead of the punctual grade $f(M)$ we usually are concerned with the grade $y(M)$ of a sample v, of a given size, shape and orientation, picked at point (M).[3]   This new variable is deducted from the previous one through an integration performed within the volume v centered in M.

$$y(M) = \frac{1}{v} \int_v f(M + h)dv. \tag{2}$$

To this variable will be bound a parameter measuring its dispersion inside V, called variance, as in classical probability calculus.   The mean

---

[2] Loc. cit., Tome III.
[3] This means that the center of gravity of v is located at point M.

163

value of the punctual variable inside V being m,

$$m = \frac{1}{V} \int_V f(M)dV$$

the variance of $y(M)$ inside V is defined as the average value within V of the square of the expression $[y(M) - m]$, let:

$$\sigma^2 = \frac{1}{V} \int_V [y(M) - m]^2 dv. \tag{3}$$

It will be noted that this notion has, at the outset, a geometrical and not a probabilistic meaning. It will not deter us from calculating these variances, in the applications, from experimental data with common statistical methods. Should they be taken according to their spatial order, as in integral (3), or previously rearranged in histograms, the same expressions $(y - m)$ are appearing, with the same weights in both the calculation procedures. But, on a conceptual ground, definition (3) has a physical content that the statistical motion has not. From expression (1) of the variogram, of (2), and the definition (3) of the variance, one may deduce, reversing the order of the integrations.

$$\sigma^2 = \frac{1}{V^2} \int_V dV \int_V \gamma(h)dV' - \frac{1}{v^2} \int_v dv \int_v \gamma(h)dv'. \tag{4}$$

Each one of these sextuple integrals has a very clear meaning: it represents the average value of the $\gamma(h)$ inside V (or v) when both the extremities of vector h sweep, each one for its own account, the volume V (or v).

If we write:

$$F(V) = \frac{1}{V^2} \int_V dV \int_V \gamma(h)dV,$$

i.e., $F(V) =$ average value of $\gamma(h)$ inside V.

One gets:

$$\sigma^2 = F(V) - F(v). \tag{5}$$

Thus knowledge of the variogram of punctual grades allows the "a priori" calculation of the variance of any sample v within any portion V of a deposit. It will be noted that this variance does not depend only upon the sizes of volumes v and V, but also upon their shapes and orientation.

Physical meaning of relation (4) is highly instructive. The variance of a macroscopic sample v, considered as the juxtaposition of a great number of microsamples dv, does not depend in any way on the number of those micro-samples nor on their variances, but only on the average value of intrinsic function $\gamma(h)$ inside the geometrical volume v. Classical statistics, considering these micro-samples as independent, should lead to a variance in terms of $1/v$. There does not actually exist any deposit in which 10 ton blasts would have a variance a thousand times lower than that of 10 kg cores. Formula (4) shows why. The grades of micro-samples are not independent at all. They are inserted into a spatial correlation lattice, the nature of which is bound to the more or less steady continuity of mineralization, and which is expressed precisely through the intrinsic dispersion law $\gamma(h)$. The grades of the micro-samples are much less different, on the average, than classical statistics would indicate, and in consequence 10 ton blasts have a much higher variance than the thousandth of the variance of 10 kg cores.

The expression of the variance in form (5) shows a law of additivity. If we consider panel V' and samples v within a field V and if $\sigma^2(V',V)$, $\sigma^2(v,V)$ and $\sigma^2(v,V')$ designate the variances of V' inside V, of v inside V and v inside V', we get:

$$\sigma^2(v,V) = \sigma^2(v,V') + \sigma^2(V',V).$$

This formula is known as *Krige's Formula*. It has been established by D. G. Krige in the case when the grades are distributed according to a (statistical) lognormal law. Its validity is actually not linked to a special statistical distribution law, but only to the existence of an intrinsic dispersion law.

Besides the variance, geostatistics introduce the notion of *covariance*. If $y(M)$ and $z(M + h)$ are the grades of two samples v and v' centered in two points M and $M + h$, covariance (inside V) of y and z is the function of h defined by:

$$\sigma_{yz} = \frac{1}{V^2} \int_V [y(M) - m][z(M + h) - m]dv.$$

It can be expressed through the variogram with a relation similar to (4):

$$\sigma_{yz} = F(V) - \frac{1}{vv'} \int_v dv \int_v \gamma(k)dv'. \tag{6}$$

The second integral represents the average value of $\gamma(k)$, when both extremities of vector k sweep, respectively, volume v and volume v', at a distance h one from the other.

Let us consider, as a particular case, the isotropic de Wijs's [4] scheme. It is defined by an intrinsic isotropic function of the form:

$$\gamma(r) = 3\alpha \ln r \tag{7}$$

in which $r = |h|$ represents the modulus of the vectorial argument h, or otherwise the distance between the two points M and $M + h$. When symbol ln represents the natural logarithm, parameter α is called *absolute dispersion*. It characterizes indeed the dispersion of grades independently from the shape and the volume of the samples and of the deposit. In the

---

[4] The starting point of development of the present theory is the original De Wijs's reasoning which is a remarkable example of transition from classical statistics to geostatistics. Reference to "Traite de Geostatistique Appliquee," where bibliographical references will be found.

particular case where the volume of the samples is geometrically similar to the volume V of the deposit, formulae (4) and (7) give:

$$\sigma^2 = \alpha \ln \frac{V}{v}. \tag{8}$$

This formula, which is the *Wijs's formula*, does express a principle of similitude. It ceases to be appliable generally as soon as the deposit is not geometrically similar to the samples. It is however possible to associate to any geometrical volume v its *linear equivalent d* devined by relation:

$$\ln d - \frac{3}{2} = \frac{1}{.v^2} \int_v dv \int_v \ln r dv'. \tag{9}$$

Formula (4) entails that sample v has the same variance in any deposit as the linear sample of length d. If D and d are the linear equivalents of the deposit and of the samples respectively, the variance may be set into the form:

$$\sigma^2 = 3\alpha \ln \frac{D}{d}.$$

The linear equivalents have been calculated and tabulated for a certain amount of geometrical figures, and, in addition, we have at our disposal some simple approximation formulae. For example for a rectangle with sides a and b we have:

$$d = a + b.$$

For a parallelogram, with sides a, b, and surface S:

$$d = \sqrt{a^2 + b^2 + 2S}.$$

For a triangle with sides a, b, c, and Surface S:

$$d = \sqrt{\frac{a^2 + b^2 + c^2}{3} + 2S}.$$

For a trapezium with basis $\quad a = \dfrac{L + l}{2}$,

$$b = \frac{L - l}{2}$$

Median: m

Surface: S

$$d = \sqrt{L^2 + l^2 + m^2 - \frac{l^2 m^2}{3L^2} + 2S}$$

For a rectangular parallelepiped with sides $a > b > c$,

$$d = a + b + \frac{c}{2}.$$

For an oblique parallelepiped with edges r_1, r_2, r_3, faces S_1, S_2, S_3 and volume V, we put up:

$$\begin{cases} R^2 = r_1{}^2 + r_2{}^2 + r_3{}^2 \\ S^2 = S_1{}^2 + S_2{}^2 + S_3{}^2, \end{cases}$$

and we obtain the following approximate equivalent:

$$d = \sqrt{R^2 + 2S + \frac{V^2 R^2}{S^3}}.$$

This notion of linear equivalent allows an easy comparison between samplings of different natures, at least in the case, common in metalliferous deposits, where the law of dispersion has the form (7).

ESTIMATION VARIANCE AND EXTENSION VARIANCE

One of the most practical problems geostatistics are supposed to resolve is the size of the possible error in the evaluation of a deposit. The general characteristics of regionalized variables indicate that this error does not only depend upon the amount of picked samples, but first of all upon their shapes, their sizes and their respective locations, in other words, on the whole, upon *the geometry of achieved mining workings*. These indications get a precise meaning through the geostatistical notion of the estimation variance. Let us suppose that, in order to estimate the real unknown grade z of a deposit or of a panel V, we know the grade x of a given net-work of mining workings Mw. The estimation error $(z - x)$ has a simple, well determined value, although unknown, for a given panel V as for the net-work Mw located preferentially. In order to make out of this error a regionalized variable, geostatistics consider the panel or the deposit to be estimated as a panel extracted from a very large fictive deposit K. This deposit is supposed to be ruled by the intrinsic dispersion law $\gamma(h)$ defined by the experimental variogram controlled in mining works Mw. We shall see that the shape and the sizes assigned to K do not actually intervene. Let us imagine that panel V which is being estimated travels across the large deposit K, drawing with its attached mining works, the error $(z - x)$ then appears as a regionalized variable with an average value equal to zero and a variance:

$$\sigma^2 = \sigma_z{}^2 + \sigma_x{}^2 - 2\sigma_{zx}. \tag{10}$$

This variance called *estimation variance* is calculated after variances $\sigma_z{}^2$, $\sigma_x{}^2$ and covariance σ_{zx} of the variables z and x inside the field K, which are themselves given by formulae of type (4) or (6). Field K interferes in the expression of $\sigma_z{}^2$; $\sigma_x{}^2$ and σ_{xz} by the simple constant $F(K)$ which is eliminated in equation (10), so that the estimation variance σ^2 is independent from the choice of K, and is calculated after the formula:

$$\sigma^2 = \frac{2}{VV'} \int_V dV \int_{V'} \gamma(h) dV'$$

$$- \frac{1}{V^2} \int_V dV \int_V \gamma(h) dV' - \frac{1}{V^2} \int_{V'} dv \int_{V'} \gamma(h) dv'. \quad (11)$$

In (11) V is the volume of the deposit being estimated and V' that of mining works Mw. The estimation variance σ^2 is calculated after integration of the intrinsic function $\dot{\gamma}(h)$ inside the geometrical volumes of the deposit and of the samples. In the same way, as the variogram could give to the concept of the influence zone of a sample a precise content, one may say that the estimation variance (11) can give a precise meaning to the "influence" of mining works over the whole deposit.

In practical calculations, formula (11) should be difficult to use. Mining works usually frame a discontinuous net-work in which the samples themselves may be picked discontinuously (for instance, groove samples cut off according to a regular grid pattern in the drifts on a vein developed at different levels). Volume V' interfering in (11) is the discontinuous volume set up by the lattice of samples actually cut off and analyzed. An influence zone is traditionally assigned to each individual sample, in the center of which it is located and supposed to represent the grade. The error usually performed in extending the grade of such an individual sample to its influence zone can be represented by a type (11) variance, where V is the volume of influence zone and V' that of the sample.

Such a variance is called *elementary extension variance* and can be calculated for a given $\gamma(h)$ in terms of geometrical parameters of the sample and its influence-zone. On condition of certain approximation hypothesis, it is possible to prove that an estimation variance of type (11) can be calculated by composing the elementary extension variances.

In practice, two cases are to be distinguished essentially. The elementary samples network, for an isotropic function $\gamma(h)$ may be isotropic[5] or not. Let us mention, as an easy example of isotropic network, the square grid pattern drilling. The errors made for an isotropic network by extending to each influence zone the grade of its central sample may be considered as independent (in other words having a geostatistical covariance equal to zero). In this case *estimation variance is obtained by dividing the extension variance $\sigma_E{}^2$ of each sample within its influence zone by the number N of these influence zones.*

$$\sigma^2 = \frac{1}{n} \sigma_E{}^2. \quad (12)$$

[5] More generally, for any given function $\gamma(h)$, the lattice may or not be adjusted to the anisotropy of function $\gamma(h)$. For questions concerning the different types of anisotropy, one should refer to the *Treatise of Applied Geostatistics*.

If, on the contrary the network is not isotropic, we are led to rearrange the samples along lines or planes of maximum density, and to compose extension variances of different natures. For example let us suppose a vein-type deposit developed by drifts and channel sampled. In the first place, we have to consider the extension variance $\sigma_{E_1}{}^2$ of a channel within the length of a drift from which it has been cut off. If N is the total number of channels, one can see that the estimation variance $(1/N)\sigma_{E_1}{}^2$ represents the error obtained by extending the grade deduced from channel samples over the mining works themselves. We consider afterwards the extension variance $\sigma_{E_2}{}^2$ of the grade (supposed to be perfectly well known) of a drift inside its influence zone. The influence zone is here the panel composed by joining both the half-levels located above and below the drift. If n is the number of developed levels, the estimation variance $(1/n)\sigma_{E_2}{}^2$ represents the error obtained by extending the average grade supposed to be perfectly well known of the mining works to the whole deposit. The resulting estimation variance becomes:

$$\sigma^2 = \frac{1}{N}\,\sigma_{E_1}{}^2 + \frac{1}{n}\,\sigma_{E_2}{}^2. \tag{13}$$

It is usually necessary to add an additional variance to this expression, representing the sampling and analyses errors. The second term in such an expression is usually broadly predominating. The greater part of the error proceeds from the extension of data from the mining works to the deposit. In particular, it would be no use to increase indefinitely the number N of samples without carrying out supplementary mining works. In fact, the estimation variance coincides very soon with the $(1/n)\sigma_{E_2}{}^2$ limit below which it cannot decrease.

Tables and graphs giving the numerical values of elementary extension variances have been established [6] for a given number of intrinsic functions (especially for type (7) of de Wijs's function). They allow a fast computation of estimation variances assigned to different drilling and underground exploration schemes.

We offer for example a vein deposit conformable to a type (7) isotropic de Wijs's scheme and developed by drifts. Let us also assume that drifts have been sufficiently well sampled as to reduce the first term of equation (13) to zero.

Let h be the raise between two consecutive levels (measured inside the plane of the vein). The extension variance of a drift of length l within an influence panel lh is proved to be:

$$\sigma_E{}^2 = \alpha\,\frac{\pi}{2}\frac{h}{l}.$$

This formula is valid only if h is small compared to l, but it may be used until $h = l$. When $h > l$, it must be replaced by a different formula. Let

---

[6] *Treatise of Applied Geostatistics*, Vol. I, for the de Wijs's functions. Vol. III for the case of a nugget effect.

us assume that lengths $l_1, l_2 \cdots l_n$ are all superior to h. The estimation variance is obtained by weighting the extension variance of each drift within its influence panel by the square of the surface of this panel:

$$\sigma^2 = \frac{l_1^2 \sigma_{L_1}^2 + l_2^2 \sigma_{E_2}^2 + \cdots}{(l_1 + l_2 + \cdots)} = \alpha \frac{\pi}{2} h \frac{l_1 + l_2 + \cdots}{(l_1 + l_2 + \cdots)^2}.$$

The explored mineralized surface being $S = h(l_1 + l_2 \cdots + l_n)$ and the total developed length being $L = l_1 + l_2 + \cdots l_n$, we obtain the following remarkable formula:

$$\sigma^2 = \alpha \frac{\pi}{2} \frac{S}{L^2}.$$

Once the estimation variance has been calculated, one has still to interpret it for practical uses under the form of *conventional error spread*. This aim is reached by allocating to this variance a probabilistic meaning. By implicit reference to a gaussian model, we shall take it that the actual average grade of the deposit is included within a 95% probability in the range $m \pm 2\sigma$, m being the estimated grade. In other cases, particularly if 2σ is not small towards m, we shall take the spread $m \exp(\pm 2\sigma/m)$, by reference to a lognormal model.

These implicit references to probabilistic models are mainly arbitrary. Actually, the notion itself of statistical distribution of an estimation error is doubtlessly meaningless. The only thing which has an objective physical meaning is the variance. This is why we speak about conventional spreads. Their practical interest resides in the fact that they draw a more intuitive picture of the possible errors than variances themselves.

KRIGING

A second application of major importance is provided by a geostatistical procedure called "kriging." It consists in estimating the grade of a panel by computing the weighted average of available samples, some being located inside others outside the panel. The grads of these samples being x_1, $x_2, \cdots x_n$, we attempt to evaluate the unknown grade z of the panel with a linear estimator z^* of the form:

$$z^* = \sum a_i x_i.$$

The suitable weights a_i assigned to each sample are determined by two conditions. The first one expresses that z^* and z must have the same average value within the whole large field V and is written as:

$$\sum a_i = 1.$$

The second condition expresses that the a_i have such values that estimation variance of z by z^*, in other words the kriging variance, should take the smallest possible value.

This is formulated with a linear equation system related to a_i, the coefficients of which are expressed with the help of the variances and covariances of the samples and of the panel. It is thus possible to tabulate, for each intrinsic function, the coefficients and the kriging variance in terms of geometrical parameters, appropriately for different configurations. Numerous drilling and underground work configurations have thus been tabulated in

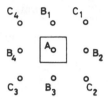

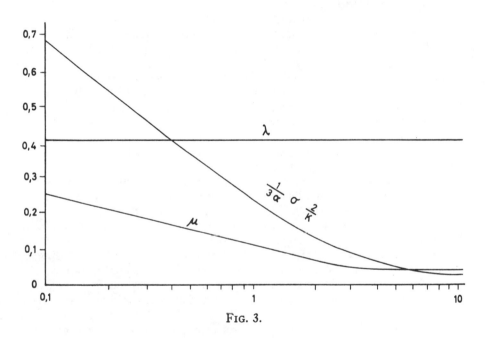

FIG. 3.

the case of an isotropic scheme of de Wijs. For information we show an example in Figure 3. The studied configuration is useful for the appraisal of a deposit explored by drilling, or for open cast selective mining. It consists, in the case of a square grid pattern drilling, in the kriging of the influence blocks of a drilling A with help of the grade of this central drilling A, and those of the 8 nearest drillings rearranged into two "aureolae" $B_1B_2B_3B_4$ and $C_1C_2C_3C_4$. Let u be the grade of A, v and w the average grades of drillings

B and C, the estimator to be used is:

$$z^* = (1 - \lambda - \mu)u + \lambda v + \mu w.$$

In Figure 3, is plotted in abscissa the ratio h/a between the width h of the formation and the size a of the mesh of the drilling grid and the numerical values of λ and μ are read on the curves as well as these of the expression $(1/3\alpha)\sigma_K^2$. The multiplication of this last expression by three times the value of absolute dispersion, 3α, yields the kriging variance.

Theoretically it is advantageous to "krige" each panel by all the samples located in the deposit, inside and outside this panel. In addition to the great complexity of computation which grows very fast, it appears in numerical examples that it is usually unnecessary to take into account remote samples. In general the one or two proximate aureolae of external samples are enough to remove practically the whole effect of remaining external samples. This is, in particular, the case of the configuration studied in Figure 3 where both aureole B and C form an almost perfect screen towards all other external drillings.

One can even notice that for high values of the h/a ratio, the weight μ of the second aureole becomes slight, so that the aureole made out of the four B drillings constitutes a screen by itself alone. This *screen effect* is a general phenomenon and plays an important part in the kriging theory.

From a practical point of view, the advantage of kriging is double. First of all, as a result of the definition itself of this procedure, it leads to achieve the best possible estimation for a given panel, that is to say the estimation with minimal variance. It can pay most appreciable services by improving, for example, the monthly output forecast for different mine-sections, and especially in the case where the mine operator is compelled to supply ores with characteristics as constant as possible.

However appreciable they are, the improvements of accuracy provided by the kriging would not always justify the amount of calculations it requires. In most cases, the major interest of the procedure does not come for the reduction of estimation variances but from its being able to eliminate the cause of systematical error. A deposit seldom happens indeed to be payable in the whole. Only some panels chosen as payable according to the grades of the samples cut off within them, are considered as payable. D. G. Krige [7] has proved that the results based only on inside samples inevitably led to over-estimating rich panels and underestimating poor ones. The geostatistical notion of kriging allows to expound this phenomenon easily and to rectify its effects. The selected panel being a rich one, the aureola of outside samples has, in general, a lower grade than that of inside samples. Yet its influence on the panel to be estimated, is not negligible, since it is allocated a weight different from zero by the kriging. Not to take in account this external aureole inevitably introduce therefore a *cause of systematical error by over-estimation* which can be eliminated by kriging.

---

[7] D. G. Krige's original reasoning constitutes a second example of an implicit passage from classical statistics to geostatistics. It is essentially based on the fact that the variance of a panel is always lower than that of its inside sampling. For references, see "Treatise of Applied Statistics."

THE NUGGET EFFECT

In the presence of a strong nugget effect the general rules outlined in the above paragraphs may suffer some apparent objections. The nugget effect has been defined in Figure 1 by a variogram characterized by a discontinuity at the origin, and corresponding to a regionalized variable that does not have the "in average" continuity. Its nature may be purely granulometrical, as in gold or diamond deposits, or, more generally, it may reveal the existence of discontinuous micro-structures. The presence of veinlets or microfractures with high-grade fillings in a stockwerk may promote such an effect. In gold deposits, the grades of two very close or even adjoining samples may be different if, by chance, one of them contains a large nugget. The smaller the samples, the more important this effect is, and it may reach a considerable magnitude for samples of several liters in volume. A translation of some millimeters only of the geometrical support of a sample is enough for it to contain or not a large nugget able to modify its grade in a proportion of 1 to 10 or 1 to 100. The possibility for a marginal nugget to be embodied, or not, inside a sample appears as an entirely random event. Actually, however, the behavior of the grade can be considered as random locally only. If it were not so, the panels of several thousand tons, on which marginal nuggets have no more detectable effect, would present almost constant grades (their variance being then a million times lower than that of samples of several kilograms). It is well known that actually, even in gold deposits, there are rich panels and poor panels. But this random effect may locally be so strong that it entirely hides the underlaying regionalization. The frequency of some expressions such as "erratic," "monstrous," or "mammoth grades" etc. . . . alluding to an hypothetical anomalous behavior of mineralization in the literature devoted to these deposits is striking. Certainly the classical statisticians were right when they noted that there was no actual anomaly, and that those monster grades, actually existing in the deposit, appeared from time to time in the sampling, with frequency determined by random laws. Historically, a clear distinction between the notions of regionalized and aleatory variables was doubtlessly hampered for a long while by the fascination aroused by this nugget effect. It appears, from the geostatistical point of view, that in fact, the ingenious terminology was not wrong while suggesting the existence of some anomaly; but the aberrant fact is not the presence of some "anomalously" high grades, but rather the locally aleatory behavior of all the grades, high or low, as well as in the deterioration of the spatial correlations grid. Those mammoth grades of the ingenious terminology are not aberrant by themselves, but the fact that they are not assorted with influence zones is so. And, on the other hand, classical statisticians were right stressing the fact that the apparitions of these aberrant grades are ruled by random laws. But they failed to note that the phenomenon can be considered as aleatory locally only.

Without trying to make a systematical statement,[8] let us show briefly how geostatistics allow us to represent a nugget effect. Let us examine the

---

[8] See *Treatise of Applied Geostatistics*, Vol. III.

$\gamma(r)$ semi-variogram representing the third type of Figure 1. We shall stick here to the case where $\gamma(r)$ is an isotropic function (in other words depending only upon the r modulus of the h vectorial argument). The C discontinuity, or jump, noticed at the origin on the $\gamma(r)$ of a variable with punctual support, is called *nugget constant*. $H(r)$ being Heaviside's function, thus defined:

$$\begin{cases} H(r) = 1 & r > 0 \\ H(r) = 0 & r = 0 \end{cases}$$

the semi-variogram may be divided into 2 components:

$$\gamma(r) = CH(r) + \gamma_1(r). \tag{15}$$

The first component $CH(r)$ represents the pure nugget effect. The second one $\gamma_1(r)$, continuous at the origin, represents the underlaying regionalization. All the variances and the covariances that have to be introduced, may then be calculated as if the variable $x(M)$ with punctual support was the sum:

$$x = x_0 + \epsilon \tag{16}$$

of a theoretical regionalized variable x_0 following the $\gamma(r)$ dispersion law continuous at the origin, and of an aleatory ϵ variable with a zero average and C variance.

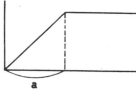

Fig. 4.

The x_0 and the ϵ are independent, and the ϵ assigned to two distinct points even very close, are independent as well. If we limit our study to the variation of the punctual grade x in the proximity of a given point, or, in other words, we consider only the small values of the distance r, $\gamma_1(r)$ will vary so slightly that it might be taken for a constant equal to C. The locally detectable variations are to be assigned almost solely to ϵ. That is what we mean when we say that the regionalized variable behaves locally as an aleatory variable. But, on a larger scale, i.e., for higher values of r, the increase of the continuous component $\gamma_1(r)$ can no longer be neglected and the regionalization of x_0 becomes perceptibly apparent.

As a matter of fact, the Heaviside function does not represent with entire satisfaction the random aspect of the behavior of a punctual variable. Unless we suppose the constant C to be infinite, the term $CH(r)$ will lose all influence over the variance of a sample of a size different from zero. It is automatically eliminated in formula (4). It means that the mean value of the ϵ independent aleatory variables, located in infinite number inside an unpunctual support, has compulsorily a zero variance.

The notion of a random variable ϵ with a punctual support has actually no physical meaning. The actual physical phenomenon will never involve a true discontinuity at the origin but a narrow transition zone in the proximity of $r = 0$. $H(r)$ must be replaced by the transition function $T(r, a)$ defined by:

$$\begin{cases} T(r, a) = \dfrac{r}{a} & \text{if} \quad r \leq a \\ T(r, a) = 1 & \text{if} \quad r > a. \end{cases}$$

The a constant, or *range*, gives the scale of the transition zone, that is to say the size of the nuggets. In the case of homogranular nuggets of same volume u it is shown that:

$$u = \frac{\pi}{3} a^3.$$

The intrinsic function $\gamma(r)$ of a punctual grade is decomposed in the following way:

$$\gamma(r) = CT(r, a) + \gamma_1(r).$$

C is still the nugget constant, and $\gamma_1(r)$ the continuous component.

The punctual grade x can be given by a sum similar to (16) in which ϵ is a regionalized variable admitting $CT(r, a)$ as its intrinsic function. Now the ϵ are only independent for distances superior to the range a. For smaller distances they are bound by a linear variogram. The nugget effect will therefore reflect itself on samples of size v different from zero. If v is large in regard to the grain size a^3 the transition zone will be diluted in the integration volume v, and the nugget effect will yield an additional variance of the type a^3/v. Indeed, let $\sigma_P{}^2$ (nugget variance) be the share of $CT(r, a)$ for the variance of sample v. According to (4) we have to compute integrals of the type:

$$\frac{C}{v^2} \iiint_v dv_1 \iiint_v T(r, a) dv_2.$$

If all sizes of v are supposed to be large in regard to a, each point inside v brings to the sextuple integral the following part:

$$C(v - \tfrac{4}{3}\pi a^3) + \frac{C}{a} \int_0^a 4\pi r^3 dr = C\left(v - \frac{\pi}{3} a^3\right).$$

This is valid only for points located at a distance superior to (a) from the boundary of v; but, when v is large, the boundary points only interfere with superior order terms. With such an approximation, the sextuple integral is equal to $C(1 - (\pi/3)(a^3/v))$.

As the integral inside V is computed in the same way, we finally have

$$\sigma_P{}^2 = C\frac{\pi}{3}\left[\frac{a^3}{v} - \frac{a^3}{V}\right]. \tag{17}$$

Practically a^3/V is negligible and the nugget variance is in terms of a^3/v, i.e., in an inverse ratio to the number of grains contained inside the sample.

Any time a nugget effect does exist, i.e., anytime a regionalized variable shows a locally aleatory behavior, an additional variance is assigned to macroscopic samples, called nugget variance, inversely proportional to their size.

The variance of those samples appears as the sum:

$$\sigma^2 = \sigma_P{}^2 + \sigma_\theta{}^2$$

of the nugget variance and of the theoretical variance $\sigma_\theta{}^2$ calculated with the continuous component $\gamma_1(r)$ of the intrinsic function.

When v is increasing, the theoretical variance is decreasing much slower than the nugget variance. In the presence of a very strong nugget effect $\sigma_p{}^2$ may happen to be widely predominating for samples of several kilograms. The underlying regionalization is almost completely hidden at the scale of these samples. If we limit the variation of the volume v in the interval of a few liters up to tens of liters, the experimentally observable variations of the variance will be those of the nugget variance effect only, and we may take the risk to conclude that the variance varies in inverse ratio of the volume.

Whereas if we consider samples of several tens of tons, the term $\sigma_p{}^2$ decreases and disappears, and the theoretical variance $\sigma_\theta{}^2$ becomes prominent. The effect of the underlying regionalization appears again and the variance is steadily decreasing as v is increasing, but much slower than $1/v$.

We have somewhat insisted upon the nugget effect in order to show, through a crucial example, how geostatistical concepts allow us to rediscover the local results that are fluently obtained from common statistical reasoning (nugget variance inversely proportional to volume) but inserting them in the general prospect of an underlying regionalization. As for the practical use of this theory, let us succcinctly mention the two following points:

In the presence of a nugget effect, the extension and the estimation variances are both increased by a term $C(\pi/3)(a^3/v)$ inversely proportional to the total volume of available samples and, therefore, in particular to the number n of those samples. In this regard, the additional estimation variance due to the nugget effect behaves itself as the sampling and analyses variances, and may be rearranged with them.

As for the kriging, the nugget effect results in partly removing all the screens. Practically, we are led to use the special forms of kriging called "aleatory kriging" which are not different from those proposed formerly by D. G. Krige himself, in connection with the gold deposit of the Rand, in which the nugget effect is probably very strong.

SEARCH FOR OPTIMUM IN MINING EXPLORATION

Geostatistics are able, through estimation variances, to provide an accurate measurement of the information yielded, by a given amount of underground workings on a deposit. Generally, these workings are expensive, and their cost must be weighed against the economic value of the provided information. Thus appears the possibility to determine the optimum amount of credits to be allocated for the exploration of a deposit, and particularly the possibility to choose the suitable moment for stopping the exploration, as well as for taking a positive or negative decision towards starting the exploitation of the deposit. These methods, permit one to solve, at least partly, one of the main problems raised by mining exploration, will be published in another connection, and cannot be treated here. Let us only, as a conclusion, stress the fact that they appear as the natural extension of geostatistics. The possibility of their adjustment was bound to the preliminary elucidation and to the thorough scientific study of the different ideas which have been summarized in this paper.

BUREAU DE RECHERCHES GEOLOGIQUE ET MINIERES,
PARIS, FRANCE,
June 10, 1963

14

Reprinted from *Mines Mag.* **58**(3):13–15 (1968)

ORE RESERVE ESTIMATION

David G. Mickle

METHODS of sampling mineral deposits and combining the assay results into ore reserve estimates have undergone very little change for nearly half a century. In recent years, however, an increasing awareness of the inadequacy of the traditional methods has resulted in extensive research efforts, much of it by the U. S. Bureau of Mines,[1] aimed at developing more satisfactory sampling and ore estimation procedures.

Reawakening of interest in the problem can be attributed in part to the continuing trend toward mining ores of lower and lower grade. As the average grade decreases, so does the allowable margin of error in estimating that grade. The uranium industry, currently staging a dramatic resurgence, has also demonstrated the shortcomings of traditional methods. The irregular and spotty nature of many uranium deposits has resulted in ore estimates that are grossly different from production results.

Wright[2] has reported an experiment which illustrates the magnitude of the problem. Five uranium deposits were selected, four of which had been mined out and the fifth was so nearly mined out that a reasonable estimate of ultimate production could be made. Maps showing drill hole locations and assay data available before mining was prepared for each deposit. The five maps were given to each of eight geologists who were asked to estimate tonnage, percent U₃O₈, and percent V₂O₅ for each deposit. Comparisons were then made of the several estimates for a given deposit, and the average of these estimates was compared with actual production records.

To anyone who believes that ore reserve estimation is a fairly exact science, the results are truly startling. For a given deposit, the maximum tonnage estimate varied from 82% to 460% greater than the minimum, the maximum U₃O₈ estimate ranged from 15% to 102% greater than the minimum, and the maximum V₂O₅ estimate was 11% to 71% greater than the minimum. One might suppose that by averaging these widely varying estimates one might arrive at a fairly accurate prediction for each deposit. The data prove otherwise, however.

Actual tonnage realized varied from 30% to 330% of the average of the estimates. Quantity of U₃O₈ produced

About the Author

David G. Mickle joined the faculty of the Colorado School of Mines in September, 1967 as assistant professor of Mining. His educational background includes B.S. degree in Mining and Geological Engineering from the University of Idaho, an M.S. in Mining Engineering from the Pennsylvania State University, and a Ph.D. in Geology from the University of Idaho. He spent three years as mining methods research engineer at the U. S. Bureau of Mines in Denver before coming to the Colorado School of Mines.

Dr. Mickle's particular areas of interest are use of statistical analysis in mineral deposit sampling and evaluation, operations research techniques, and the application of computers to mining problems.

ranged from 20% to 240% of the average, and quantity of V₂O₅ ranged from 30% to 350% of the average. Gross value of the deposits ranged from 20% to 230% of the gross value estimated from the average of the tonnage and grade estimates.

For three of the five deposits, tonnage produced exceeded the greatest tonnage predicted by any of the estimators; for one deposit the tonnage was less than the smallest estimate. Percentage of U₃O₈ exceeded the highest estimate in one case and was less than the lowest estimate in three cases. V₂O₅ exceeded the highest estimate twice and was less than the lowest estimate twice.

By any yardstick, the estimates must be classed as very unsatisfactory. It is small wonder that interest in improved estimation methods runs particularly high in the uranium industry.

Before examining some recent developments in estimation procedures, it will be instructive to review briefly the traditional approaches and point out some of their shortcomings. Nearly all of the traditional calculation procedures fall into one of the following two categories:

(1) **Area of influence** methods, which assume that some area or volume of ore surrounding a sample has the same assay value as the sample. Usually the area of influence is assumed to extend half way to each adjoining sample.

(2) **Gradual change** methods, which assume a gradual linear change in ore grade from one sample location to the next.

Figure 1 illustrates the distribution of ore grade assumed by the area of influence methods. The abscissa represents distance along an exposed face of ore in a vein deposit, and the ordinate shows grade of ore at each location along the face. Changes in grade are assumed to occur as discrete jumps located half way between sample locations.

The fallacy of this assumption is obvious. In the first place, it is very doubtful that any mineral deposit exhibits the well-behaved pattern of grade changes shown in the figure. But the area of influence calculation procedures require in addition that samples be taken at exactly those locations which will cause the discrete grade changes to occur half way between the samples. The joint occurrence of these two improbable events is indeed very unlikely.

Exactly the same argument applies to the assumption of gradual changes **(Figure 2).** Both the assumed pattern of grade change, and the fortuitous selection of sample locations precisely at the changes in slope of the grade curve are highly improbable.

Numerous examples could be given demonstrating the failure of ore deposits to obey the assumptions underlying the area of influence and gradual change methods of ore estimation. **Figure 3** shows the variation in grade along a portion of a diamond drill hole, as reported by Hazen.[1] Each two-foot length of core was assayed, and **Figure 3** shows these assays plotted at the footage of the center of the sample interval. If the deposit had been sampled every ten feet rather than every two feet, one might have obtained the assays indicated by the circles. Obviously, these assay values do not extend half way to the adjoining samples, nor does the grade change along a straight line from one sample to the next. Of course, even the lines joining the assays at two-foot intervals are imaginary.

One must also keep in mind that assay values do not necessarily equal the actual metal content of the deposit at the sampled location. Variation is introduced during sample reduction and assaying, so that the assay is only an estimate of the grade at the place sampled. The amount of variation depends upon the nature of the ore, the sample preparation procedure used, and the assaying technique.

Rickard[3] reports the results of splitting each of 31 samples into two halves and assaying them separately. Designating one half as the "original" and the other as the "duplicate," it was found that the difference between the two assays ranged from 3 percent to 400 percent of the "original" assay, with an average difference of 72 percent.

A more recent example of duplicate assaying is quoted by Wolfe.[1] Presumably identical splits from eight samples were sent to each of three analytical laboratories. In no instance did the results check to the first significant figure after the decimal, and for three of the eight samples one laboratory differed by an order of magnitude from the other two. If by some chance the grade distribution in the mineral deposit had obeyed the area of influence or gradual change assumptions, the assay results would certainly give a very imperfect representation of that distribution.

It is significant to note that, from the very beginning, there was a recognition of the failure of ore deposits to meet the assumptions of the traditional ore reserve calculation procedures, but these procedures were advocated nonetheless because no better method was known. For example, in 1905 Crosley[5] recommended the gradual change method in preference to the area of influence method but had to admit that the assumptions underlying both were erroneous: "That this even and gradual rise or fall in value of the reef, of course, is not supposed to exist except for the purpose of calculation; but we are as much entitled to use the supposition as to say the value at one point exists unchanged, half way to the next, or as far as the next, and we must make the supposition if we are to use numbers and mensuration with any approach to accuracy."

Since the mining industry, for more than 50 years, has been in the curious position of using calculation procedures based upon admittedly false assumptions, how is it that the results, for the most part, have been accepted as satisfactory? Probably the answer is that we have been saved by the law of averages. Assays and weighting factors are not consistently biased in one direction but contain random errors, which tend to compensate for one another when a large number of assays are averaged. For example, it has been observed that ore reserves computed by the polygonal method, which assumes an area of influence extending half way to adjoining samples, often are reasonably accurate in terms of average grade even though the ore mined within any polygon may differ markedly from the assumed grade of the polygon.

Two serious problems result from the use of the traditional calculation procedures. **One is that they provide**

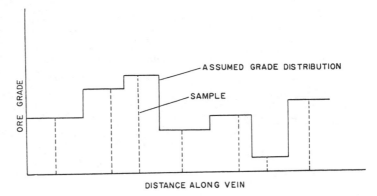

FIGURE I. AREA OF INFLUENCE ASSUMPTION

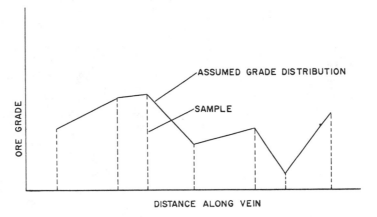

FIGURE 2.   GRADUAL CHANGE ASSUMPTION

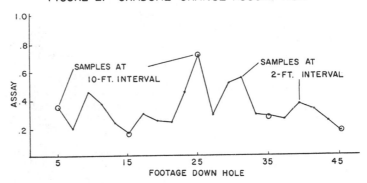

FIGURE 3. ASSAYS FROM A DRILL HOLE

no estimate of the reliability of the calculation. It is recognized that the accuracy of an ore reserve estimate depends upon the natural variability of the deposit, the number and size of samples taken, and the procedures used to prepare and analyze the samples. But the traditional methods furnish no estimate of the number or size of samples that should be taken if one wishes to evaluate a given deposit with a specified degree of accuracy.

The engineer must estimate the required number of samples on the basis of past experience, compute the ore reserves hoping that the number of samples was sufficient, and accept the result as valid because no information to the contrary is available. If the deposit differs markedly from ones previously investigated, sampling may be excessive or inadequate. The incorrect estimates for the uranium deposits previously discussed probably resulted from the failure of the estimation procedure to give warning that the deposits were erratic and required more sampling if a reliable estimate were to be obtained.

The second problem with the traditional methods is their tendency to conjure up a false conception of the distribution of ore in the deposit. For example, there is a widespread practice of drawing contour maps of ore grade by carefully interpolating between assay values using the assumption of a gradual linear change of grade between sampled points. If that assumption is false, the contour map will have little validity and is largely a waste of time.

Another example is the practice of calculating the grade of an ore block by carefully measuring the area of each partial or complete polygon that it intersects and weighting the assays at the polygon centers by these areas. This calculation, of course, assumes that the grade of ore throughout a polygon has the same grade as the sample at its center. Since the assumption is incorrect, the validity of the calculation is questionable.

An improved ore estimation procedure is required which will overcome the shortcomings of the traditional methods. In brief it should

(a) use a more reasonable assumption about continuity of grade between samples, or require no such assumption at all,

(b) furnish an estimate of the probable accuracy of ore reserve calculations,

(c) provide a means for determining the number of samples required to evaluate a given deposit within specified limits of accuracy, and

(d) aid visualization of actual grade distribution both in the deposit and in the various stages of sample reduction and analysis.

In recent years it has been recognized that well-established techniques of statistical analysis fit these requirements remarkably well. No assumption is made about continuity or trends between sample locations. Observed variability in the assay results can be used to estimate the range within which the true average grade of the deposit probably lies. Similarly, one can estimate the number of samples required from a particular deposit in order to narrow this range to a specified size. Analysis can be made of the amount of assay variability due to sample preparation and assaying rather than to the deposit itself, thus pinpointing practices which need improvement. Finally, statistical analysis recognizes the entire process of ore estimation as involving chance or probability at every step, rather than giving the misleading implication that an exact answer can be obtained.

It is not the intention here to review the techniques of statistical analysis which can be applied to ore reserve estimation. For an excellent summary, the reader is referred to the previously mentioned publication by Hazen.[1]

Statistical analysis is the most promising basis upon which to build improved methods of ore estimation, but the problem is not as simple as might appear at first glance. Several of the requirements of statistics, which textbooks usually assume are easily met in any sampling problems, are rarely achieved in ore deposit sampling. One is the requirement that samples be random, so that every time a decision is made to take a sample, every possible sample in the deposit has an equal chance of being selected. Theoretically one could achieve this by choosing three numbers from a random number table, letting them represent the X, Y, and Z coordinates of a point somewhere in the deposit, and taking the sample at this randomly-selected point. In practice, of course, this scheme would be excessively costly and wasteful. A drill hole or drift would be required to reach each sample site, resulting in an enormous cost per sample, and all the information acquired in reaching that point must be ignored because it does not meet the requirement of randomness.

Various suggestions for achieving random sampling of ore deposits have been made, but none seems to be completely satisfactory. Moreover, there are often sound reasons for not sampling at random even if it were practical, such as concentration of samples in areas critical to geological interpretation or the pinpointing of ore boundaries. It appears that in most cases we must accept the fact that sampling will not be random. We need to devise modifications of standard statistical procedures in order to gain the most information possible under this non-ideal situation.

Another requirement of many statistical techniques is that the data be distributed according to the bell-shaped normal curve. Few, if any, mineral deposits yield assay data that have a normal distribution. In some cases the distribution of logarithms of the assays is approximately normal, but this is not universally true. Highly skewed and multi-modal distributions are the rule rather than the exception.

Obviously, confidence limits and probabilities derived from the normal courve cannot be applied directly to such data. However, the use of stratified sampling, transformations, and sample grouping will often solve this problem and should be more widely adopted in ore reserve estimation.

Another point of divergence is that mineral deposit sampling is of a special type called **bulk sampling,** in which choice of the size, shape, and orientation of the sampling unit can have a profound effect upon the shape of the resulting assay frequency distribution. Under certain assumptions one can show that sample variance should be inversely proportional to sample volume, but non-random distribution of grade in the deposit and sample preparation procedures can complicate this simple relationship. The complexities of bulk sampling, particularly as related to ore deposits, provide a fertile area for significant research contributions.

In summary, we can say that methods of estimating ore reserves have long been overdue for a revision that would put them on a sound engineering basis. Although furthr research is needed to overcome certain difficulties, statistical analysis holds great promise for becoming the standard procedure of the future.

REFERENCES

1. Hazen. Scott W.. Jr.. 1967. Some Statistical Techniques for Analyzing Mine and Mineral-Deposit Sample and Assay Data: U. S. Bureau of Mines Bull. 621.

2. Wright. Robert J.. 1959. An Experimental Comparison of U_3O_8 Estimates vs. Production Eng. and Mining Jour.. v. 160. n. 11. p. 100-102. 188.

3. Rickard. T. A.. 1904. The Sampling and Estimation of Ore in a Mine: Eng. and Mining Jour.. New York.

4. Wolfe. John A.. 1962. Geostatistics and the Exploration Economy: Symposium on Mathematical Techniques and Computer Applications in Mining and Exploration. Univ. of Arizona. p. H1-H28.

5. Crosley. William. 1905. The Computation of Assay Values: Trans. Inst. of Mining and Met.. v. 15. p. 90.

ERRATUM

David G. Mickle received the Ph.D. in geology from the University of Arizona, not the University of Idaho as stated in "About the Author."

Reprinted from *Eng. and Mining Jour.* **180**(10):106–116, 119–123 (1979)

THE VOLUME-VARIANCE RELATIONSHIP: A USEFUL TOOL FOR MINE PLANNING

Harry Parker

Knowledge of the amount and average grade of ore in a deposit is vital to the engineer who plans a mining operation, a metallurgical facility, or downstream fabrication plant. Because this knowledge cannot be obtained directly, inferences must be made about the distribution of ore within the deposit from samples obtained by sampling programs. Geostatistical methods provide powerful insights into the prediction of average grade and tonnage.

All mineral deposits are made up of mixtures of ore and waste rock. Various mining methods may be employed to extract ore and waste, with varying degrees of efficiency. If a deposit is mined by hand, the miner can ensure a high degree of segregation of ore and waste material, resulting in a very high-grade, high-unit-cost product at a very low rate of production. On the other hand, if a deposit is mined using large draglines, bucket-wheel excavators, or power shovels, there may be little segregation of ore and waste, and a low-grade, low-unit-cost product will be produced at a high rate of production.

Obviously, an optimum mining method exists for each deposit—a method that yields a product of required average grade for downstream processing at a reasonable cost and at a suitable rate of production. After a mining engineer has analyzed equipment and labor requirements, he can estimate costs and productivity and, using a geostatistical method known as the volume-variance relationship, he can estimate average grade and tonnage that a given mining method will produce. Knowing the average grade and tonnage, he can use discounted cash flow analysis to select a mining method and production rate that maximize the net present worth or return on investment. The volume-variance relationship is also useful for mining and metallurgical planners confronted with the problem of assessing the periodic variability of ore grades in a mining scheme.

VOLUME-VARIANCE RELATIONSHIP

Common to all mining methods is the notion of the selective mining unit—the smallest practical volume that can be classified as ore or waste. For a manual mining operation, this unit is obviously a muck-car load, while for a mechanized open-pit operation it might be an individual truck load or even an entire shift's production from a face. To estimate ore reserves, an engineer must know the frequency distribution of the selective mining units.

For most mines, these data are not directly available. Samples are comprised of much smaller volumes, so the problem is to predict the frequency distribution of selective mining units by examining the distribution of appropriate samples.

This is done by assuming that each selective mining unit is comprised of the aggregation of many sample-sized volumes (Fig. 1). Implicitly, the selective mining units will have the same average grade as the sample-sized volumes, without noting at this stage whether they are ore or waste. Schurtz[1] recognized that the variance of the frequency distribution—a parameter that measures the tendency of selective mining units to differ from the mean—will be large if the selective mining units are small in volume. Conversely, the variance will be small if the selective mining units are large in volume. Indeed, the variance vanishes if the selective mining unit is as large as the whole deposit.

The inverse relationship between volume and variance can be seen intuitively by observing that small selective mining units tend to be uniform in grade throughout—some units are all high grade and others all low grade. Thus, strong deviations from the mean may occur. For larger selective mining units, there tends to be a mixture of high and low grade material, and average grades tend to be nearer the average for the whole deposit.

VARIANCE OF THE SMU

The variance of selective mining units (smu) is found using a simple formula:

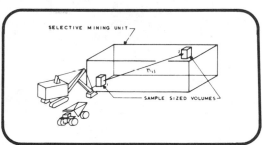

Fig. 1 — A selective mining unit is composed of the aggregation of sample-sized volumes, two of which are shown here. The vector h_{ij} separates them.

$$\sigma^2_{smu} = \overline{\gamma}_D - \overline{\gamma}_{smu} \qquad (1)$$

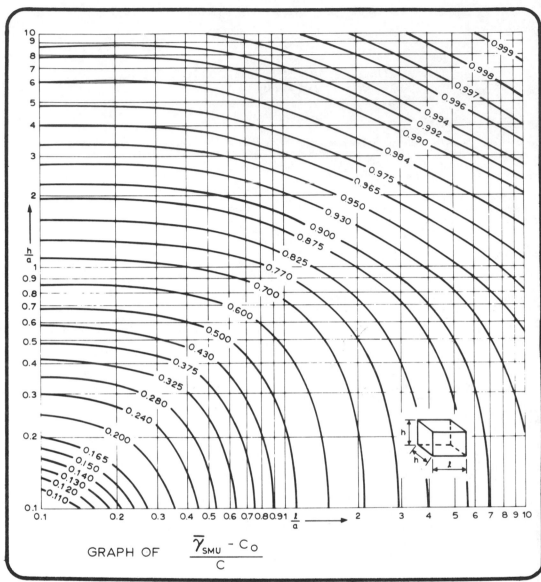

GRAPH OF $\dfrac{\overline{\gamma}_{\text{SMU}} - C_0}{C}$

Fig. 2—Graph of $(\overline{\gamma}_{\text{smu}} - C_0)/C$, for parallelepiped blocks (David[2]). C is the sill value of the variogram, C_0 is the nugget effect, and a is the range. The dimensions of the block are h and l.

$\overline{\gamma}_D$ is the average value of the variogram function within the deposit and $\overline{\gamma}_{\text{smu}}$ is the average value of the variogram function within the selective mining unit. To compute these values, a vector must be drawn between all possible pairs of points lying within the deposit or the selective mining unit, respectively. For each pair, the variogram value is noted, and the average of all of these is computed. In these computations, the vector between points A and B is not assumed to be the same as that between points B and A, even though the variogram values will be the same. Vectors of zero length, such as $A \longrightarrow A$ and $B \longrightarrow B$, also must be included and assigned a variogram value equivalent to the nugget effect.

Computation of the average variogram values is tedious if done by hand. Frequently the estimator assumes that $\overline{\gamma}_D$ is equivalent to the variance of sample grades (or variogram sill and nugget effect), provided the range of the variogram is short compared with the dimensions of the deposit. To obtain $\overline{\gamma}_{\text{smu}}$, simple computer programs can be written that do the calculations for a representative number of points, usually 50, either on a grid or scattered at random within a unit. David[2] discusses this process in more detail and provides a computer program for the purpose. If the grid approach is used, the grid should be defined so that the outermost points are half a grid spacing from the boundary of the selective

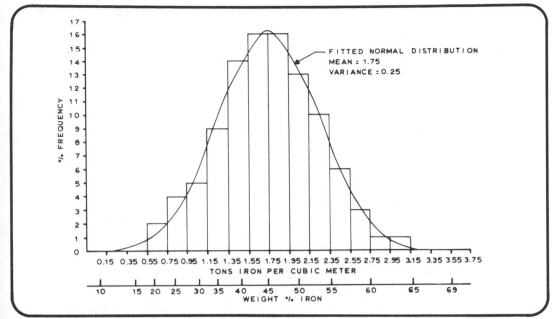

Fig. 3—Normal distribution of samples for an iron ore deposit.

mining unit (Clark[3]).

David[2] provides charts to be used if the shape of the selective mining unit is simple and the variogram is of a standard type. One of these charts, reproduced in Fig. 2, is a graph of:

$$\frac{\overline{\gamma}_{\text{volume}} - C_0}{C} \tag{2}$$

where C_0 is the nugget effect, and C is the sill for a spherical variogram. For example, a block 15 m high, 15 m deep, and 30 m long, with the range a of the variogram* equal to 60 m, has a graph value of 0.325, corresponding to $h/a = 15/60 = 0.25$, $l/a = 30/60 = 0.5$. If C is 0.8 and C_0 is 0.3, $\overline{\gamma}_{\text{volume}} = \overline{\gamma}_{\text{smu}}$ has a value of 0.56. Now if the deposit has dimensions 300 x 300 x 600 m, h/a is now 5.0, and l/a is 10.0. Thus we obtain a graph value of 0.998, and $\overline{\gamma}_{\text{volume}}$ $(= \overline{\gamma}_D)$ is now equal to 1.098, or practically equivalent to the sill plus nugget effect of 1.10. The variance of selective mining unit grades is then $\overline{\gamma} - \overline{\gamma}_{\text{smu}} = 1.10 - 0.56 = 0.54$.

SMU FREQUENCY DISTRIBUTION

Estimation of the frequency distribution of selective mining units requires statistical inference. For many mineral deposits, the sample grade distribution approximates a normal or lognormal distribution. If the sampling distribu-

tion is normal, then the distribution of selective mining unit grades, which represent a linear combination of the grades of sample-sized units within the selective mining units, must be normal. Where the sampled grade distribution is lognormal, it cannot be proven that the distribution of selective mining units must also be lognormal; however, Switzer and Parker[5] and David[2] have shown that this is empirically true.

For more complicated sampling distributions, Parker and Switzer[6] discuss a case having a mixed lognormal and normal population. In another approach, Marechal[7,8] and Dagbert and David[9] use functions that transform the sample histogram into a normal distribution, find the distribution of selective mining units for the normal variable, and then retransform to obtain the distribution of the selective mining units themselves.

In the following paragraphs, two cases are discussed. The first involves a normal distribution and the second a lognormal distribution.

Example 1—a normally distributed iron ore deposit

Fig. 3 illustrates the normal distribution of samples taken from an iron ore deposit, the variograms for which are shown in Fig. 4. Since the density of iron ore varies with grade, any averaging must include weighting by density, which is difficult since density is a nonlinear function of grade. Therefore the units that have been used are density-free tons of iron per cubic meter. The following formula adapted from Parker[10] has been used for conversion:

$$\frac{\text{Tons Fe}}{\text{m}^3} = \frac{A\rho_0\rho_g Z}{100 \, (A\rho_0 - Z \{\rho_0 - \rho_g\})} \tag{3}$$

where A = percentage of iron in ore mineral, in this case 69.9% Fe in hematite; ρ_0 = density of ore mineral, in this case 5.26 gm/cc; ρ_g = density of gangue minerals, in this

---

*David's chart assumes that the variogram is comprised of dimensionless "point" samples. This assumption is generally acceptable if sample dimensions are less than ⅕ of the corresponding dimension of the block. If sample dimensions are greater, the sample variogram must be transformed to that of a point sample using a process known as regularization (David[2] and Clark[4]). Alternatively, one may use other charts. For instance, if the sample length is equivalent to the height of the block, but the sample dimensions are small enough to be considered as a point in the horizontal dimension, $\overline{\gamma}_{\text{smu}}$ may be found from another graph in David[2]. In complex cases, the computer method indicated above is preferred.

183

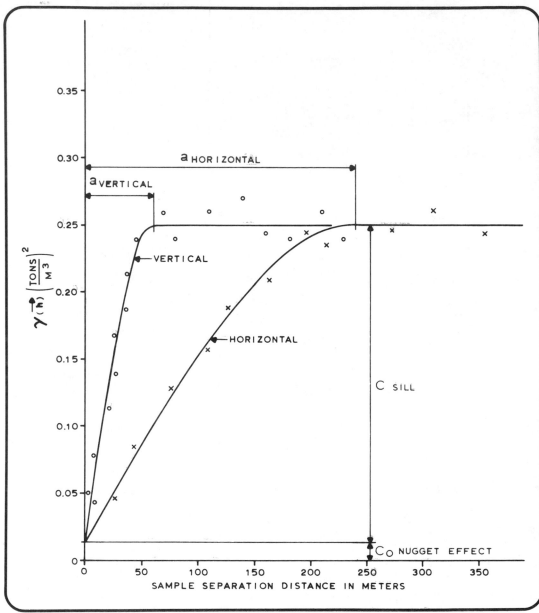

Fig. 4—Variogram for tons of iron per cubic meter.

case 2.65 gm/cc; and Z = weight percent iron.

Using W to represent tons per cubic meter in this example:

$$W = \frac{974.34Z}{36767 - 261Z} \qquad (3a)$$

and, conversely:

$$Z = \frac{36767W}{974.34 + 261W} \qquad (3b)$$

Table 1 shows the variances and cutoff grades for selective mining units that might typically be associated with operations of various sizes. The variances were derived from a computer program employing the spherical models fitted to the variograms in Fig. 4.

Table 1—Geostatistical parameters for the iron ore example

	Production rate (tpd)	Loading unit size (cu m)	Dimensions of selective mining unit (m) length x width x height	Variance, σ^2	Cutoff grade,* % Fe	Proportion of tonnage above cutoff	Average grade above % Fe cutoff	Proportion of total metal contained in ore selective mining units	Profitability per ton	Total profitability
Case 1	500	1.5	10 x 5 x 6	0.225	67.6	0.001	68.8	0.001	1.195	0.001
Case 2	1,500	2.5	15 x 7 x 9	0.217	55.0	0.103	57.9	0.132	2.952	0.303
Case 3	10,000	4	20 x 10 x 12	0.210	44.1	0.580	50.6	0.65	6.470	3.754
Case 4	20,000	9	25 x 10 x 12	0.209	43.0	0.631	50.0	0.700	6.978	4.403
Case 5	40,000	14	25 x 15 x 15	0.203	42.7	0.648	49.7	0.715	7.076	4.587

*Cutoff grade selected for each case reflects order of magnitude economic studies.

Assuming that the W values, or tons of iron per cubic meter, are normally distributed, then the mean of the deposit can be written as μ tons per cubic meter and the standard deviation as σ tons per cubic meter. The distribution of selective mining units will also be normal, with a mean of μ tons per cubic meter and a variance of σ^2_{smu}/n tons per cubic meter, which can be read from the appropriate column in Table 1. The usual procedure when dealing with a normal distribution is to convert to the so-called Standard Normal, for which tables are widely available. For any W value, the standardized normal value corresponding to it is found by computing:

$$X = \frac{W - \mu}{\sigma}$$

for "point" samples, or

$$X_{smu} = \frac{W - \mu}{\sigma_{smu}}$$

for selective mining units. Given a cutoff value expressed in tons per cubic meter (W_c) and writing

$$X_c = \frac{W_c - \mu}{\sigma_{smu}}$$

then the volume proportion of the deposit containing values above the cutoff is

$$V_c = 1 - \Phi(X_c) \qquad (4)$$

where $\Phi(\ )$ is the cumulative probability function for a standard normal variable, which is available in most statistics tables and text books.

The average value (tons per cubic meter) of the proportion of the deposit above cutoff is

$$\overline{W}_c = \mu + \frac{\sigma_{smu}}{V_c} f(X_c) \qquad (5)$$

where $f(X_c)$ is the height of the normal probability density curve at X_c and may be calculated by

$$f(X_c) = \frac{1}{2\pi} \exp\left(-\frac{1}{2} X_c^2\right)$$

The average grade, in %Fe by weight, can be found by setting $W = \overline{W}_c$ in equation (3b).

$$\overline{Z}_c = \frac{36767 \overline{W}_c}{974.34 + 261 \overline{W}_c} \qquad (6)$$

The tonnage proportion of the deposit that is above cutoff is given by

$$T_c = \frac{V_c \rho_c}{\rho_\mu} \qquad (7)$$

where $\rho_c = 100\ \overline{W}_c/\overline{Z}_c$ is the average density of the ore above cutoff, and $\rho_\mu = 100\mu/\overline{Z}_0$ is the density of the whole deposit.

The proportion of metal in the deposit contained in the selective mining units is

$$R = \frac{\overline{Z}_c T_c}{\overline{Z}_0} \qquad (8)$$

The difference between the average grade of selective mining units above the cutoff grade and the cutoff grade is a measure of the profitability per ton of production:

$$E = \overline{Z}_c - Z_c \qquad (9)$$

Finally, a concept of total profitability over the operating life is

$$E_T = E T_c \qquad (10)$$

These variables are graphed as a function of cutoff grade in Figs. 5-9. It is important to observe that the curve based on samples is nearly identical to that based on selective mining units. From Table 1, it is clear that the variance of selective mining units does not vary markedly with their size; hence, only the curve for case 3 (10,000 tpd) has been plotted on the graphs. The reason for this relative lack of variance is that many sedimentary iron ore deposits show strong continuity of grade on the scale of selective mining units, meaning that the range of the variogram is extremely large with respect to the block size. The averaging effect that should reduce the variance is therefore weak.

Curves are also shown that relate cutoff grade to produc-

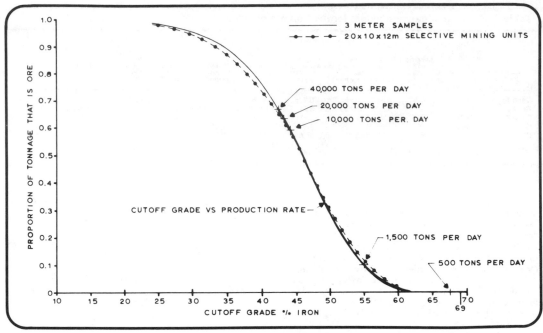

Fig. 5—Cutoff grade vs. tonnage for an iron ore deposit.

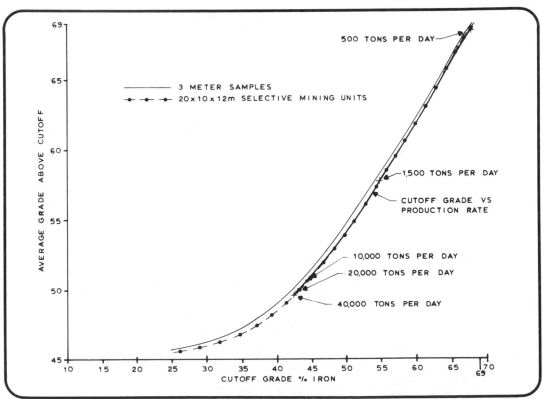

Fig. 6—Cutoff grade vs. average grade for an iron ore deposit.

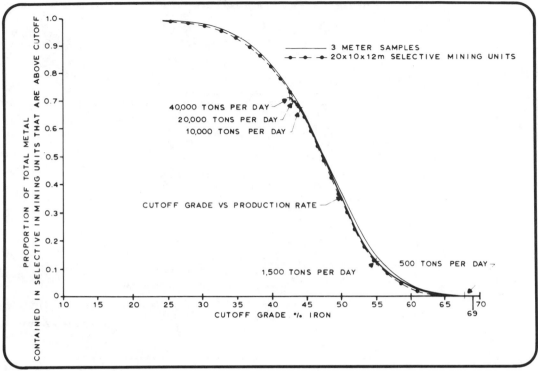

Fig. 7—Cutoff grade vs. metal recovery for an iron ore deposit.

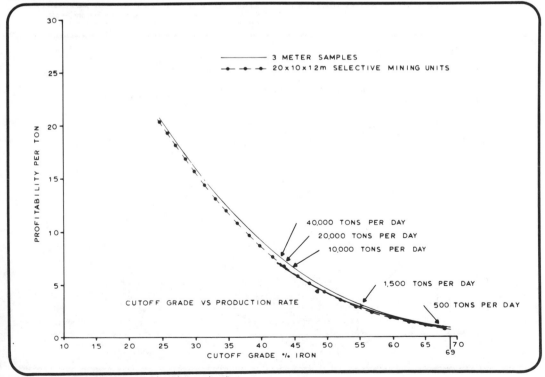

Fig. 8—Cutoff grade vs. profitability per ton for an iron ore deposit.

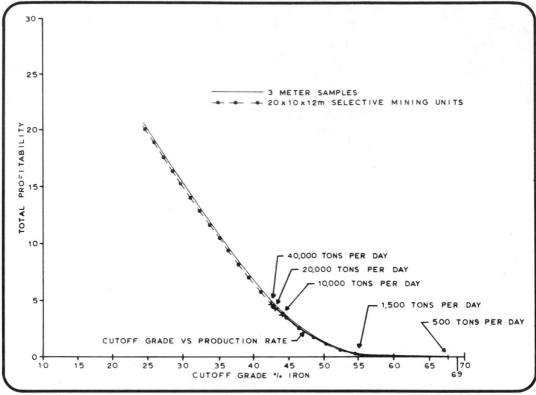

Fig. 9—Cutoff grade vs. total profitability for an iron ore deposit.

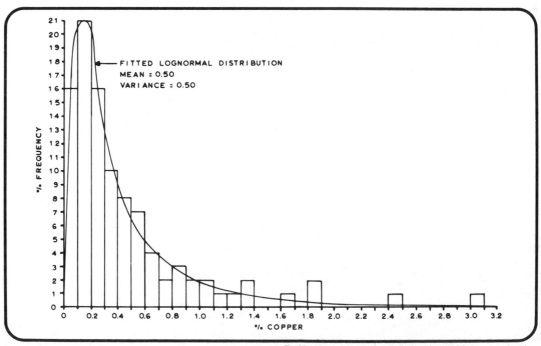

Fig. 10—Lognormal appearing distribution of samples for a copper deposit.

Table 2—Geostatistical parameters for the copper example

	Production rate (tpd)	Loading unit size (cu m)	Dimensions of selective mining unit (m) length x width x height	Variance, σ^2	Cutoff grade,* % Cu	Proportion of tonnage above cutoff	Average grade above % Cu cutoff	Proportion of total metal contained in ore selective mining units	Profitability per ton	Total profitability
Case 1	500	2	10 x 5 x 6	0.356	1.64	0.042	2.58	0.214	0.936	0.039
Case 2	1,500	3	15 x 7 x 9	0.305	1.40	0.055	2.19	0.240	0.787	0.043
Case 3	10,000	5	20 x 10 x 12	0.259	0.49	0.346	0.97	0.672	0.482	0.167
Case 4	20,000	15	25 x 10 x 12	0.242	0.36	0.495	0.80	0.791	0.439	0.217
Case 5	40,000	19	25 x 15 x 15	0.212	0.33	0.555	0.75	0.822	0.411	0.228

*Cutoff grade selected for each case reflects order of magnitude economic studies.

tion rate and to selective mining unit size. A proposed mine must lie on these curves. The graphs indicate that an operation with the lowest cutoff grade and highest production rate will recover the most metal and will be the most profitable, although once the mining rate reaches 10,000-15,000 tpd, a law of diminishing returns appears to apply. In the western hemisphere, current mining practice follows this premise, with nearly all open-pit iron mines producing more than 20,000 tpd and some more than 40,000 tpd.

Some orebodies will not be large enough and in some areas downstream demand at steel mills will not be great enough to justify the most efficient, high-production-rate operation. The graphs demonstrate that in such cases the largest possible operation will be the most profitable.

Example 2—a lognormally distributed copper deposit

The lognormal distribution of sample values representative of a disseminated copper deposit are shown in Fig. 10 and the variograms for the deposit in Fig. 11. The variances of selective mining units were determined using a computer program (Table 2). For this example, the density does not vary markedly over the range of the distribution; hence, transformation to density-free units is unnecessary.

For a lognormal distribution, calculations are made using the mean and standard deviation of the logarithms of the grades, not the grades themselves. Writing μ and σ for the mean and standard deviation of the sample grades, then

$$\alpha = \ln \mu - \frac{1}{2}\beta^2$$

$$\beta^2 = \ln \left(\frac{\sigma^2}{\mu^2} + 1 \right)$$

where α and β are the mean and standard deviation of the logarithms of the grades. Assuming that the selective mining units are also lognormally distributed, the mean μ will remain constant, but the standard deviation will become σ_{smu}. Therefore both α and β will change with block size.

$$\alpha_{smu} = \ln \mu - \frac{1}{2}\beta^2_{smu}$$

$$\beta^2_{smu} = \ln \left(\frac{\sigma^2_{smu}}{\mu^2} + 1 \right) \qquad (11)$$

Sichel[11] and Link, Koch, and Schuenemeyer[12] provide tables for estimating these parameters if the number of samples is small.

As in the previous example, the distribution (now of the logarithms) is converted to the standard normal. For any grade Z, the standardized value is

$$X_{smu} = \frac{\ln Z - \alpha_{smu}}{\beta_{smu}}$$

for a specified cutoff

$$X_c = \frac{\ln Z_c - \alpha_{smu}}{\beta_{smu}}$$

Since the density of the ore is assumed to be constant, the proportion of the deposit that is above cutoff is

$$T_c = 1 - \Phi(X_c) \qquad (12)$$

The average grade of this ore is

$$\overline{Z}_c = \frac{\mu}{T_c} \left[1 - \Phi(X_c - \beta_{smu}) \right] \qquad (13)$$

The proportion of total metal that will be mined from ore units is

$$R = \frac{\overline{Z}_c T_c}{\mu} \qquad (14)$$

The profitability factor for mining is

$$E = \overline{Z}_c - Z_c \qquad (15)$$

Finally, the measure of total profitability is, as before

$$E_T = ET_c \qquad (16)$$

In Figs. 12-16, these variables are graphed as a function of cutoff grade for the cases listed in Table 2. Overlain is a curve for cutoff grade versus size of selective mining unit typically associated with a given production rate.

In this example, the variance for selective mining units is significantly lower than the variance for samples, and predictions made on the curve for samples may be seriously in error. For instance, if the cutoff grade is 0.4% Cu, the samples predict 38% of the deposit will be ore, and the average grade will be 1.04% copper. However, at a 40,000-tpd production rate, the ore tonnage will be 46% of the deposit, and the average grade will be only 0.82% copper.

The graphs indicate that while large-scale operations recover the greatest proportion of the total metal in the

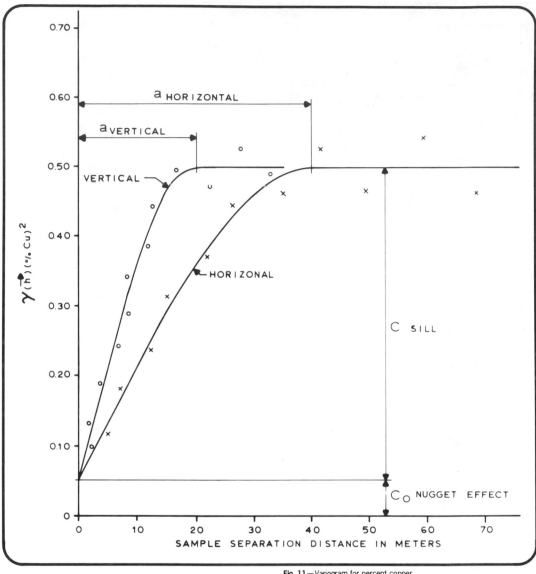

Fig. 11—Variogram for percent copper.

deposits and have the greatest total profitability, profitability per ton is least for high production rates. Clearly, if down-stream demand for product is low, or if there are a large number of deposits available to satisfy the demand, high-cutoff-grade, low-production-rate operations will be most profitable in the short term. In the longer term, lower-cutoff-grade, higher-production-rate operations will recover more metal and have greater total profitability.

Until recently, most mining companies have tended to follow the latter course. However, during the last five years, the costs of mine and metallurgical facilities have risen to the extent that profitability translated into net revenue is sometimes incapable of repaying them within a reasonable time. As a result, many new projects have scaled down production rates.

More sophisticated analysis requires construction of multidimensional economic models, employing grade-tonnage curves for each production rate. Then, a calculation of net present worth or return on investment as a function of grade, tonnage, production rate, metal price, and percent-loan financing can be made. The values of these parameters that maximize net present worth or return on investment can then be determined and project development planned accordingly. A recent case study was presented by Recny.[13]

The previous discussion has assumed that the mineralized area, here called the "deposit," does not change shape from case to case. Examples would be deposits where ore is roughly equally accessible to extraction, regardless of cutoff grade, such as near-surface deposits of uranium, bauxite, or nickel laterite and many underground deposits. For deposits

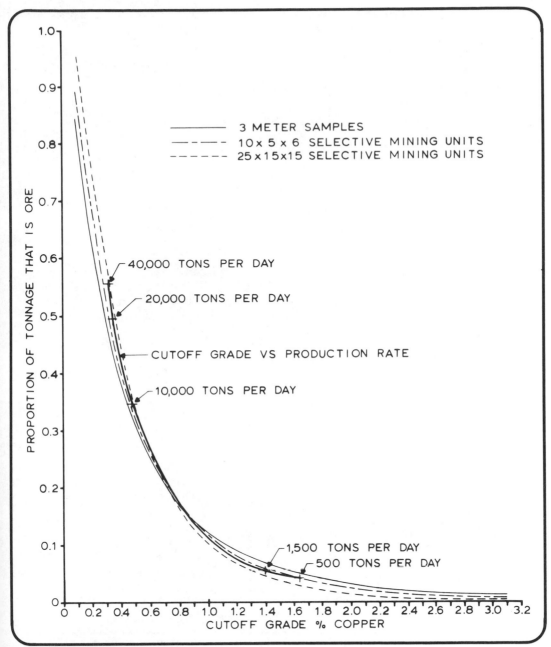

Fig. 12—Cutoff grade vs. tonnage for a copper deposit.

with zoned mineralization and/or deep overburden, such as porphyry coppers, the frequency distribution and variogram must be calculated for each case to reflect only those samples lying within a proposed mining area.

EXTENDING THE V-V RELATIONSHIP

In production planning, it is often necessary to predict the variability of selective mining units within an area of a deposit that is considerably smaller than the deposit as a whole. For instance, we may wish to know the variability of daily production units within a month's time. The variance in mill feed grades over a month's time is

$$\sigma^2_{smu/V} = \overline{\gamma}_{\substack{\text{volume} \\ \text{of} \\ \text{month's} \\ \text{production}}} - \overline{\gamma}_{smu} \tag{17}$$

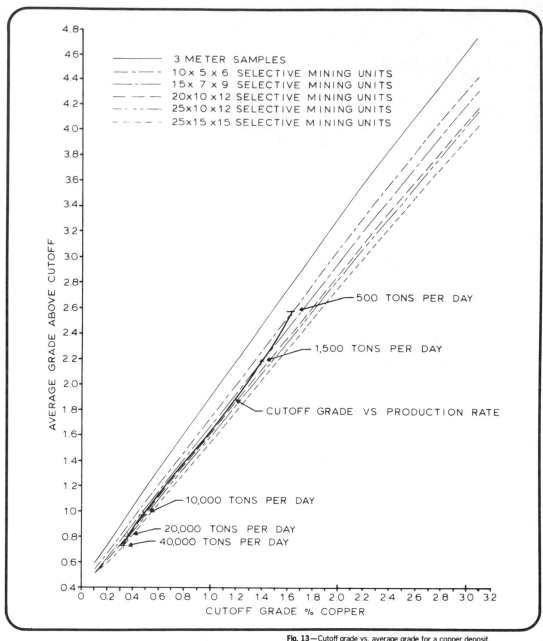

Fig. 13—Cutoff grade vs. average grade for a copper deposit.

Assume a spherical variogram with a range of 60 m, sill of 0.8, and nugget effect of 0.3. In one month, a volume of 30 x 30 x 225 m is mined, giving $h/a = 0.50$, $l/a = 3.75$. The graph value is 0.87, and $\bar{\gamma}$ for a month's production is 1.00. If selective mining units are as before (15 x 15 x 30 m), $\bar{\gamma}_{smu} = 0.56$.

$$\sigma^2_{smu/V} = 1.00 - 0.56$$
$$= 0.44$$

The standard deviation of the daily grades produced over the month is $\sigma_{smu}/V = 0.66$.

For some deposits, this may be too great. If we assume that ore is stockpiled a week before milling* and thorough mixing occurs during reclaim, then the selective mining unit

*More complicated models (Elbrond[14]) take into account the continuous nature of loading and reclaiming of stockpiles.

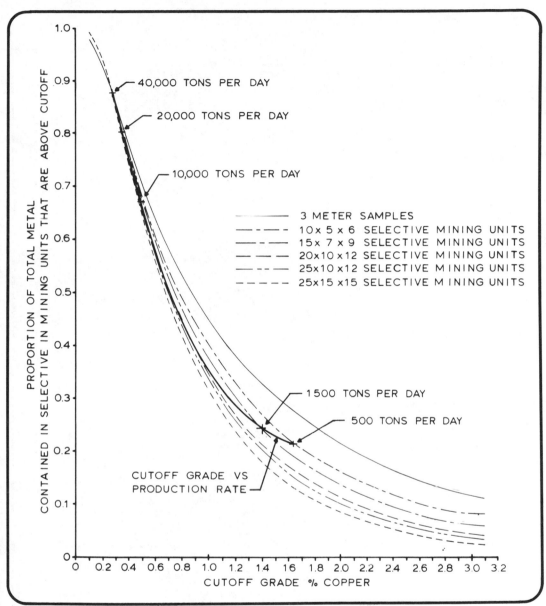

Fig. 14—Cutoff grade vs. metal recovery for a copper deposit.

becomes 15 x 15 x 225 m, assuming a shovel works its way along a bench during the week. In this case

$$\sigma^2_{smu/V} = 1.00 - 0.97$$
$$= 0.03$$

and the standard deviation is 0.17.

If a second shovel is working in another area of the pit far enough away to be outside the variogram range, then the variance of average daily mill feed grades over the life of the

operation (with one 15 x 15 x 30 m volume from each bench and no stockpiles) is

$$\sigma^2_{smu/V} \text{ or } 0.44/2 = 0.22$$

For n shovels, it will be σ^2_{smu} /n.

Another question often asked is, "What will be the daily fluctuation—the difference between yesterday's grade and today's grade?" To obtain this quantity, we must find

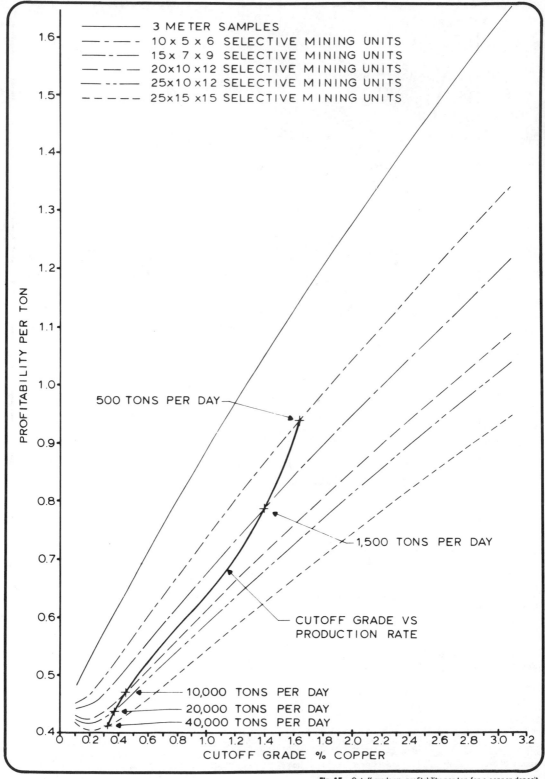

Fig. 15—Cutoff grade vs. profitability per ton for a copper deposit.

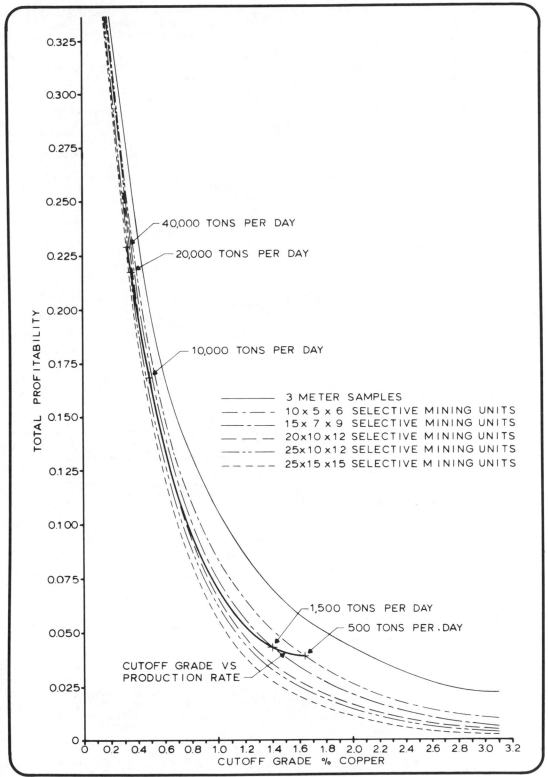

Fig. 16—Cutoff grade vs. total profitability for a copper deposit.

$$\sigma^2_{\substack{daily \\ fluctuation}} = 2\ (\sigma^2_{sml} - \overline{\gamma}_D + \overline{\gamma}_{ij})$$

The calculation of $\overline{\gamma}_{ij}$ is made by averaging the variogram values for vectors between points i constrained to be in yesterday's unit and points j in today's unit. Usually, this calculation is done by a simple computer program, calculating the variogram values between about 50 points within each unit respectively. For the case under consideration, the variance of differences between days is 0.62, and the corresponding standard deviation is 0.78.

CONCLUSION

The volume-variance relationship implies that for each mining method there exists a unique frequency distribution of selective mining units. Using this distribution, average grade and tonnage of ore for various cutoff grades can be predicted and, in turn, an economic analysis can determine a suitable mining method and cutoff grade.

The relationship can be used to predict the variability of grade or other metallurgical variables for selective mining units within small areas of a deposit, providing a guide to design of metallurgical treatment plants and blending facilities.

Like any tool, the volume-variance relationship has its limitations, in particular an inability to locate within the deposit those selective mining units that are ore and those that are waste. To accomplish this, a local estimation technique, kriging, must be used, as discussed in previous articles in this series. Kriging, if properly applied (Dagbert and David,[9] Parker and Switzer[6]), will yield a distribution of estimated grades that is close to that predicted using the volume-variance relationship.

Fluctuations in grades within an orebody for daily, monthly, or other time periods can be studied using a technique called conditional simulation (Journel[15]). Using the variogram and actual sample data, a model of the deposit can be made that simulates the grade of each selective mining unit. These models are very useful if a deposit shows zonation of grades and/or marked comingling of ore and waste selective mining units.■

REFERENCES

1) Schurtz, R. F., "The Electronic Computer and Statistics for Predicting Ore Recovery," MINING ENGINEERING, Oct. 1959, pp 1035-1044.

2) David, M., Geostatistical Ore Reserve Estimation, Elsevier, Amsterdam, 1977.

3) Clark, I., "Some Practical Computational Aspects of Mine Planning," In Guarascio, M., et al., editors, Advanced Geostatistics in the Mining Industry, D. Reidel, Dordrecht, Netherlands, 1976, pp 391-402.

4) Clark I., "Regularization of a Semivariogram," COMPUTERS & GEO-SCIENCES, Vol. 3, 1977, pp 341-346.

5) Switzer, P., and H. M. Parker, "The Problem of Ore Versus Waste Discrimination for Individual Blocks: The Lognormal Model," In Guarascio, M., et al., editors, Advanced Geostatistics in the Mining Industry, D. Reidel, Dordrecht, Netherlands, 1976, pp 203-218.

6) Parker, H. M., and P. Switzer, "Use of Conditional Probability Distributions in Ore Reserve Estimation, A Case Study," Proceedings of 13th APCOM Symposium, Clausthal-Zellerfeld, W. Germany, October 1975, pp MII: 1-16.

7) Marechal, A., "Forecasting a Grade-Tonnage Distribution for Various Panel Sizes," Proceedings of 13th APCOM Symposium, Clausthal-Zellerfeld, W. Germany, October 1975, pp EI: 1-18.

8) Marechal, A., "The Practice of Transfer Functions: Numerical Methods and their Applications," In Guarascio, M., et al., editors, Advanced Geostatis-tics in the Mining Industry, D. Reidel, Dordrecht, Netherlands, 1976, pp 253-276.

9) Dagbert, M., and M. David, "Predicting Vanishing Tons Before Production Starts or Small Blocks are No Good for Planning in Porphyry Type Deposits," AIME Convention, Atlanta, Ga., March 1977, Preprint 77-AO-83.

10) Parker, H. M., "The Geostatistical Evaluation of Ore Reserves Using Conditional Probability Distributions: A Case Study for the Area 5 Prospect, Warren, Maine," Unpublished Ph.D. thesis, Stanford University, 1975.

11) Sichel, H. S., "The Estimation of Means and Associated Confidence Limits for Small Samples from Lognormal Populations," Proceedings of symposium on Mathematical Statistics and Computer Applications in Ore Valuation, South African Institute of Mining and Metallurgy, 1966, pp 106-123.

12) Link, R. F., G. S. Koch, and J. H. Schuenemeyer, "Statistical Analysis of Gold Assay and Other Trace Element Data," US Bureau of Mines Report of Investigation 7495, 1971.

13) Recny, C. J., "Assessment of Economic Scale of Operations Mine Planning," Unpublished Masters thesis, Stanford University, 1978.

14) Elbrond, J., "Evaluation of Blending Performance," Proceedings of seminar on Sampling, Blending, and Proportioning of Bulk Solids, University of Pittsburgh School of Engineering, Pittsburgh, Pa., 1973.

15) Journel, A., "Geostatistics for the Conditional Simulation of Ore Bodies," ECONOMIC GEOLOGY, Vol. 69, 1974, pp 673-687.

CORRECTIONS

The following are corrections for Dr. H. Parker's paper, "The Volume-Variance Relationship," presented as Geostatistics Part 5 in the October 1979 issue of E&MJ:

■ Page 110, line 6—The test indicates a variance of σ^2_{sml}/n. The correct term is simply σ^2_{sml}.

■ Page 110, bottom left—The equation should read:

$$f(X_c) = \frac{1}{\sqrt{2\pi}}\ \exp\left(-\frac{1}{2}X_c^2\right)$$

■ Page 120, bottom right—The equation should read:

$$\frac{\sigma^2_{smu}/V}{2}\ \text{ or } 0.44/2$$

■ Page 123, top left—The equation should read:

$$\sigma^2_{\substack{daily \\ fluctuation}} = 2\,(\sigma^2_{\substack{daily \\ production}} - \overline{\gamma}_D + \overline{\gamma}_{ij})$$

In Dr. I. Clark's paper, "The Semivariogram—Part 1," published in the July 1979 issue of E&MJ, the equation at the bottom left part of p 92 should read:

$$\gamma^*(h) = \frac{1}{2\,(N-h)}\ \sum_{i=1}^{N-h}\ (g_i - g_{i+h})^2$$

16

Reprinted from *Chemical, Metall., and Mining Soc. South Africa Jour.* **53**(2):47–64 (1952)

A STATISTICAL ANALYSIS OF SOME OF THE BOREHOLE VALUES IN THE ORANGE FREE STATE GOLDFIELD

D. G. Krige

SYNOPSIS

It is shown that the lognormal frequency curve can be applied to the observed distribution of borehole values in the Main Sector of the Orange Free State Goldfield. On this basis statistical estimates of the average recovery grade, with limits of error, and the total mill tonnage are prepared. General conclusions are drawn regarding the confidence to be placed on estimates of individual mine grades and regarding the additional confidence to be gained, for valuation purposes, from regular deflections of boreholes.

In the statistical section it is shown that the distribution of the Maximum Likelihood Estimator for the mean of a lognormal population tends to lognormality as 'n' increases and that this estimator and its standard error, as based on random sampling theory, will be biassed if applied to a stratified random sample.

INTRODUCTION

The extensive information made available by the unique drilling programme in the Orange Free State Goldfield has formed the basis for important conclusions regarding geological formations, reef correlations, faulting, mining depths, stoping widths, geothermic gradient and a number of other factor which will govern mining conditions in this field.

The gold values disclosed in the large number of boreholes which have intersected reef, have naturally provided the confidence for reaching decisions on the establishment of mines and for raising the capital funds necessary for financing these mines. This confidence in going ahead on the indications from the limited number of boreholes drilled on any one mining property emanates mainly from the tenor of the values encountered and from the experience gained on the Witwatersrand in the exploitation of reefs which have proved persistent gold carriers over very extensive areas. There is also a general belief that due to unavoidable core losses in intersections on a friable reef such as the Basal, borehole values will, if anything, understate the actual gold values in a mine.[1, 2, 6] Attention has also been drawn

to the comforting fact of the high percentage of boreholes in which payable values have been encountered.[2, 6] Without the aid of the powerful tool of applied statistics the confidence which is generally placed in the future of this goldfield can, however, not be defined or analysed on a scientific basis.

It is the object of this paper therefore to analyse the available borehole results on a statistical basis in order to draw conclusions regarding the possible economic tonnages and grades of ore to be mined with indicated limits of error, the extent of additional confidence to be gained from deflections, etc. This investigation has been confined to the block of ten mines lying to the west of the Welkom fault and forming what may be termed the Main Sector of the field. The three mines to the east of this fault are not contiguous to the ten mines concerned and their valuation presents a separate problem.

As this paper is intended primarily for mining engineers, it has been divided into a Valuation and a Statistical Section, the latter forming the background to the general conclusions discussed in the former. Except where specifically indicated, only Basal Reef values have been considered.

I. VALUATION SECTION

1. *Basic concepts*

In a previous paper[4] the author discussed in some detail the general basic considerations involved in the application of statistics to mine valuation on the Witwatersrand. In the statistical evaluation of borehole data the fundamental concept on which all deductions are based, is that the few available borehole cores represent only an infinitesimal part of the entire reef body, in fact, only 1 part in some 5,000 million in the case of first intersections on the Basal Reef in the Main Sector. Similarly, the known borehole values form an infinitesimal percentage of the virtually infinite number of borehole values which could be obtained (theoretically) by repeated drilling. In statistics the available values are termed the 'sample' and the infinite number of possible values is called the 'population' or 'universe,' from which the 'sample' was obtained.

The application of statistical theory involves the estimation of the essential characteristics of the unknown population from the known sample with the accompanying inferences as to the possible errors inherent in such an estimation. In the case of borehole values, it enables *inter alia* an estimate to be made of the mean gold value of the

91 first intersections on Basal Reef in the Main Sector* in the form of a histogram (or step-diagram) and shows the outline of a lognormal frequency curve which has been fitted to the observations. A distinctive feature of the lognormal curve is that it can be transformed into the symmetrical Normal Curve of Error by plotting the abscissae values on a logarithmic scale. On Diagram No. 2, therefore, the 91 observed values are shown classified on a logarithmic scale with the Normal Curve (corresponding to the lognormal curve on Diagram No. 1) superimposed on the step-diagram.

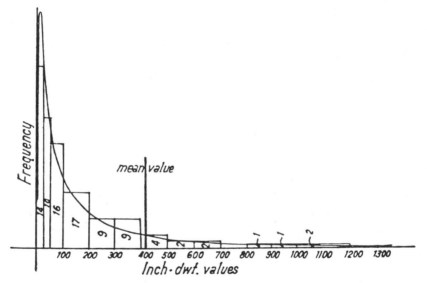

Diagram No. 1—Showing frequency histogram of 91 first intersection values on Basal Reef and the corresponding lognormal curve (4 values in excess of 1,200 inch dwts not shown)

entire reef area concerned (this naturally corresponding to the mean value which would be obtained from the infinite number of possible boreholes), and the qualification of this estimate by the appropriate limits of error.

2.   *Application of the Lognormal Curve*

Investigations carried out on a number of Witwatersrand mines[4, 5, 7] have indicated that individual underground gold values in a specific area form a distinctive frequency distribution which can be represented satisfactorily by the lognormal frequency curve.

Diagram No. 1 represents the frequency distribution of the values obtained in the

The 'goodness-of-fit' of the curve on Diagram No. 2 has been tested with the Chi-square test and the normality of the histogram by Fisher's and Geary's tests (see Section II) and it can be stated that the observed deviations from the theoretical model can be attributed reasonably to the operation of chance in the selection (by drilling) of the 91 values concerned.

It is, however, generally contended that first intersections are usually incomplete and that deflections are made in order to obtain better and if possible complete core recoveries. If this contention is correct, *and*

* As listed in Section II

there is no bias in the human decisions to either deflect or not to deflect,* the values obtained in all the last intersections of the 91 boreholes in question should be more representative of the true population of values than the first intersections. Diagram No. 3, therefore, illustrates the distribution of the 91 last intersections† on the same basis as Diagram No. 2. In this case the step-diagram is highly irregular in shape and its departure from the theoretical curve is evident. The statistical tests referred to above confirm the fact that this departure cannot be attributed reasonably to chance

negligible effect on the lognormal trend of their distribution. The main deductions made in this paper are therefore based only on these first intersection values.

It may be mentioned that on certain of the mines concerned, it appears to have been the regular practice to deflect every hole at least once, regardless of the value encountered in the first intersection. In such a case there appears to be no reason why the deflection values should be biassed and why these may not be taken into account in an estimate for the specific mine in question.

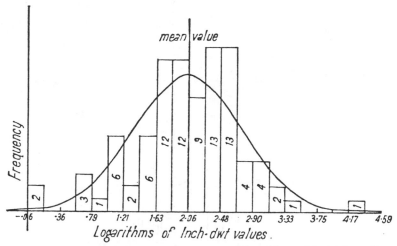

Diagram No. 2—Showing frequency histogram of logarithms of first intersection values on Basal Reef and the corresponding normal curve

alone. If, therefore, the lognormal curve is accepted as a suitable representation of the distribution of gold values, it is clear that *for the Main Sector as a whole* only the first intersection values can be regarded as unbiassed. The fact that some of these values are admittedly below normal on account of core losses has apparently had a

* There will be no bias if the examination of every borehole core before it is split and assayed for gold, is the *only* factor influencing the classification of a recovery as either unsatisfactory or satisfactory, and the consequent decision to deflect or not to deflect respectively, i.e. if such decisions are made regardless of the assay values disclosed subsequently.

† Where no deflection was made the borehole value has been regarded both as a first and a last intersection.

3. *The extent of core losses.*

The apparent bias present in the distribution of the last-intersection values in the Main Sector as a whole, makes it impossible to compare these values with the first intersection values on a proper basis in order to estimate the average extent of core losses. It is, therefore, not possible to state what proportion of the difference of nearly 30 per cent between the average values from first and last intersections, respectively, in the 62 boreholes where deflections were made, can be accounted for by actual core losses.

4. *Valuation of the main sector.*

The mean of the lognormal curve which has been fitted to the 91 first intersection values on the theory of maximum likelihood is 411 inch-dwts which can, therefore, on available procedures be accepted as the best estimate of the actual overall mean value of the 10 mines concerned. Further analysis shows that there is an estimated chance of only 1 in 100 of the actual mean value being less than 239 inch-dwts and a similar chance of its exceeding 750 inch-dwts. In other words, it is estimated that the actual mean value of these 10 mines will almost certainly lie between 239 and 750 inch-dwts, the most likely value in this range being 411 inch-dwts.

this value is not out of keeping with the rest.

This borehole estimate may also be compared with the following shaft intersection values already available :—

Welkom : 187 and 420 inch-dwts.
President Steyn : 235 and 966 inch-dwts.
President Brand : 470 inch-dwts.
Western Holdings : 378 and 652 inch-dwts.
Free State Geduld : 635 inch-dwts.
Freddies South : 162 and 234 inch-dwts.
Freddies North : 151 inch-dwts.

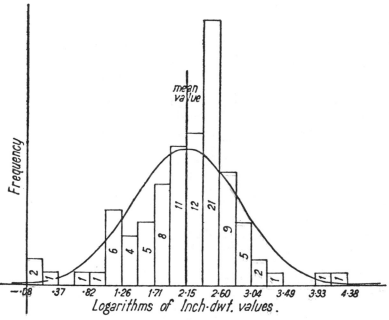

Diagram No. 3—Showing frequency histogram of logarithms of last intersection values on Basal Reef and the corresponding normal curve

This statistical estimate may be compared with the arithmetical average of the 91 borehole values, i.e. 518 inch-dwts, which is subject to limits of error far in excess of those indicated above. The definite advantage of the above statistical estimate over the usual arithmetic average is referred to briefly in Section II. Reference to Diagram No. 2 will also make it evident that on the statistical basis the effect of the one extremely high value (i.e. Geduld No. 1—23,037 inch-dwts) is more than counterbalanced by the 2 low values at the extreme lefthand side of the step-diagram, and that

If these 11 values are assumed to constitute a random selection from the Main Sector, the statistical 't' estimate for the overall mean value is 408 inch-dwts and the estimated 1 in 100 limits are 258 and 672 inch-dwts. The two estimates are, therefore, in close agreement, but have not been combined because the shaft values are not representative of all the mines in the sector. It is, however, evident that there is, as yet, no confirmation of the general popular belief that the underground disclosures are or will be on the whole of a higher order than the borehole values.

An analysis of the relative variations between the borehole values within individual mines and within the deflection areas around the original intersections respectively has, by interpolation, also enabled an estimate to be made of the relative variation to be expected between the values of ore blocks within the Main Sector. In other words, for any value accepted as the overall mean gold value of the total reef body, it is possible to estimate the likely distribution

Diagram No. 4* has been prepared on this basis and on the borehole estimate of 411 inch-dwts. The solid line *AB* indicates the likely average payable grade and percentage payability for any specified pay limit. The confidence to be placed in this estimate is reflected by the 4 dotted lines parallel to *AB*, viz. the 1 per cent and 5 per cent confidence limits. For example :—

At a pay limit of, say, 150 inch-dwts, the grade of payable ore is likely to be 560

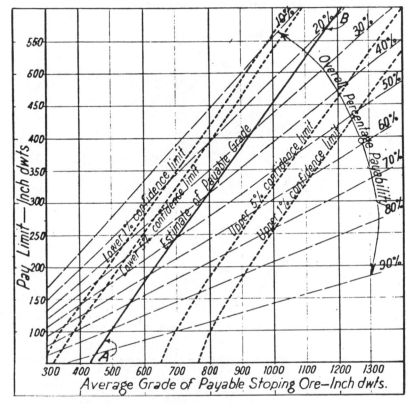

Diagram No. 4—Showing estimated average grade of payable stoping ore and the overall percentage payability corresponding to different pay limits; also the estimated 1 per cent and 5 per cent confidence limits for these estimates

of block tonnages in the various grade categories, and therefore to determine the average grade of the payable ore above any specified pay limit, as well as the corresponding percentage payability.

*This diagram is based on the principle of the *Ore Graduation Graph* as originated by the late B. J. Kloppers, ex Inspector of Mining Leases, Department of Mines, and further developed by F. W. J. Ross in his unpublished master's thesis, University of the Witwatersrand, 1950.

inch-dwts, and the percentage payability 68 per cent. There is a chance of only 1 in 100 of these 2 figures being lower than 420 inch-dwts and 48 per cent respectively (lower 1 per cent confidence limit) and a chance of 1 in 20 of being lower than 450 inch-dwts and 54 per cent respectively (lower 5 per cent confidence limit). Also there is a chance of only 1 in 100 of these 2 figures being in excess of 860 inch-dwts and 85 per cent respectively (upper 1 per cent confidence limit).

A further analysis of Diagram No. 4 will reveal the interesting feature that, whatever value is accepted as the overall mean value of the Main Sector, about 80 per cent of the total gold content of the Basal Reef is expected to be contained in that 40 per cent of ore blocks falling in the highest grade categories.

unpayable *individual* values (as represented by boreholes or underground sample sections) which are found in the payable ore blocks usually by far exceed the number of payable *individual* values to be found in unpayable ore blocks. The position can, however, under special circumstances be reversed, e.g., when the pay limit is well in excess of the overall mean value.* In the present case the position is that the per-

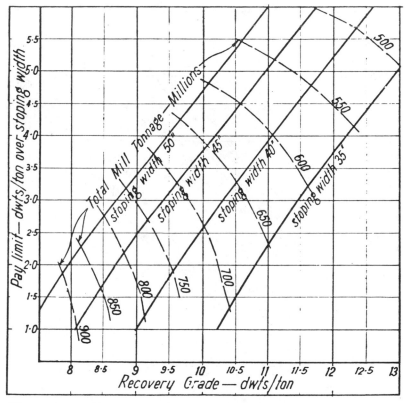

Diagram No. 5—Showing statistical estimates of recovery grade and total mill tonnage for the Main Sector and corresponding to different pay limits and stoping widths (confidence limits not shown)

Attention is also drawn at this stage to the wrong practice of estimating the percentage overall payability of a mine and the average value of the payable ore from the percentage of payable holes and on the average value of such holes respectively. This practice will in many cases lead to a serious under-estimation of the percentage payability and an even more serious over-estimation of the grade of payable ore. This is due to the fact that the number of

centage payability if estimated directly from individual borehole values would be 44 out of 91, i.e. 48 per cent (as against 68 per cent above) and the average value of payable ore would be the average of the 44 pay values, i.e. 1,013 inch-dwts (as against 560 inch-dwts above).

A comparable position exists in the case of the customary practice on many mines of

* This will be evident from a critical examination of Formula No. (10) in the Statistical Section.

publishing the percentage payability and average payable value for current reef development on the basis of individual sampling sections. In relation to the payable ore which will eventually be stoped, such development payability and payable grade figures will in general be under- and over-statements respectively.

The conversion of the average grade of payable ore and the percentage payability

ing; the mine call factor; and the mill recovery factor. The variation in these factors from mine to mine is appreciable and provides considerable scope for research. Preliminary investigations on a number of mines indicate, however, that the overall ratio between the recovery grade and the average underground sampled grade of payable ore from stopes, when plotted against the mine's percentage payability, follows a

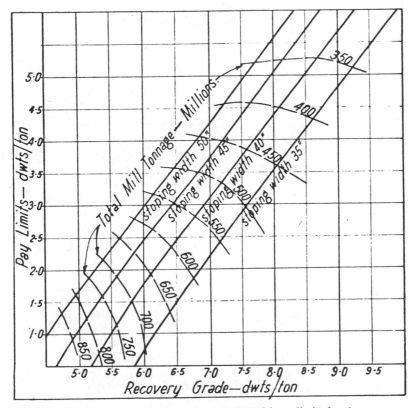

Diagram No. 6—Showing the lower 1 per cent confidence limits for the recovery grade and mill tonnage estimates shown on Diagram No. 5, i.e. there is an estimated chance of only 1 in 100 of the actual average position in the Main Sector being worse than that shown above

of a mine into the corresponding recovery grade and mill tonnage, respectively, involves a number of uncertain factors which depend not only on physical features but also on mining policy and the nature of the control exercised on mining and milling operations. The main factors are stoping width; the extent of dilution of payable stoping ore by development tonnages and by the stoping of unpayable ore; the surveyor's shortfall or excess; the degree of waste sort-

distinctive trend in spite of individual deviations. This trend is linear and indicates a ratio of about 0·70 corresponding to a 40 per cent payability, increasing to 0·85 at a 100 per cent payability.

Similarly, the ratio of mill tonnage to tonnage of payable stoping ore underground appears to follow a linear trend when plotted against percentage payability of the mine, the ratio being approximately 1·285 at a

40 per cent payability and 1·095 at a 100 per cent payability.

These two indicated average trends will naturally not apply during the initial stage in a mine's life when there is a considerable amount of gold absorption in the mill as well as an abnormally high percentage of development ore being milled.

On the basis of these trends the estimates shown on Diagram No. 4 have been converted into the recovery grades and mill tonnages reflected on Diagrams No. 5 and 6. Tonnages have been determined on the following additional factors :—

Total reef area = 51,115 claims.
Allowance for losses due to faults and dykes = 15 per cent.
Allowance for reef area extracted by development = 6 per cent.
Average dip of reef = 17°.

Diagram No. 5 indicates the position for 4 different stoping widths ranging from 35 inches to 50 inches and Diagram No. 6 shows the lower 1 per cent confidence limits for the estimates shown on Diagram No. 5. For example :—

Assuming a pay limit of, say, 3·5 dwts/ton and a stoping width of, say, 45 inches, the indicated recovery grade and mill tonnage when conditions are normal are 9·8 dwts/ton and 680 million, respectively (Diagram No. 5) which at an annual milling rate of, say, 12 million tons per annum, provides for a life of 57 years and a gold output of nearly 6 million ounces per annum. Further, there is an estimated chance of only 1 in 100 of these figures being less than 6·8 dwts/ton, 500 million mill tons (Diagram No. 6), 42

years life and 4 million ounces per annum, respectively.

There appears, therefore, to be ample justification for the general confidence placed in the future of this sector and of the Orange Free State Goldfield as a whole.

5.   Conclusions regarding individual mines

As in the case of the sector as a whole, it is also possible to estimate the relative variation between block values within a mine and hence to estimate the mill tonnage and recovery grade corresponding to any specified pay limit. Table I has been prepared on this basis, and on the following factors :—

Mine area = 5,112 claims.
Dip of reef = 17°.
Losses due to faulting = 15 per cent.
Stoping width = 46 inches.
Reef area extracted by development = 6 per cent.
Milling rate = 1,200,000 tons per annum.
Pay limit = 150 inch-dwts = 3·26 dwts/ ton.

From Table I it appears that, particularly on account of the life aspect and the high capital structure of the Orange Free State mines, the economic position of a mine will be somewhat precarious if the overall mean value of the mine turns out to be less than, say, 150 inch-dwts.

On average there are, however, only 9 borehole values available per mine on which an estimate of the overall mean value can be based. From the knowledge of the likely relative variation between borehole values

TABLE I

| Overall mean value of mine, inch-dwts | Payable ore from stopes | | Mill Tonnage, millions | Recovery grade, dwts/ton | Life, years |
	Value, dwts/ton	Tonnage, millions			
100	5·46	15·9	21·4	3·52	18
150	6·21	30·1	39·1	4·28	33
200	6·98	42·2	53·2	5·06	44
250	7·78	51·8	63·5	5·83	53
300	8·62	59·4	71·2	6·64	59
400	10·37	69·5	80·6	8·31	67
600	14·12	79·4	89·4	11·71	75
800	18·10	83·4	92·6	15·20	77
1,000	22·22	85·3	94·3	18·90	79

within a mine, the errors to which such an estimate based on 9 values may be subject, can be determined. The probability of the true mean value of a mine being less than 150 inch-dwts can then be guaged from the estimate of the mine's mean value based on the 9 known borehole results. For the statistical 't' estimate discussed in this paper, these probabilities are reflected in Table II.

TABLE II

Statistical mean of 9 observed borehole values (inch-dwts)	Probability of mine's true mean value being less than 150 inch-dwts
	Per cent
100	58
150	36
200	22
250	14
300	9
400	4
600	1
800	0·4
1,000	0·1

E.g., if the Statistical Mean of the 9 available values is, say, 300 inch-dwts, the chance of the mine's overall mean value being less than the critical limit of 150 inch-dwts is about 1 in 11.

The author has for obvious reasons preferred to leave to the discretion of the reader the determination of the statistical means of individual mines and hence of the confidence to be placed in the successful exploitation of any one mine.

6. *Additional confidence gained from deflections*

It is possible to estimate, from the observed relative variations between borehole values within a deflection area and within the Main Sector the extent to which the errors inherent in valuing the Main Sector as a whole on, say, 100 boreholes. will be reduced if any specific number of deflections are made at each hole. It is assumed for this purpose that all boreholes will be deflected as a regular routine and that core losses can be disregarded.

A suitable basis for gauging the extent to which the overall estimate will be improved by deflections is to indicate the number of single borehole intersections required to

yield as reliable a result as a 100 boreholes each with a stipulated number of deflections. Table III has been prepared on this basis and indicates that considerable improvement is obtained from 2 deflections at each hole, but that any further substantial increase in improvement requires an increase in the number of deflections per borehole which will probably not be warranted.

TABLE III

No. of deflections per borehole	1	2	3	5	Maxm
No. of single intersections required to yield as reliable a result as 100 boreholes with deflections ...	127	143	153	166	200

Similarly, the position has been estimated for the theoretical case where the deflection area is increased to one with a radius of, say, 50 feet, and is reflected in Table IV.

TABLE IV

No. of deflections per borehole ...	1	2	3	5	Maxm
No. of single intersections required to yield as reliable a result as 100 boreholes with deflections ...	136	164	184	212	305

It is interesting to note that the above conclusions agree very closely with those arrived at previously by the author in practical test calculations based on underground sample values which were regarded for that purpose as analogous to borehole values.[5]

7. *Reefs other than the Basal Reef*

A statistical investigation of the borehole results obtained from the Leader Reef in the Main Sector indicates a likely payability of only some 15 per cent and this reef is therefore a problematical source of mill tonnage. It is, however, possible that it will be exploited to a limited extent on at least some of the 10 mines in question.

The available values from the *A* and *B* reefs are confined almost exclusively to the northernmost 4 mines in the sector and are therefore not suitable for estimates applicable to the sector as a whole.

II. STATISTICAL SECTION

1. *The lognormal frequency distribution*

The properties of this distribution with detailed references are listed in Reference 4, and for the purpose of this investigation, therefore, only certain of these are listed again.

The lognormal frequency distribution is represented by

$$\psi(z)dz = \left[\sqrt{2\pi} \cdot \sigma \right]^{-1} exp \left[\frac{\sigma^2}{2} - \xi - \frac{1}{2\sigma^2} \right.$$

$$\left. (\log_e z - \xi + \sigma^2)^2 \right] dz, \text{ with the variable}$$

$$= z, \text{ mean} = \theta = exp \left[\xi + \frac{\sigma^2}{2} \right]$$

and variance

$$= \theta^2 \left[exp(\sigma^2) - 1 \right] \quad \dots \quad \dots \quad \dots \quad \dots \quad (1)$$

The normalised form is:

$$f(x)dx = \left[\sqrt{2\pi} \cdot \sigma \right]^{-1} exp \left[- \frac{1}{2\sigma^2} (x - \xi)^2 \right] dx$$

for the variable $x = \log_e z$; mean $= \xi$
and variance $= \sigma^2 \dots \quad \dots \quad \dots \quad \dots \quad \dots \quad (2)$

The maximum likelihood estimators for θ are:
For σ^2 known *a priori*:

$$t'' = exp \left[\bar{x} + \frac{n-1}{2n} \sigma^2 \right] \quad \dots \quad \dots \quad \dots \quad (3)$$

$$= exp (\bar{x}) exp(\frac{\sigma^2}{2}) exp(-\frac{\sigma^2}{2n}) \quad \dots \quad \dots \quad (3a)$$

and variance $= \theta^2 \left[exp(\frac{\sigma^2}{n}) - 1 \right] \quad \dots \quad \dots \quad (4)$

$$\doteqdot t''^2 \left[exp(\frac{\sigma^2}{n}) - 1 \right] \quad \dots \quad (4a)$$

For σ^2 unknown:

$$t = e^{\bar{z}} \left[1 + \tfrac{1}{2}V + \frac{n-1}{2^2 2!(n+1)} V^2 \right.$$

$$\left. + \frac{(n-1)^2}{2^3 3!(n+1)(n+3)} V^3 + \dots \right] \dots (5)$$

where $\bar{x}$ = mean of natural logs of observed values in a sample of size n, and V = variance (unadjusted) of natural logs of observed values.

For n large, t approaches

$$t' = exp (\bar{x} + \frac{V}{2}) \dots \quad \dots \quad \dots \quad \dots \quad \dots \quad (6)$$

The estimated variance for t

$$= t^2 \left\{ e^{V(n-1)} \left[1 + \frac{V^2}{2(n-1)} \right. \right.$$

$$\left. \left. + \frac{V^4}{2!2^2(n-1)(n+1)} + \dots \right] - 1 \right\} \dots (7)$$

which for n large becomes

$$t'^2 \left\{ exp \left[\frac{V}{n-1} + \frac{V^2 n}{2(n-1)^2} \right] - 1 \right\} \dots (8)$$

the actual variance being

$$= \theta^2 \left[exp(\frac{\sigma^2}{n} + \frac{\sigma^4}{2n}) - 1 \right] \quad \dots \quad \dots \quad (8a)$$

The frequency of z values above value z_i

$$= \frac{1}{\sqrt{2\pi}} \int_{w_i}^{\infty} exp(-\frac{w^2}{2}) dw \quad \dots \quad \dots \quad \dots \quad (9)$$

where $w_i = \frac{1}{\sigma} (\log_e z_i - \xi)$

$$= \frac{1}{\sigma} (\log_e z_i - \log_e \theta + \frac{\sigma^2}{2}) \quad \dots (10)$$

The average of all z values above value z_i

$$= \theta \frac{\int\limits_{w_i - \sigma}^{\infty} exp(- \frac{w^2}{2}) dw}{\int\limits_{w_i}^{\infty} exp(- \frac{w^2}{2}) dw} \quad \dots \quad \dots \quad (11)$$

2. *The application of the lognormal curve to borehole values*

The suitability of the lognormal curve for representing the distribution of borehole values from the Basal Reef in the Main Sector of the Orange Free State Goldfield can be gauged from the fit of the corresponding normal curve to the logs of these values.

The following is the usual χ^2 test as applied to the values obtained from the 91 first intersections on Basal Reef*:—

The probabilities of exceeding the above two values are 33·6 per cent and 58·6 per cent respectively.

A more effective test for kurtosis (Geary's ratio)† shows

TABLE V

(Estimates made from observed values : Mean = 2·05586 and Variance = 0·49682.)

Category limits—common logs of inch-dwt values	Observed frequencies F_1	Theoretical frequencies F_2	Difference F_1-F_2	Difference² $\dfrac{}{F_2}$
	6	6·079	−0·079	0·001
———— 0·99851 ————				
	8	10·670	−2·670	0·668
———— 1·42145 ————				
	6	8·207	−2·207	0·594
———— 1·63292 ————				
	12	9·813	+2·187	0·487
———— 1·84439 ————				
	12	10·730	+1·270	0·150
———— 2·05586 ————				
	9	10·730	−1·730	0·279
———— 2·26733 ————				
	13	9·813	+3·187	1·035
———— 2·47880 ————				
	13	8·207	+4·793	2·799
———— 2·69027 ————				
	8	10·670	−2·670	0·668
———— 3·11321 ————				
	4	6·079	−2·079	0·711
Totals	91	90·998	—	7·392

Degress of freedom = 10 − 3 = 7 ; ·5 > P > ·3.

Application of Fisher's test for symmetry and kurtosis to these individual observations shows :—

$V_1 = 2·05586$ $k_1 = 2·05586$

$V_2 = 0·49137$ $k_2 = 0·49682$

$V_3 = -0·07572$ $k_3 = -0·07828$

$V_4 = 0·67745$ $k_4 = -0·03366$

$g_1 = -0·24308$ $Sg_1{}^2 = 0·06381$

$$\frac{g_1}{Sg_1} = -0·96230$$

$g_2 = -0·13638$ $Sg_2{}^2 = 0·06258$

$$\frac{g_2}{Sg_2} = -0·54519$$

$$a = \frac{\text{Average deviation (from mean)}}{\text{Standard deviation}}$$

$$= \frac{0·53434}{0·704934}$$

$$= 0·758$$

From Geary's tables for $n = 90$, and $a = 0·758$,

$P = 0·04$.

This indicates a critical level for the kurtosis, but such that the hypothesis of normality is not rejected at the 1 per cent level.

A similar examination of the distribution of the values obtained from the last intersections from the same set of boreholes clearly rejects the hypothesis of lognormality :—

χ^2 test : Observed $\chi^2 = 18·3$; Degrees of freedom = 7

---

* These 91 values are as follows : (inch dwts).
1, 1, 5, 5, 5, 8, 11, 12, 12, 14, 14, 15, 24, 24, 33, 34, 37, 39, 39, 41, 43, 45, 47, 48, 51, 53, 53, 54, 54, 56, 59, 64, 79, 80, 83, 85, 88, 92, 94, 96, 104, 108, 109, 109, 129, 143; 149, 150, 157, 160, 166, 170, 180, 188, 191, 195, 198, 201, 210, 222, 227, 227, 238, 241, 244, 261, 310, 312, 327, 336, 349, 383, 376, 388, 400, 405, 421, 437, 439, 518, 546, 665, 678, 890, 906, 1009, 1085, 1747, 1893, 2898, 23037,

†Ref. 8, p. 481.

$P = \cdot01$ (approximately).

Symmetry :

$k_1 = 2 \cdot 15024, \quad k_2 = 0 \cdot 55034,$

$k_3 = -0 \cdot 68814, \quad k_4 = 5 \cdot 51415,$

$g_1 = -1 \cdot 6855 \quad \dfrac{g_1}{Sg_1} = -6 \cdot 672, \, P < \cdot001.$

Kurtosis :

$g_2 = 18 \cdot 21, \quad \dfrac{g_2}{Sg_2} = 72 \cdot 78, \quad P < \cdot001$

Note.—The application of the above tests for normality naturally presupposes that the 91 values were selected at random, whereas due to their geographical distribution, these values actually approximate to stratified sampling. Also as the variance of the transformed variates tends to increase with increasing size of reef area concerned (see subsection 4 below), these tests are, if anything not rigid enough.

It will be evident by reference to Diagram No. 3 that the rejection of the hypothesis of lognormality in the case of the last intersections is due mainly to the abnormal concentration of values in the category 236 to 394 inch-dwts. This concentration may have been produced as a result of a bias in the human decisions to either deflect a borehole (again) or not to. In view of the possibility of this bias, only first intersections are used in subsequent analyses, except in the examination of the ' variance—size of area ' relationship which is not appreciably affected by any such bias.

3.   *Extent of core losses*

A further result of the departure from lognormality in the case of last intersections is that a valid test cannot be applied to determine the extent, if any, to which the difference between the values from first and last intersections can be attributed to core losses in drilling.

4.   *Analysis of variance (logs of observed values)*

Within mines

The following are the calculated variances for the 10 individual mines in the Main Sector of the field in respect of Basal Reef first intersections only. (These variances are based on common logs and require multiplication by $5 \cdot 3019$ for conversion to natural logs.)

TABLE VI

Mine	No. of inter-sections	Calculated unadjusted variance (common logs)
Welkom	14	0·756852
President Steyn ...	13	0·313464
President Brand ...	9	0·146932
Western Holdings	7	0·130088
Free State Geduld	9	0·593728
Freddies South ...	7	0·352047
Freddies North ...	6	0·136186
St. Helena ...	7	0·362907
Loraine	8	0·131359
Jeannette ...	11	0·447464
	91	

The pooled estimate of variance per mine based on 81 degrees of freedom = 0.420275.

The L test for homogeneity of the variances* within mines shows :—

$L_1 = 0 \cdot 82793 \qquad k = 10$

Harmonic mean of degrees of freedom per mine = $7 \cdot 39$

5 per cent limit for $L_1 < 0 \cdot 802$ from Nayer's Tables*.

The hypothesis of homogeneity of variances is therefore not rejected at the 5 per cent significance level.

The analysis of variance between and within mines is as follows :—

TABLE VII

	Degrees of freedom	Sum of squares	Mean square
Between mines ...	10–1	10·6722	1·1858
Within mines ...	91–10	34·0423	0·4203
Whole sector ...	90	44·7148	0·4968

Ratio $\dfrac{1 \cdot 1858}{0 \cdot 4203} = 2 \cdot 821, \, n_1 = 9, \, n_2 = 81.$

Referring to the ' F ' tables, the differences in mean values from mine to mine are, therefore, significant even at the 1 per cent level. (F at 1 per cent for $n_1 = 9$ and $n_2 = 80$, is $2 \cdot 64$.)

It has, however, been shown previously[4, 5] that a combination of lognormal subpopulations with identical coefficients of variation

* Ref. 3, pp. 83–86 and 366.

and with lognormally distributed means (corresponding to a combination of normal subpopulations with identical variances and normally distributed means) forms a parent lognormal distribution with mean equivalent to the overall mean of the subpopulation means and with parameter

$$\sigma_p{}^2 = \sigma_s{}^2 \text{ (of subpopulations)} + \sigma_m{}^2 \text{ (of means of subpopulations)} \quad \dots \quad (12)$$

Homogeneity of the variances of the transformed subpopulations and nonhomogeneity of the means of these subpopulations are therefore not in conflict with the concept of a parent normal population in the transformed form and a parent lognormal population in the untransformed form.

' *Variance—Size of area* ' *relationship*

The variances of the logs of borehole values have now been estimated within the Main Sector as a whole, within individual mines and within a deflection area. It is also obvious that the variance within an area represented by the cross-section of half of a borehole core (which forms the ore sample assayed for gold) will be zero. The trend of the relationship between these estimated variances and the corresponding size of reef areas can be analysed suitably by plotting the variance against log of area as shown on Diagram No. 7, prepared from the following data (Table IX):—

TABLE IX

	Area (horizontal)	Area on Reef Plane sq. ft.	Log sq. ft.	Variance
Main Sector	51,115 claims	3.422 million	9·5343	0·4968
Mine	5,112 claims	342 ,,	8·5343	0·4203
Deflection Area	20 sq. ft.	20·91 ,,	1·3204	0·1824
Half Core Section	0·0072 sq. ft.	0·0075 ,,	− 2·1249	0·0000

Within deflection area:

In order to obtain an estimate of the variance within the reef area covered by borehole deflections, ie. some 20 square feet, all the individual intersections (141) at holes where deflections were made (62), were stratified according to boreholes to give the following analysis of variance :—

TABLE VIII

	Degrees of freedom	Sum of squares	Mean square
Between strata ...	62–1	51·3911	0·8425
Within strata ...	141–62	14·4086	0·1824
Whole sample ...	140	65·7997	0·4700

Ratio $\dfrac{0·8425}{0·1824} = 4·619$, $n_1 = 61$, $n_2 = 79$.

Referring to the tables for ' F,' it is evident that the differences in value between individual deflection areas are highly significant at the 1 per cent level. (' F ' at 1 per cent for $n_1 = 61$ and $n_2 = 79$ is approximately 1·73.)

The trend appears to be linear or very nearly so and suggests that linear interpolation between observed points may be resorted to. Thus by interpolation between the variances for a mine and a deflection area, the variance within an average size ore block (say, 73,200 sq.ft. ; log = 4·8645) is estimated at 0·2992.

The differences between any two variances are also very significant, e.g., the difference between 0·4968 and 0·2992 (i.e. 0·1976) represents the variance (of logs) associated with the distribution of ore block values within the Main Sector (assumed to be lognormal—see No. (12) above and Ref. 4.)

Similarly, if the area covered geographically by the current ore reserves (including unpay) in a mine at any stage is taken at, say, 1,000 claims (66,951 million sq. ft. on plane of reef; log of area = 7·8258) the variance of borehole values (logs) within this area is interpolated to be 0·3968. The distribution of ore blocks within the ore reserve area will, therefore, be associated with a variance (on the log basis) of 0·3968 − 0·2992 = 0·0976. On the natural log basis this latter variance becomes 0·5175 which is almost indentical with the corresponding average variance (σ^2) associated with ore reserve distributions on the Witwatersrand, viz. $0 \cdot 5^4$.

It seems, therefore, as if the relative variability of block values in the Basal Reef will be very similar to that encountered on the Witwatersrand.

5.   Valuation of the Main Sector
Estimate of overall mean

For the 91 observed values from first intersections on Basal Reef, $n = 91$, $\bar{x} = (2 \cdot 302585 \times 2 \cdot 05586)$ and $V = (5 \cdot 3019 \times 0 \cdot 49137)$. Interpolation from the tables published in Reference 7 puts the t estimate, No. (5) above, at 410 inch-dwts, whereas the solution to Equation No. (5) yields 411 inch-dwts.

The t' estimator, No. (6) above, is 424 inch-dwts and therefore shows a positive bias exceeding 3 per cent. This is due to the extreme skewness indicated for the parent lognormal population.

Standard error

The estimated variance of the t estimator from No. (7) above reduces to

$t^2(e^{0 \cdot 0667} - 1) = 11,656$ inch-dwts^2,
with S.E. $= t(e^{0 \cdot 0667} - 1)^{\frac{1}{2}}$
$= 108$ inch-dwts

The corresponding indicated variance of the t' estimator is $t'^2(e^{\cdot 0671} - 1)$. If the unbiassed t is substituted for t', however, the S.E. becomes 108 inch-dwts which is the same as above.

Effect of stratification

Due to the fact that the parent population is stratified with a more or less even geographical spread of the boreholes, the 91 available values can be regarded as approximating a stratified sample with 1 value selected at random per stratum. The average size of a stratum amounts to some 37,607,000 sq. ft. on the plane of the reef (log $= 7 \cdot 5753$) and interpolation in Table IX puts the average (log) variance within the strata at $0 \cdot 3887$ (or $2 \cdot 0606$ on a natural log basis) which is considerably lower than that for the Main Sector as a whole (i.e. $0 \cdot 4968$). This difference not only reduces the standard error of the mean (estimated above on random sampling theory), but due to the logarithmic transformation of the

variable can also be shown to introduce a bias into all the maximum likelihood estimates of the untransformed population mean if these are based on random sampling theory.

Take *firstly* the case of the maximum likelihood estimator t'' in No. (3a) above, where the last factor, i.e. $\exp(-\frac{\sigma^2}{2n})$ is the correcting term for size of sample employed. In the case of a stratified sample with random selection of one unit within each stratum, the variation in the first term in (3a) above will be governed by the within strata variance σ_s^2, and the correcting term for size of sample then becomes

$\exp\left[-\dfrac{\sigma_s^2}{2n}\right]$ and the t'' estimator

$$= t''_s = \exp(\bar{x}) \exp\left(\frac{\sigma^2}{2}\right) \exp\left(-\frac{\sigma_s^2}{2n}\right) \ldots \text{(13)}$$

If No. (3a) is applied to a stratified sample, therefore, where σ_s^2 is less than σ^2, a negative bias factor will be introduced in the t'' estimate equivalent to

$$\dfrac{\exp(-\dfrac{\sigma^2}{2n})}{\exp(-\dfrac{\sigma_s^2}{2n})} = \exp\left(\frac{\sigma_s^2 - \sigma^2}{2n}\right) \quad \ldots \quad \ldots \quad \text{(14)}$$

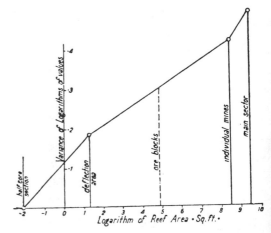

Diagram No. 7—Showing 'Variance—Size of area' relationship

With σ^2 and σ_s^2 of the order of $(0 \cdot 4968 \times 5 \cdot 3019)$ and $(0 \cdot 3887 \times 5 \cdot 3019)$ respectively, and $n = 91$, this factor will amount to $0 \cdot 998$ and can therefore be disregarded.

Further analysis shows that the maximum bias will, in the case of a ' variance—size of area ' relationship as shown on Diagram No. 7, occur with a sample of size 2, (i.e. with 1 value selected at random from each of the 2 strata in the population), and will only be of the order of 2 per cent.

The estimated variance of t''_s will, as compared with No. (4a) above, be

$$= t''^2_s \left[\exp\left(\frac{\sigma_s^2}{n}\right) - 1 \right] \quad \ldots \quad \ldots \quad \ldots \quad (15)$$

which indicates an appreciable reduction in the case of the population now under consideration.

Consider *secondly* the case where σ^2 and σ_s^2 are unknown and have to be estimated from the sample itself, but where n is large. For stratified sampling as above (6) then becomes

$$t'_s = \exp\left[\bar{x} + \frac{Vn}{2(n-1)} - \frac{V_s'}{2n} \right] \quad \ldots \quad (16)$$

where $\dfrac{Vn}{2(n-1)}$ is the estimate of the factor $\dfrac{\sigma^2}{2}$ in No. (13), $\dfrac{V_s'}{2n}$ the estimate of the factor $\dfrac{\sigma_s^2}{2n}$ in No. (13), and V'_s represents an unbiassed pooled estimate of σ_s^2.

Comparing No. (16) with No. (13) it is obvious that the bias introduced in applying No. (6) instead of No. (16) will approach that defined in No. (14) as n increases and V and V_s' approaches σ^2 and σ_s^2 respectively. The estimated variance of t_s' as compared with that of t' presents a difficult problem in view of the fact that both V and V'_s will be estimated from the same data and their sampling errors will therefore, not be independent. When n is large, however, the correction for stratification in No. (16) may be disregarded, with a reversion to No. (6). In No. 6 the variance of $\bar{x}$ will then reduce to $\dfrac{\sigma_s^2}{n} \doteqdot \dfrac{V'_s}{n}$ and the variance of the factor $\dfrac{V}{2}$ will also be reduced on account of stratification since a stratified sample will, under the specified conditions, provide a . more efficient estimate of the parent population variance than a random sample. It is therefore evident that the variance of t'_s

will be less than that of t'. The determination of the extent of the reduction in the variance of the factor $\dfrac{V}{2}$, however, falls outside the scope of this investigation and as a first approximation, therefore, the variance of t'_s will be regarded as

$$t'^2 \left[\exp\left(\frac{V'_s}{n}\right) \exp\left\{ \frac{V'^2 n}{2(n-1)^2} \right\} - 1 \right] \quad \ldots \quad (17)$$

The use of No. (17) will result in an overstatement of the error variance of the estimate of the population mean from a stratified random sample, but will be closer to the true error variance than that indicated by No. (8).

In the *third* case where n is not large, the extent of the bias introduced into the t estimator and into the estimate of its variance, as defined by Nos. (5) and (7), on account of stratification of the sample, is unknown, and presents a problem outside the scope of this paper.

For purposes of the present estimate, however, n is large (91) and it is, therefore, intended to accept the t estimator for the population mean and to use No. (17) as a first approximation of its error variance.

Thus $t = 411$ inch-dwts.

With variance $= 411^2(e^{0.0604} - 1) = 10,510$

and standard error $= 103$ inch-dwts.

It is evident from the insignificant difference between this standard error and that calculated on the basis of random sampling theory that in the present case it is largely an academic question as to whether the sample of 91 values can or should be regarded as a random or a stratified random sample.

Confidence limits

As the sample size increases the t estimator approaches the t' estimator[7] i.e. $\exp(\bar{x} + \frac{V}{2})$. Now, for large samples the independent distributions of V and $\bar{x}$ can both be regarded as normal and therefore also the function $(\bar{x} + \frac{V}{2})$. Hence the limiting distributions of t and t' as the size of sample is increased. ie. of $\exp(\bar{x} + \frac{V}{2})$, are lognormal.

In the present case $n = 91$, and the assumption of lognormality of the t distribution seems a justifiable approximation. A further approximation is, however, necessary in defining confidence limits for t, i.e. assumption of *a priori* knowledge of the variance parameter σ^2 for the t distribution, because of the fact that the ' Student's ' distribution for the lognormal case is unknown, and presents a problem involving awkward mathematical theory (see also discussion by Dr D. J. Finney to Ref. 4).

On the assumption of lognormality and an actual variance for the t estimator of $\theta^2(e^{\cdot 0604} - 1)$, the 5 per cent and 1 per cent confidence limits will be as follows :—

For $\sigma^2 = 0\cdot0604$, No. (10) reduces to

$$\log_{10}\frac{z}{\theta} = \frac{w - 0\cdot12289}{9\cdot36885}$$

and $\theta = 0\cdot6879z$ or $1\cdot5442z$ for 5 per cent confidence limits

and $\theta = 0\cdot5819z$ or $1\cdot8257z$ for 1 per cent confidence limits.

The estimated confidence limits for the mean are therefore :—

283 and 635 inch-dwts at the 5 per cent level and

239 and 750 inch-dwts at the 1 per cent level.

Arithmetic mean

It is interesting to note that the arithmetic mean of the 91 observed values is 518 inch-dwts with a calculated standard error of 254 (based on random sampling theory). The advantage of the t estimator is, therefore, evident.

Reference to Nos. (1) and (7) indicate that on the t basis 91 values are on average equivalent for confidence purposes to more than twice this number on the usual arithmetic mean basis.

Value of payable ore from stopes

From Section 4 above, the variance parameter σ^2 associated with the distribution of the values of ore blocks within the Main Sector is estimated at $0\cdot1976 \times 5\cdot3019 = 1\cdot0475$.

Accepting the pay limit for ore in the Main Sector at, say, 150 inch-dwts, and solving in Nos. (10) and (11) yields :—

For $\theta = 411$:- Payable ore = $68\cdot2$ per cent at 562 inch-dwts.

For $\theta = 239$:- Payable ore = $47\cdot7$ per cent at 417 inch-dwts.

For $\theta = 750$:- Payable ore = $85\cdot6$ per cent at 861 inch-dwts.

Diagram No. 4 was prepared on this basis by solving these equations for different pay limits and for the relevant θ values.

6. CONCLUSIONS CONCERNING INDIVIDUAL MINES

The variance parameter σ^2 for borehole values within a mine has been estimated in Subsection 4 above at $(0\cdot4203 \times 5\cdot3019)$, i.e. $2\cdot2284$. The average number of borehole values available per mine is 9 and the t estimator applied to an individual mine will therefore have a variance of the order defined by No. (8a) above,

i.e. $t^2(e^{0\cdot5235} - 1) = t^2(0\cdot688)$

and Standard Error $= t(0\cdot8295)$.

Substitution of $\sigma^2 = 0\cdot5235$ in No. (10) on the assumption of lognormality in the t distribution* yields :—

5 per cent confidence limits : $\dfrac{z}{\theta} = 0\cdot23$ and $2\cdot53$.

1 per cent confidence limits : $\dfrac{z}{\theta} = 0\cdot14$ and $4\cdot14$.

(*Note.*—The effect of any stratification on these limits will be slight for $n = 9$.)

This suggests that if the lower limit of recovery value corresponding to the economic flotation of a mine is known and the corresponding overall minimum mean value limit for the mine can be calculated, then the t estimate of the mine's mean value will have to be about four times this minimum mean value limit, in order to be almost certain that the mine will not turn out to be an uneconomic venture.

It is on this basis that the general conclusions concerning individual mines were arrived at in the Valuation Section.

For this purpose the variance parameter σ^2 for ore blocks within an individual mine was accepted as

$5\cdot3019(0\cdot4203 - 0\cdot2992) = 0\cdot642$ (see Subsection 4 above).

* For n small the distribution of t will not be lognormal, but as a first rough approximation for valuation purposes this assumption seems justified.

It should be pointed out that, if the hypothesis of homogeneity of the variance parameters σ^2 for individual mines as tested in Subsection 4, is accepted, the standard error of the pooled estimate of a mine's variance parameter will be considerably less than that based on only 9 values for a specific mine. In this case the *a priori* knowledge of a mine's σ^2 within narrow limits could be employed in a revised form of No. (3) above to yield an estimate of the population mean with standard error approaching that defined by No. (4) above. Such an estimate is particularly suitable in the case where the calculated variance based on the logs of, say, 9 borehole values, appears from *a priori* knowledge to be unrealistically high due to the presence of either an exceptionally high or low value in the sample. If the sample variance is accepted in such a case and is actually considerably in excess of the true population variance, the *t* estimator could yield an unrealistically high estimate of the population mean.

7. ADDITIONAL CONFIDENCE TO BE GAINED FROM DEFLECTIONS

Consider now the Main Sector as a whole on the basis of random sampling by, say, 100 boreholes. The variance of the *t* estimator can be regarded as the same as that of the *t'* estimator, which from No. (8a) above for $n = 100$ can be accepted as $t'^2(e^{0.0609} - 1)$ $= 0.0628t'^2$.

The estimated variance parameter σ^2 for values within a deflection area $= 0.1824 \times 5.3019 = 0.9671$. With 1 deflection at every borehole, which can in conjunction with the original intersection be regarded as approximating a random sample of 2 from the deflection area, the variance parameter τ^2 of the *t* estimator for each deflection area will, on the assumption of lognormality be 0.59617 (see No. (7) above amended for σ^2 instead of V). The parameter σ^2 for the distribution of the *t* estimators from all the deflection areas within the Main Sector would then reduce to $(2.6342 - 0.9671)$ $+ 0.59617 = 2.26327$, and the variance of an estimate based on 100 boreholes, each with one deflection then becomes $t'^2(e^{0.04824} - 1)$.

In order to yield this variance the *t'* estimator based on boreholes without deflec-

tions will require a total of 127 values. Table III in the Valuation Section was prepared on this basis.

Now consider the possibility of increasing the deflection area to one with, say, a 50 ft radius (i.e. log area $= 3.8951$). Interpolation in Table IX puts the variance parameter σ^2 for this size area at 1.3992, and proceeding as above, Table IV in the Valuation Section was calculated to indicate the additional confidence gained from such deflections.

8. THE LEADER REEF

The results of a similar investigation carried out in respect of the 77 first intersections on Leader Reef in the Main Sector are briefly :—

t estimate $= 88$ inch-dwts
χ^2 test for fit of curve : $0.2 > P > 0.1$
Variances (natural logs) :
  Main Sector : 1.848
  Mines : 1.429
  Deflection Area : 0.173
  Ore Blocks : 0.790
  Ore Blocks within Field : 1.058
Payable Ore within Field (Pay limit 150 inch-dwts)
$= 15$ per cent at 293 inch-dwts.

Although it appears from these figures that the Leader Reef is an extremely doubtful source of milling ore, the confidence limits attaching to the above estimates are wide and it is therefore possible that workable tonnages will be found in a number of the ten mines concerned.

CONCLUSION

It should be noted that the basic borehole information on which this paper has been based, is available to the public ; also that the deductions made are purely personal and should not be construed as representing the conclusions of any party with which the author is or has been associated in an official capacity.

REFERENCES

1. DE KOCK, DR W. P. The Carbon Leader on the Far West Rand, with special reference to the proportion of the gold content of this horizon recovered in drilling. *Proceedings of the Geological Society of South Africa*, Vol. 51 (1948).
2. JEPPE, PROF. C. B. Shaft sinking and development in the Orange Free State Goldfields—

Optima, a quarterly review published by the Anglo-American Corporation of South Africa, Ltd., Vol. 2 (September 1951).

3. JOHNSON, P. O.   Statistical methods in research.   Prentice-Hall Inc., New York.   1949.

4. KRIGE, D. G.   A statistical approach to some basic mine valuation problems on the Witwatersrand.   *Journal of the Chemical, Metallurgical and Mining Society of South Africa* (December 1951). Discussion (March 1952).

5. KRIGE, D. G.   A statistical approach to some mine valuation and allied problems on the Witwatersrand.   Master's Thesis, University of the Witwatersrand (1951).

6. PAPENFUS, E. B.   The importance of borehole results in prospecting and developing a new goldfield.   *Optima*, a quarterly review published by the Anglo-American Corporation of South Africa, Vol. 1, No. 1 (June 1951).

7. SICHEL, H. S.   New methods in the statistical evaluation of mine sampling data.   *Transactions of the Institution of Mining and Metallurgy*, London (March 1952).

8. SMITH, J. G., and DUNCAN, A. J.   Sampling statistics and application, 1st edition.   McGraw Hill, New York.

[*Editor's Note:* The Discussion has been omitted.]

17

MAGNETITE TREND SURFACE ANALYSIS: A GUIDE FOR OPEN PIT QUALITY PRODUCTION CONTROL AT THE SMALLWOOD MINE, NEWFOUNDLAND

D. P. Harris[1] and E. L. Zodrow[2]

ACKNOWLEDGMENT

The authors wish to express their gratitude to the Iron Ore Company of Canada, notably to Mr. D. Selleck and Mr. W. Campbell, Superintendent of Engineering and Chief Mining Engineer, respectively, for their kind permission to publish this paper at the Seventh Symposium on Operations Research and Computer Applications in the Mineral Industries.

ABSTRACT

Reliable grading is one basic means of upholding quality control and minimizing production costs. Complex geologic and mineralogic relationships in the ore body of the Smallwood Mine cause excessive error in grading of magnetite by methods of simple weighted averages. Trend surface analysis assists the grading of magnetite by expressing the interrelationship of the geologic and statistical characteristics of the ore body.

INTRODUCTION

Location

The Iron Ore Company of Canada is Canada's largest producer of combined pellets and concentrate and direct shipping ore; the Carol Lake Project, one part of this company's operation, produces pellets and concentrate at Wabush Lake, Labrador. This area is situated in northeastern Canada near the boundary of the Labrador portion of Newfoundland and Quebec (approximately 200 miles north of Sept-Iles, Quebec (fig. 1). In terms of geotectonics, this part of the Labrador geosyncline is described as the Normanville geological sub-province, which is part of the Grenville province (fig. 1). [6]

---

[1]*Assistant Professor of Mineral Economics, Department of Mineral Economics, The Pennsylvania State University.*
[2]*Mine Geologist, the Iron Ore Company of Canada at Labrador City, Newfoundland.*

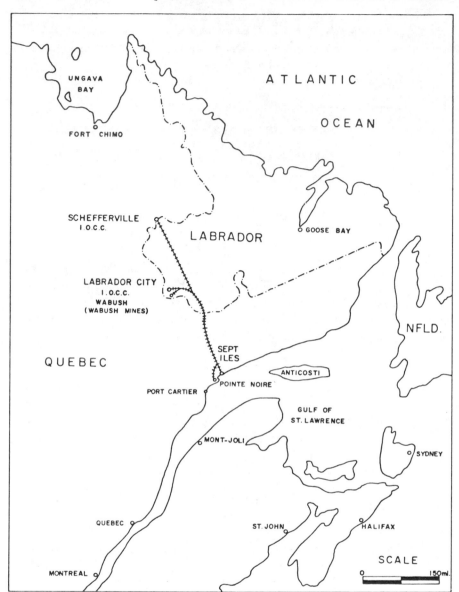

Figure 1.   Location of the mining operations of the Iron Ore Company
of Canada (I.O.C.C.).

Legend:
            Interprovincial boundary   — · — · —
            Railway   +++++++

In this paper our discussion is limited to the Smallwood Mine, one of the two open pit mines operated by the Iron Ore Company of Canada at Wabush Lake.

GENERAL STATEMENT

Crude iron ore quality control before production blasting is based upon computerized block grading.[8] After blasting, the muckpiles are graded by muckpile sampling; that is, sampling of the topographical surfaces of the muckpiles; this is essentially a short range grade prediction not exceeding one week. Daily crude grades are predicted by toe sampling. These samples are collected along the muckpile ahead of and behind the producing shovels. In both types of muckpile sampling the sample volume is held constant, as advocated by Hazen (p. 35),[3] and grade estimation is based on arithmetic means.

Block grading estimates possess minimum-variance as compared to the estimates from both muckpile and toe samples. However, these grade estimators are all seriously questioned as unbiased estimators on the basis of theoretical consideration and grading experience (p. 167-178).[5]

The Smallwood Mine ore body contains chiefly specular-hematite with minor quantities of magnetite as recoverable iron ore minerals. The magnetite component is by far the most influential in the autogenous grinding process with respect to grinding rate and generation of fines (minus 200 mesh particles); the grinding rate decreases and the quantity of fines increases, ceteris paribus, with an increase in magnetite. Moreover, unforeseen high magnetite-ore shipments to the mill unduly tax the efficiency and the capacity of the magnetic separation plant. Thus, potentially recoverable magnetic iron units are lost when there is an excessively high content of magnetite. The problems in milling are made acute because of the unreliable prediction for blocks of ore (prior to blasting) of their magnetite content.

PREVIOUS STATISTICAL WORK

Prior to the present analysis, an attempt was made to isolate geological causes responsible for the magnetite variation across the Smallwood Mine. This was done to assist interpretation of the trend surface relative to the most probable geological influences so that the trend surface could be reconciled with structural relationships and co-variations emanating from the sum total of natural forces. Factor analysis identified metamorphic influences as one cause mainly responsible for the formation of magnetite.[7] Since metamorphism implicitly implies time and space dependency (metamorphic advance) and since most of the problems in grading stem from the magnetite variability, a trend surface analysis was made on the magnetite to examine the relationship of the geographic distribution of magnetite to general geology and to the ore body.

STRUCTURAL GEOLOGY AND VALUABLE MINERAL DISTRIBUTION

General Structural Geology

This ore body is roughly canoe-shaped bulging out in the central portion (fig. 2). The basic structural form is a doubly plunging, inverted syncline which is structurally superimposed on an anticline. The former structure is associated with the Grenville deformation (orogeny), while the latter is part of the Labrador geosyncline orogeny. The traces of the axes of these features make an angle of intersection of approximately 30°. The folding of the inverted syncline is locally isoclinal but varies considerably throughout this mine (fig. 3). At the northeast part of the ore body the bedding dips consistently at about 45° SE, but on the southeast part of the strucural influence of the steeply plunging anticline together with more prevalent secondary folds combine to give very erratic attitudes.[1]

Valuable Mineral Distribution

Geological mapping and sampling indicate a rather uniform distribution of specular-hematite. Statistically, this may be interpreted as approximation of spatial randomness in this mineral. On the other hand, the distribution of magnetite appears nonuniform. Based on observable magnetite mineralogy and statistical interpretation, three possible controls for magnetite are proposed:

(1) Sedimentary.
(2) Metamorphic.
(3) Surface oxidation under normal temperature and pressure.

The first two origins are inferred from factor analysis;[7] field evidence supports this hypothesis. The last origin of magnetite is substantiated by observation in the mine near the quartz-carbonate rock and iron formation contacts where the magnetite, occurring as bands of massive and very fine-grained material, is sometimes associated with earthy manganese dioxide, limonite-goethite replacements, and leaching on a large scale.

Summarizing, the authors conclude that the magnetite distribution over the mine is a function of all three origins each of which imposes its own

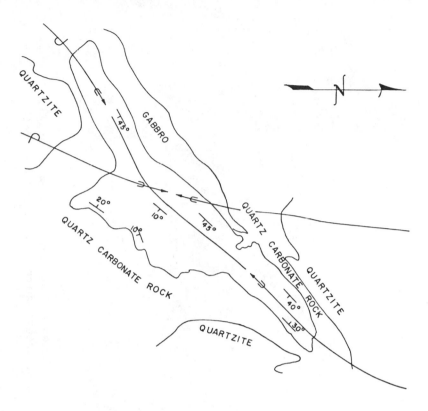

Figure 2. Structural interpretation of the Smallwood Mine,
Labrador. The plunging overturned anticline (1)
is associated with the Labrador geosyncline orogeny, while the overturned plunging syncline (2)
is part of the Grenville deformation system, i.e.
(2) is superimposed upon (1).

Legend:

Strike and dip of bedding planes

Overturned syncline with plunge
direction

Overturned anticline with plunge
direction

Geological boundary mapped and
inferred.

Scale: 1" to 2000'

219

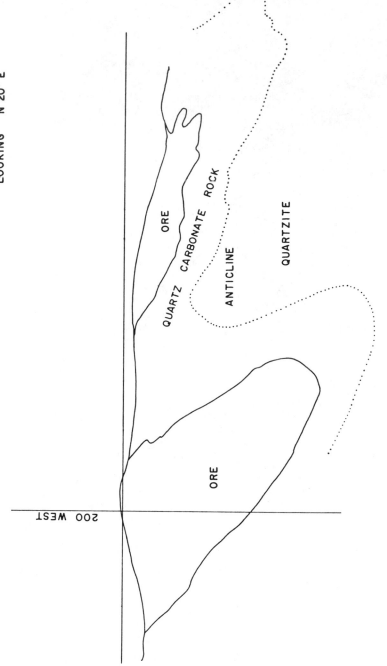

LOOKING N 20° E

200 WEST

ORE

QUARTZ CARBONATE ROCK

ORE

ANTICLINE

QUARTZITE

Figure 3. Schematic cross-section of the Smallwood Mine, Labrador. The designation ANTICLINE represents that fold structure which is associated with the Labrador geosynclinal orogeny.

Scale: 1" to 500'

characteristic control at specific locations in this mine and contributes preferentially, depending on the dominant type of origin, to the nonuniform spatial distribution.

TREND SURFACE ANALYSIS OF MAGNETITE

BACKGROUND

This analysis uses as basic input data 150 drill hole coordinates on the 2055 bench with reference to a local grid established on the Smallwood Mine (fig. 4). The drill holes are all vertical, and drill hole deviation from the vertical is assumed negligible.

While trend surface analysis may have many somewhat different applica-

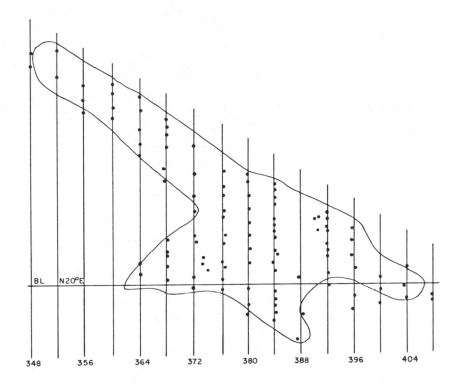

Figure 4.   Bench 2055, Smallwood Mine, Labrador.   Showing in part the
locations of the diamond drill hole locations used in this
analysis.   The cross-sections are 400' apart.

Scale: 1" to 1000'

tions in mineral industry studies, most applications can be viewed as falling within two major categories: (1) As a smoothing technique, where the data are subject to local and spurious variatioȧ, and where one is primarily interested in the overall more stable component (trend), and (2) as an intermediate step in seeking local anomalies which are difficult to discern because of the overpowering regional effect (trend). Category (2) is typical of the application of trend surface analysis as applied to geophysical and geochemical measurements. It is category (1) that is given attention in this paper.

The general method of trend surface analysis can be briefly described as statistical examination of the relationship of the mineral value to its location (in two dimensional space, the X, Y, plane). More complex analyses consider the joint variation of the mineral value with some other measurement as well as its coordinates (4, p. 323-346; 2). The first objective in a trend analysis, especially if it is of the type one category, is to determine if there is a significant trend. If a significant trend exists, then interest focuses on the form of and the parameters of the equation relating the mineral value to its location. Given a sound equation of a well defined trend, it may be employed in various ways, one of which might be to generate trend values for points between drill holes as well as for those data points that went into the analysis. After plotting the trend values, the trend map may then be used to control ore production grades and to assist in mine planning where grade consideration is of primary importance.

Generation of Trend Equation

Equations that express two-dimensional trends may be computed by one of three ways:

(1) Orthogonal polynomials.
(2) Standard polynominals.
(3) Polynomials with heterogeneous terms.

The method of (1) requires either that the original data be taken on an equispaced grid, or that new data be generated from the sample data for points on an equispaced grid. The method employed here is that of type (3) for three reasons: (a) Data are not on an orthogonal grid, (b) there is nothing sacrosanct about conventional polynomials, especially as they apply to a complex system, and (c) a computer program is available which allows parsimonious elimination of terms of an equation that contribute little to explaining the distribution of the mineral value.

In order to eliminate the generation of excessively large numbers (in the computer storage), it was necessary to scale the coordinates of locations as follows:

W1 = the northings,
W2 = the eastings,

then X = W1/10,000
Y = (W2 + 4,000)/2,000

Each of these coordinates was raised to all powers from 1 to 10, and all combinations (of the powers) of the two coordinates were computed. Thus for each of the 151 data points there existed the quantity of magnetite and all terms in X and Y required for a 10th degree polynomial of X and Y. These data were then submitted for correlation and regression analysis on the IBM 360-67 of The Pennsylvania State University Computation Center. Several computer runs were made to explore the following relationship:

(1) Raw measurments of magnetite with the coordinate terms,
(2) The natural logarithmic value of magnetite, ln (magnetite + 10), and the coordinate terms,
(3) The exponential value of magnetite, exp (magnetite/10), and the co-ordinate terms.

The exponential value of magnetite gave the strongest relationship.

A problem in trend surface analysis is determining the degree of equation to select as an expression of the trend. An equation with many terms will always be a better fit to the data base than one with less terms. Of course, a near perfect fit can be obtained by employing an equation of very high degree, but such an equation is not desired for it merely duplicates the raw data. On the other hand, equations of too few terms may not fully define the trend.

The approach used in this study, to select the trend equation, was to test sequentially the additional variance in magnetite explained by the inclusion of one more term. Although the fit of the equation to the data always improves by including an additional term, eventually the incremental increase in goodness of fit is not statistically significant. That is to say, such an increase could arise from random events and cannot be associated specifically with that term.

The above approach yielded an equation with nine terms:

$$
\begin{aligned}
_e(\text{magnetite}/10) \ = \ & \overset{(10.4236)}{729.3872} - \overset{(0.1059)}{23.5239X^3} + \overset{(167.3687)}{0.2289X^6} - 575.6397Y \\
& \overset{(130.2908)}{+\ 336.9631Y^2} - \overset{(44.9321)}{111.3109Y^3} + \overset{(0.8449)}{2.1461Y^6} - \overset{(0.0033)}{0.0091Y^{10}} \\
& \overset{(2.2493)}{+\ 6.1499X^4Y} - \overset{(0.4902)}{1.3332X^5Y}
\end{aligned}
$$

where the number in brackets above each term is the standard error of the coefficient of that term. Taking the natural logarithms on both sides, the above equation becomes:

$$
\begin{aligned}
\text{Magnetite} = \ln\ (& 729.38729 - 23.52399X^3 + 0.2289753X^6 - 575.6397Y \\
& + 336.9631Y^2 - 111.3109Y^3 + 2.146149Y^6 - 0.009168Y^{10} \\
& + 6.14999X^4Y - 1.33321X^5Y)\,10
\end{aligned}
$$

This equation explains 20 percent of the variance in magnetite across the area of the 2055 bench of the Smallwood Mine. A greater amount of the variance in magnetite was explained with more terms in the equation; however, the increment of explained variance was not significant at the 0.05 percent level. This implies that there is greater than one chance in 20 that the seemingly better fit is a result of the combination of random influences and cannot be attributed to the additional term. In the above equation, the standard errors of the coefficients are one-half to one-third as large as the coefficients; however, in the ten term equation determined, the coefficient of the variable added to the nine variables above is 0.0020157 and its standard error 0.0018949, the standard error being nearly as large as the coefficient.

Although the amount of variance in magnetite explained by the trend equation is rather low, 20 percent, the null hypothesis that there is not statistically significant trend is rejected at the 0.05 and 0.01 percent levels; that is, the probability of obtaining as good a fit as this when there really is not any relationship between magnetite and its location, as defined in the equation, is considerably less than one chance in 100. Specifically for the degrees of freedom $n_1 = 9$ and $n_2 = 141$, the F-distribution statistics are as follows:

$$
\begin{aligned}
F_{\text{trend equation}} &= 3.86 \\
F_{.05} &= 1.94 \\
F_{.01} &= 2.53
\end{aligned}
$$

Thus, the equation can be accepted as an expression of the overall trend of magnetite across the Smallwood Mine ore body.

INTERPRETATION OF STATISTICAL RESULTS

Although the value of the trend surface as a grading model, at this stage of evaluation, can only be regarded as qualitative in nature; it has important implications to production scheduling, crude grade control, and geologic interpretation.

According to geological surface and bench mapping, company geologists theorized that the magnetite content of the ore increases throughout the mine when approaching the ore/waste contacts. Contradicting this evidence is production muckpile sampling which shows that this theory does not hold for the entire ore body. For, as mining proceeded west on benches, immediately above the 2055 bench (the central portion of Smallwood Mine), magnetite grades remained remarkably constant on approaching the ore/waste contact. Further, as mining proceeded north from the central portion, muckpile samples became richer in magnetite. Simultaneously magnetite values increased closer to the contacts west and east. In the south end of the mine where mining follows the narrow portion, fluctuating magnetite values were reported, but noticeable increases were evident further south.

The above distribution of magnetite is generally valid above the 2055 bench for which controlled mine sampling exists. Thus, a good agreement exists between actual observations, based on extensive sampling in this mine over 4 years, and this trend surface.

We observe then that this trend surface serves as a qualitative guide to magnetite control in mine planning. Magnetite fluctuations in the crude ore are inevitable, because magnetite distribution is a function of its formations. To compensate for these fluctuations (if metallurgically desirable), the authors propose ore blending by benches rather than by individual production blasts. That is to say, sequential development of new benches, starting from the eastern part of the central portion, should coincide with ore shipments from the central portion, and ores from the central portion with ores from the northern and southern end of the mine. Of course, other factors must be considered simultaneously in this type of blending to ensure the most profitable crude ore shipments to the mill.

The geologic structure of this mine can also be well interpreted on the basis of this analysis. Figure 5 reveals that the position of interpreted traces of structural axes in magnetite concentration generally coincide with the geologically interpreted synclinal and anticlinal traces of these axes. One

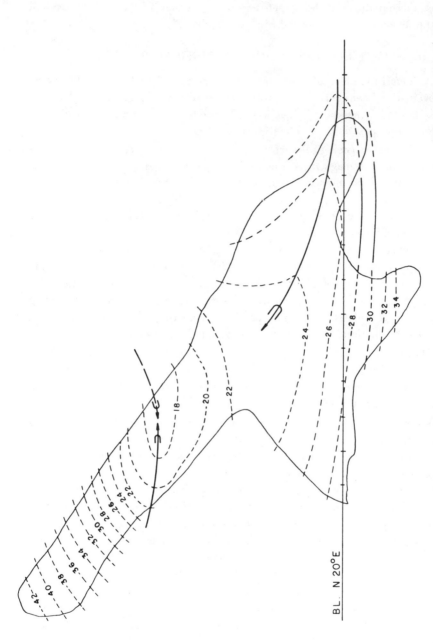

FIGURE 5.—Magnetite trend surface of bench 2055, Smallwood Mine, Labrador. Trend interval is in 2 per cent magnetite. The structural features of magnetite values are according to trend surface interpretation.

exception is noted in the southern portion where geological evidence indicates a plunging anticline (fig. 2), whereas this analysis suggests a plunging syncline of magnetite values.

In summary, a statistically significant and geologically meaningful trend of magnetite distribution exists in this ore body. Because of the moderately low degree of the trend equation for the area considered, the trend is quite smooth and exhibits only the large features. Inasmuch as a statistically acceptable equation of higher degree could not be obtained, it is proposed that figure 5 exhibits that component in the magnetite values that can be attributed to the overall trend for the area.

The deviations (residuals) of the measured magnetite from the trend values generated by the equation are quite large in some instances. In a statistical sense, these deviations contain variations which may reflect local erratic (unsystematic variations) as well as meaningful microtrends; however, in geological terms, the residuals may be considered a function of the different origins of magnetite. Therefore, meaningful trend may yet exist in the following two forms (the proposition that meaningful trend may yet be present in the residuals may seem to contradict the statement that higher degree equations were not justifiable statistically; this is not necessarily true in a geological sense):

(1) Trend which might be determined by analyzing smaller areas, say dividing the area in two parts, one being the central wide portion and the other the narrow extension to the south.
(2) Trend which is associated with the interaction of magnetite and its coordinates with a mineral component.

In general, the larger the area the greater the distance between data points the less influence microtrend has in determining the form of the equation, and consequently, the less the microtrend is described by the equation. Enlarging the model to include one or more mineral components would in effect allow a more meaningful expression of geologic-mineralogic relationships. It is believed that further research along these lines might be rewarding.

REFERENCES

1. Hamilton, C. G., 1967, Structures in the Wabush Formation west of Wabush Lake with special reference to the Smallwood mine: M.Sc. thesis, Carleton Univ., Ottawa, p. 82.
2. Harbaugh, J. W., Application of four-variable trend hyper surfaces in oil exploration, in Computers in the Mineral Industry, Stanford Univ., School of Earth Sciences, v. 9.

3. Hazen, S. W., Jr., 1967, Some statistical techniques for analyzing mine and mineral-deposit sample and assay data: U.S. Bur. Mines Bull. 621, p. 223.
4. Krumbein, W. C., and Graybill, F. A., 1965, An introduction to statistical models in geology: New York, McGraw-Hill Book Co., Inc., 475 p.
5. Mood, A. M., and Graybill, F. A., 1963, Introduction to the theory of statistics, 2d ed., New York, McGraw-Hill Book Co., Inc., 443 p.
6. Stockwell, C. H., and Williams, H., Age determinations and geological studies, Part 2: Canadian Geol. Survey Paper 64-17, p. 29.
7. Zodrow, E. L., and Harris, D. P., 1967, An ore grading model for the Smallwood mine: Mining Eng., p. 70-73, Aug.
8. Zodrow, E. L., 1967, Block grading models in iron ore mines: Canadian Mining Jour., p. 41-44, Dec.

18

Space Models of Inclined Limestone Deposits

V. Nemec

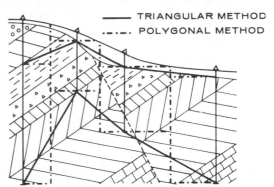

—— TRIANGULAR METHOD
.—.—. POLYGONAL METHOD

FIGURE 1—Blocks of reserves in cross section, based on triangular and polygonal methods.

ABSTRACT

A new approach to ore reserve estimation and quality control for cement raw material deposits has been developed. Space models allowing for the estimation of chemical composition at any point have been constructed, using moving averages, contouring and regression analysis. Three systems of blocks are used simultaneously for computing reserves: microblocks, macroblocks and megablocks. The described methodology has been used successfully in more than fifteen Czech deposits.

INTRODUCTION

EXPLORATION AND EXPLOITATION of industrial minerals has not been one of the more interesting and exciting areas of mining activity. The geological and technical problems of these deposits have been overshadowed by problems in the more popular areas of metallic ores and petroleum.

Evaluating industrial minerals for cement factories is not an easy matter. The complete chemical composition of the raw materials should be known with satisfactory reliability. The six properties estimated are: CaO, MgO, SiO_2, Al_2O_3, Fe_2O_3 and loss on ignition. Other important parameters are derived from them, such as the silica module $Ms = SiO_2 : (Al_2O_3 + Fe_2O_3)$, the alumina module $Ma = Al_2O_3 : Fe_2O_3$ and the coefficient of saturation (according to Lea-Parker) where $S = 100\ CaO : (2.8\ SiO_2 + 1.18\ Al_2O_3 + 0.65\ Fe_2O_3)$. The clinker usually is a mixture of two, three or four components and its final composition is prescribed within narrow limits for particular oxides and modules.

In most situations, undesirable local deviations in the parameter values for the deposit may be corrected easily by changing the mixture ratio or by temporarily using, if unavoidable, imported correction material (in a larger quantity than would be usual otherwise or admissible for economic reasons). Of course, a completely wrong design of the technical dimensions and economic profitability based on false assumptions of the geological evaluation may lead to serious problems, especially from the point of view of economic stability and viability of the plant. It is evident that such problems cause administrators to have an increased interest in the evaluation of industrial minerals.

A short case history of the evaluation of some Czech limestone deposits is used as an example of how the methodological approach was developed.

CZECH LIMESTONE DEPOSITS EVALUATION

About 15-20 years ago, exploration of limestone deposits for new cement plants was undertaken in Czechoslovakia. The density of drill holes permitted a study of tectonic features of particular deposits, and the numerous cores obtained presented excellent material for paleontological and stratigraphic research, especially in the famous Barrandian area of central Bohemia. Simple classical methods for computing reserves of deposits were used for estimating quantity and quality of reserves. At first, triangular and polygonal methods were used, without special regard to mining conditions (levels). Results of chemical analyses of material obtained from drill holes also provided the basis for computing reserves.

Figure 1 shows an example of the classical methods, which generally resulted in a discrepancy between computed reserves and the requirements for extraction because no mining levels were considered (Nemec, 1965). Thus, without further comment, another phase of evaluation methodology can be introduced in which either geological cross sections or geological blocks (based on arithmetic means of quantitative and qualitative parameters) are used with respect to mining conditions. The deposit is divided into a system of blocks, more or less independent, at different levels representing proposed (or already established) exploitation levels. It may be seen in cross section (*Figure 2*) how individual blocks are constructed. The average chemical composition of each block was computed using:

(1) a weighted average of all sample segments analyzed in the space of the block; or

(2) the average chemical composition for different stratigraphic units (or petrographic sub-units) in the deposit or for large segments thereof for particular blocks weighted by the estimated percentages of the units (derived from cross sections or geological maps).

A series of new deposits were analyzed on this basis, quarries opened and factories constructed. Geologists thus had an opportunity to check the results of their evaluation; unfortunately, discrepancies were found in many situations. Additional exploration work was then carried out and an attempt was made to find the source of the errors. It became evident that some errors had been caused by defects in drilling technology and in the chemical analyses. However, the approach to computing reserves was also found to be unsatisfactory. Neither of the two systems of estimating the chemical composition of the deposit survived a critical analysis.

(1) It may be seen easily in *Figure 2* that using only samples taken within the space of the computed block left the majority of this block unsampled; the system

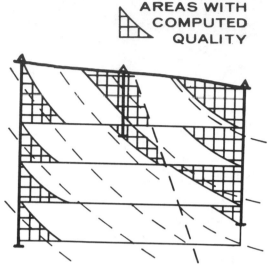

FIGURE 2—Cross section showing areas with computed quality.

could be valid within certain reservations, although only in the case of horizontal beds, which are the exception in the Barrandian area.

(2) Although the other system (mentioned above, sub 2) seemingly is more satisfactory, some important aspects are not taken into consideration:

    (a) the chemical composition of different beds is variable and the variability can be expressed in some trends showing heterogeneity and anisotropy even if these cannot be seen macroscopically; and

    (b) the geological processes operative during the formation of the deposit are not simple — a schematic model of the process (*Figure 3*) shows how many factors are involved.

An example of one of these factors is the tectonic force constantly working in accordance with the dynamic equilibrium of the Earth and obeying a simple law of regular structural pattern which is expressed by (Nemec, in preparation):

$$y_x = 2^{-x}\,D$$

where
y = distance between structural lines of a given order.
x = order of structural lines (= 0, 1, 2, 3, ...) and
D = original constant-diameter of the Earth (approximately 12,754 km).

It was therefore necessary to find methods for analyzing the mass of new data to obtain better decisions on problems involved in evaluating deposits (Nemec, 1966). Many methods were already in use in the world in 1966 for mathematical modelling of deposits. However, a new tailor-made methodology had to be developed, consisting of two main parts: (1) space modelling of quality parameters; and (2) a logical system of blocking.

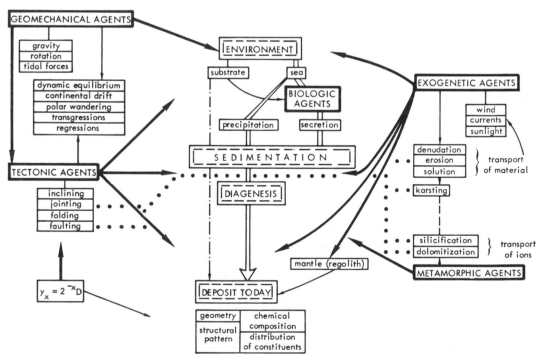

FIGURE 3—Schematic model of geological processes involved in the formation of a limestone deposit.

MODELLING DEPOSITS

Moving Average

The moving average is a valuable tool for treating data (Nemec, 1968) and is used:

(1) to express basic structural features of the deposit as controlled by sedimentational processes;
(2) to express some basic mining parameters of future exploitation; and
(3) to reduce bias introduced in the original sample of the population.

The sedimentation process often results in numerous thin beds with different qualitative characteristics, which are to a certain extent expressed by chemical analyses in a sequence of individual samples. For mining purposes, only average parameters are needed, representing chemical composition for particular spots or segments of mining floors. Moving averages, as used for our modelling, take these practical needs into consideration.

Sequences of chemical analyses referring to individual samples taken in exploratory holes are cumulated into series of higher units, which represent average qualitative parameters for a cumulative thickness equal (or at least similar) to the thickness of a mining floor (in the example given in *Figure 4*, this thickness, a, = 24 m). Mean values of qualitative parameters are computed as weighted averages using thicknesses corresponding to individual samples (or — in case of border samples which are subject to division — a respective part of their thickness) as weights for numerical operations.

Another constant expresses the calculation step (in the given example, this constant distance between adjoining average values, b, = 4 m).

All numerical operations are carried out with the aid of computers, based in part on the following formulae. The mathematical formulation for the first midpoint of intervals under consideration is

$$d_1 = nb \geq \left(\frac{a}{2} + c_o \right) > (n-1)b$$

where
- a = the interval for average values (constant thickness),
- b = step of numerical operations (constant distance between adjoining average values or midpoints),
- c_k = thickness of individual samples (or section of them), whereas c_o = thickness of unassayed waste,
- d_k = depth of the midpoint of the k-th interval, and
- n = a whole number.

Further intervals form a series:

$$d_2, d_3, \ldots d_{m+1}, d_m$$
where $d_k = d_1 + (k-1)b$,
for $k = 2, 3, \ldots, m$, until

$$d_m = (d_{max}) \leq \overset{max}{\underset{0}{\Sigma}} c_k - \frac{a}{2}.$$

In the foregoing computer program, constants a and b are to be given as common input data; all other input data consist of individual chemical analyses systematically arranged on punched cards. In other words (using the practical example in *Figure 4*), the computer examines a series of samples of a drill hole and checks for every 4 m if an average qualitative characteristic can be computed which would represent the given thickness of the mining floor (24 m). The first theoretically possible average can be computed for the interval 0 - 24 m, but with regard to the thickness of overburden, c_o = 3 m, the first average can be elevated for the interval 4 - 28 m (representing derived quality parameters referring to the depth of d_1 = 16 m).

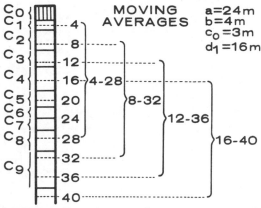

FIGURE 4—Formation of moving averages from drill-hole information.

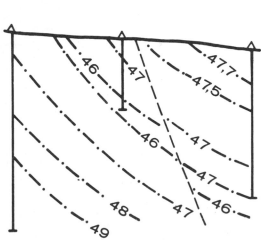

FIGURE 5—Contours of CaO content in cross section.

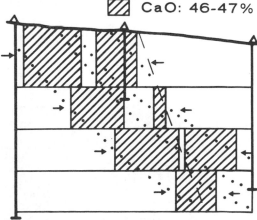

FIGURE 6—Content of CaO, 46-47 per cent, at different mining levels.

A set of moving averages is a set of new regionalized variables leading mostly to more reliable and easily applicable results (when compared with the original set of individual samples). The advantages of using moving averages will be demonstrated further in the text.

Contours of Quality Parameters

Contours may be constructed based on moving averages that represent basic quality parameters for each mining level. As shown in *Figure 5*, the contours conform with the stratigraphic and petrographic layers. If the position of the mining levels is known, the actual quality of the deposit in each spot of the floor will be equal to the value represented by the center of the floor height (*Figure 6*). In a vertical sense, differences of different strata are replaced by a continuous series of data. In practice, this enables the geologist to look deeper into the sedimentation process than allowed by the usual macroscopic description of cores or individual chemical analyses from samples representing the core segments. In a horizontal sense, the contours may be interrupted by tectonic phenomena and, depending on geological conditions, a different discipline may be accepted for such zones. Data based on samples taken in such zones are then to be eliminated before contouring. Of course, this does not mean that only "good" and "reliable" data will enter into the final evaluation. Because the zones represent different conditions from those regulating primary sedimentation, special steps need to be taken before computing the final results.

Preliminary contouring in cross section permits the construction of contours on respective maps representing horizontal planes passing through the center (mid-height) of proposed mining floors (*Figure 7*). The transfer of contours into horizontal maps is not a tedious job if the sections are based on careful geological considerations. The construction of contour maps for each of the mining levels in the proposed quarry serves as a valuable control for geological ideas.

It would be a tedious job to contour each of the six constituents needed in a final evaluation; however, by the use of correlation analysis, the number of primary constituents represented may be limited.

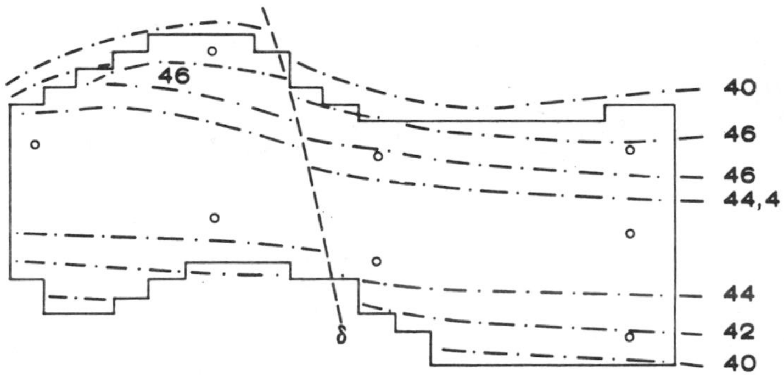

Correlation Analysis

Correlation analysis is a well-known tool for mathematical statistics. The interdependency of data is estimated in the process of our modelling and redundant information eliminated. Coefficients of correlation are usually

FIGURE 7—Contour map of a mining level.

TABLE 1— Changes in Correlation Coefficient by Reducing Original Samples (atypical data eliminated)

Layers	Set of Samples	Number of Observations	Average Values %			Correlation Coefficients	
			CaO	MgO	SiO₂	CaO − SiO₂	(CaO + MgO) − SiO₂
A	orig.	470	43.19	1.38	12.98	−0.518	−0.509
	red.	89	43.35	1.37	12.16	−0.943	−0.958
B	orig.	82	44.62	3.61	8.03	−0.091	−0.535
	red.	52	46.04	1.28	9.09	−0.908	−0.918
C	orig.	96	44.37	6.47	3.42	−0.221	−0.862
	red.	54	46.81	4.65	3.59	−0.382	−0.896

TABLE 2—Changes in Correlation Coefficient by Selecting from Original Samples and by Using Moving Averages

Set of Samples	Number of Observations	Correlation Coefficients			
		CaO − L.O.I.	(Cao + MgO) − L.O.I.	CaO − SiO₂	(CaO + MgO) − SiO₂
original	470	0.342	0.904	−0.191	
selected	220	0.587	0.945	−0.342	
moving averages	31	0.584	0.847	−0.734	−0.953
moving averages reduced	22	0.757	0.829	−0.902	−0.957

Constituents	Input Data for Computations		
	CaO, Ms, MM	CaO, Ms, Fe₂O₃	CaO, Fe₂O₃
	1	2	3
CaO............	contours		(1)
MgO............	$MgO = a_i\,CaO + b_i$		(1, 2)
Loss on Ign........	$LOI = c_i\,CaO + d_i\,MgO + e_i$		(3)
SiO₂............	$SiO_2 = f_i\,CaO + g_i\,MgO + h_i$ (4)		$SiO_2 = (f_i\,CaO + k_i\,Fe_2O_3 + m_i):2$
Al₂O₃...........	$Al_2O_3 = \dfrac{SiO_2 \cdot MM}{Ms\,(Ms + MM)}$	$Al_2O_3 = \dfrac{SiO_2}{Ms} - Fe_2O_3$	$Al_2O_3 = n_i\,Fe_2O_3 + p_i$
Fe₂O₃...........	$Fe_2O_3 = \dfrac{SiO_2}{Ms + MM}$	contours	

$MM = Ms \cdot Ma$
i = system of regression equations
(1) In case of increased dolomitization, contours for MgO and for (CaO + 1.39 MgO) are constructed, and CaO is derived as the difference between these values.
(2) In certain cases: $a_i = 0$.
(3) For MgO < 3%:$d_i = 0$, for Mgo ≥ 3%:$d_i = c_i$.
(4) For MgO < 3%:$g_i = 0$, for MgO ≥ 3%:$g_i = f_i$.

high for the relations between CaO and SiO₂, CaO and loss on ignition, and Al₂O₃ and Fe₂O₃.

The samples from a deposit may be categorized into different classes with regard to geographic location, stratigraphical origin, petrographic unit, degree of change by tectonic or other agents, etc. Table 1 is an example of results. The reduced samples give, in all three situations, a higher quality of limestone which shows a decomposition of the previous noisy mixture of these limestones with some impurities; the latter have been mostly eliminated in the reduced versions. Correlation analysis gives a summary of the interrelation of variables. If the population is smoothed by moving-average values (selected to avoid duplication of original data), larger coefficients of correlation are usually obtained, as seen in Table 2. It is not possible to determine which constituents should be evaluated by contouring and which by correlation analysis. The final selection depends on:

(1) geological conditions;
(2) purpose of computation (including degree of reliability necessary); and
(3) quantity and degree of reliability of data which are available.

Generally, CaO content is important, but in particular situations, due to an increased MgO content, it is useful to separately contour MgO and the sum of CaO + 1.39 MgO (the coefficient 1.39 expressing the volume weight of MgO and enabling us to transform this constituent and to replace it with CaO on the same volumetric basis). The CaO content then may be computed easily as the difference between the two sets of data.

The content of SiO₂ is one of the easiest to contour, but in some instances (e.g. if high-quality products are required), the Fe₂O₃ seems to be the most important constituent. The Al₂O₃ may be important in situations where cement clinker is composed from various sources; to achieve the narrow limits given by the industrial conditions, this constituent seems to be the most delicate. Differences found between evaluated and real SiO₂ or Fe₂O₃

may be corrected easily by small changes in portions of sand or iron ingredients, but the Al₂O₃ content, having such an important role in modules used for evaluation, cannot be changed practically and small discrepancies may cause serious technical or economic troubles in production. Sometimes, the modules seem most useful for contouring and respective oxides may be derived easily from them. The saturation coefficient is excellent for contouring, but its utilization for computing particular components would practically be impossible due to serious differences in orders of coefficients used in its formula. Table 3, therefore, gives some examples of how the various constituents were determined.

Regression coefficients used to estimate constituents not expressed in contours are to be chosen carefully from data obtained by the correlation analysis. It is beyond the scope of this paper to demonstrate in detail all the steps, but it should be mentioned that only linear regression has been used. It has been shown that the effect of increasing correlation coefficients by using various nonlinear regressions according to special tests would not yield any improvements of practical significance. The reliability of regression equations is checked by comparing the original data of moving averages (other than those used for correlation analysis) with those computed by regressions. The results of several deposits are shown in Tables 4 and 5 and demonstrate a good accord. Serious discrepancies, if encountered, usually demand a complete revision of the analysis, resulting in subsequent territorial subdivision of the deposit. Let us only remark that effects of closed arrays of data demonstrated by Chayes (1960) and Vistelius-Sarmanov (1961) have not been observed in the retrieval of our data.

Correlation analysis is also used for estimating constituents other than the six previously mentioned; e.g. NaO, K₂O, MnO, P₂O₅, etc. The number of samples in this situation is usually small, but the results may be transformed easily in accordance with the final representation of the basic constituents.

233

Practical Advantages

The described modelling gives both detailed and generalized ideas concerning the distribution of different constituents within the deposit. Despite many fields in which this system has been proved as useful and efficient, our attention is concentrated on mainly one point of view — that of reliable reserves. A set of contours, representing basic qualitative parameters in cross sections and in horizontal maps, permits a reliable estimate of the chemical composition of each "materialized" point of the deposit if the point is located at the mid-height of the mining floor. This advantage is utilized in the computation of reserves. At this stage, the geologist and technologist have concluded their work and the mining specialist and economist can begin (*Figure 8*).

COMPUTING RESERVES

System of Blocking

Three systems of blocks are used simultaneously for computing reserves:

(1) microblocks,
(2) macroblocks, and
(3) megablocks.

Microblocks are fundamental units; they have a constant area and all necessary input data are derived from maps (topographic, overburden, quality parameters). For each microblock, input data are estimated at its center. Similar basic units have also been used elsewhere (e.g. Pana, 1965; Weiss, 1965). The system of microblocks is an adaptation of regular grids which have been used for various purposes in ore reserve computations. The dimension of a block is usually 10 x 20 meters, and this has been found suitable (compared to smaller units of up to 5 x 5 meters). This size is practical from several points of view because it fits general mining conditions, such as final slopes in different levels of the quarry, minimum width of the quarry at its deepest level, etc. Also, the block can be utilized for recalculating reserves after its mining activity has progressed, and the ease of application of this approach is advantageous. The microblocks are numbered by a system of coordinates so that they can be located easily on a map. Input and output data for computing reserves in microblocks are given in Table 6.

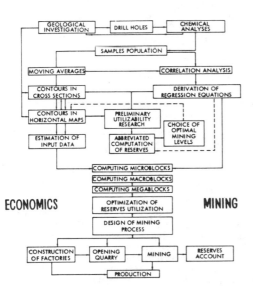

FIGURE 8—Flow chart for modelling and computing reserves.

TABLE 4—Comparing Real and Derived Average Basic Chemical Constituents

Deposit	Number of Checked Data	SiO₂ % a	SiO₂ % b	Al₂O₃ % a	Al₂O₃ % b	Fe₂O₃ % a	Fe₂O₃ % b	MgO % a	MgO % b	L.O.I. % a	L.O.I. % b
Hvizdalka	39	15.44	15.59	3.31	3.32	1.41	1.41	1.26	1.27	34.78	34.58
Prachovice	121	12.68	12.84	4.34	4.37	input	input	2.05	2.04	35.26	35.21

a: real data;   b: derived data

TABLE 5—Accord of Real and Derived Chemical Constituents

Kind of accord		SiO₂	Al₂O₃	Fe₂O₃	MgO	L.O.I.
Very Good	A(1)	33.3%	84.6%	71.8%	38.5%	48.7%
Good	B	53.8%	15.4%	28.2%	33.3%	48.7%
Poor	C	10.3%	—	—	28.2%	2.6%
Unsatisfactory	D	2.6%				
Total		100.0%	100.0%	100.0%	100.0%	100.0%

(1) A: results practically equal
    B: differences do not exceed the tolerated errors in chemical analyses
    C: differences exceed tolerated errors
    D: substantial differences

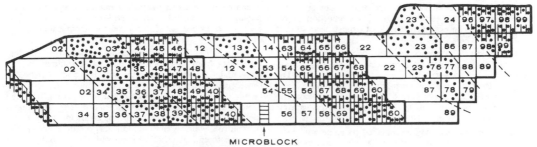

FIGURE 9—Microblocks and macroblocks in cross section.

Macroblocks represent a summation of microblocks in larger blocks; the objectives considered in this summation may differ, and usually a clash of interests may occur. A certain technological or geological homogeneity is called for primarily, but a mining homogeneity also is a factor to be respected. The example in *Figure 9* may give some idea of how to express simultaneously geological and technological criteria on the one hand and mining criteria on the other. Three main layers of limestones are represented — two of them contain high-calcium strata for primary use in lime production; the other, with interbedded shales, supplies a large cement factory. Exploitation of a deposit having five levels is enough to satisfy an equilibrium of limestone and shale (sand and iron additions forming the other parts of the clinker). A quasicoordinate system may be seen in *Figure 9;* e.g., in the third layer of limestone (on the right-hand side), number 23 is utilized for blocks containing high-calcium limestone, and numbers 22 and 24 express other limestones — the extraction of all of them is unaffected by any necessity to exploit or remove the adjacent shales. Another system of macroblocks serves to find the necessary equilibrium; numbers 77, 78 and 79 are used for high-calcium limestones, 88 and 89 for other

limestones, and 96 to 99 for shales; the last cipher shows the respective "mining-antisystem". Extraction of limestone in blocks 77 or 87 requires, at first, the extraction of shales in block 97 and naturally all blocks with smaller numbers in this series, such as 96. Even the condition of minimal level width is respected as shown; e.g., block 87 in the 4th level, where a separate block of limestone is numbered 22, cannot be formed just for this reason.

An easily understandable conception of *megablocks* now can be introduced. The larger units summarizing macroblocks serve as the basis for searching for a final optimal exploitation variant containing the so-called balance reserve only. An illustrative example of all three systems in a map is given in *Figure 10.*

Microblocks form a large population and preparing input data is a tedious job. However, up to the present, no system has been developed, with respect to the numerous geological and other factors (see *Figure 3*), for reading in the necessary data from the respective maps. Perhaps remote sensing may be utilized for this purpose, as well as an automated analysis of textures developed by J. Serra (1967).

TABLE 6—Input and Output Data for Microblocks

Data:	Input:	Output:
POSITION	Number of block	Number of block
INSTRUCTION	constant area; system of coefficients for regression equations	
QUANTITY	Overburden — thickness (1) Deposit — thickness (1)	Overburden — volume Deposit — tonnage
QUALITY: basic chemical constituents	selected parameters (see Table 3)	L.O.I. $\%$ SiO$_2\%$ Al$_2$O$_3\ \%$ Fe$_2$O$_3\ \%$ CaO $\%$ MgO $\%$
additional qualitative parameters		M's Ma Ms. Ma Saturation

(1) Only in case of microblocks exposed at the surface.

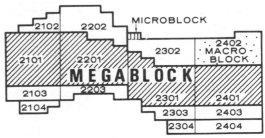

FIGURE 10—Microblocks, macroblocks and megablocks in map of mining level.

TABLE 7—Example of Number of Blocks in Different Systems and their Interrelationship

System	Number of Blocks	Ratio in Systems	
		All 3 systems	2 highest systems
Microblocks	7183	100.0$\%$	—
Macroblocks	317	4.4$\%$	100$\%$
Megablocks	16	0.2$\%$	5$\%$

Macroblocks and megablocks are used to obtain a generalized view of the entire deposit. The proportion of various blocks in an example of a large deposit near Prague is shown in Table 7.

All computations are effected on computers. The data of microblocks, printed in tables, also are stored on magnetic tape, which serves as input for macroblocks. The same method is used for megablocks. A computer, such as the Minsk 22, needs only 5 minutes to calculate and process data on hundreds or even thousands of variants if a convenient system of numbering megablocks is used.

Optimization work may be replaced by some automatic process, perhaps simulation of mining, which will be based on microblocks as input units. The development of such a program does not at present seem to be necessary where results of hundreds of computer variants are compared visually. Restrictions on which blocks are available would be necessary in the program to account for the distribution of variants in the model. Obviously, only those blocks at the surface or on the mining face could be used as short-term units on which any optimization mining model is to start.

Reliability of Results

It is easy to express the reliability of results computed by the above approach mathematically. Taking those results, however, which may be based on a limited number of data — although mathematically correct — might be misleading. The best control is the mining process and this is where thoroughly reliable tests are to be sought.

After the methodology of computing had been used for the first time, the results for two upper mining levels were checked by using additional data gathered by the mining organization (fire holes sampled and analyzed). Data were obtained from holes drilled on a small grid. The information served as a basis to check the differences. Table 8 shows the good agreement of the results of this test. Systematic errors may be explained by the fact that, in the additional holes, the surface mantle was not separated as in exploration holes.

SPECIAL PROBLEMS

Space is not available in this paper to explain many special problems in the practical application of modelling and computing reserves. These problems are mostly connected with

(1) processes used for modelling and computing,
(2) geological phenomena, or
(3) the complexity of morphology.

A short comment can perhaps demonstrate that the methodology as developed — if carefully applied — is flexible enough to assure satisfactory results from many points of view.

The tedious job of contouring and especially of reading input data from contour maps may cause some practically unavoidable deviations in individual data and — if two or three quality parameters are used as input data — it may result in sums of chemical constituents deviating from the most probable ones. A series of special tests therefore is included in the computer program, arranging for automatic control, which eliminates all suspect data and allows for adjustment. It has been proved that deviations are normally distributed. The blocks taken as a series representing the width of mining level (often multiplied by the number of levels actually under exploitation) is interesting for various mining and production purposes, and it is natural that the deviation from the normal will obey laws known in mathematical statistics.

Geological problems to be considered in the modelling and computing exercises are connected mostly with the basic tectonic structure of the deposit under consideration. The system of joints, fissures and faults provokes or facilitates other features, such as karst topography, silicification or dolomitization. Data derived from contours, representing mostly the primary sedimentation, cannot be used in microblocks containing these phenomena. They are usually studied on a statistical basis, covering the entire deposit or large parts of it separately.

The construction of contours derived from moving-average data presumes a constant thickness (height) of all floors. Therefore, some deviation may be caused in the case of microblocks exposed at the surface where the top mining floor has considerable relief. In practical tests, it has been shown that with differences of about 40 per cent in thickness, no errors are expected provided that the contour maps of the top floors are constructed not in a horizontal plane but at a plane intersecting all centers of the floor. In the case of a smaller thickness (where bigger deviations are to be expected), the respective microblocks probably will give some incorrect qualitative parameters, but as weighted averages are used for summarizing all microblock data into macroblocks this approximation cannot cause much trouble in the final result.

FINAL REMARKS

Taking into consideration the complexity of the geological processes involved in the formation of a limestone deposit (as shown in *Figure 3*), and all problems connected with the final results of exploration from not only the geological, but technological, chemical, technical, hydrogeological, mining, economic, social and political points

TABLE 8—Comparing Average Quality Parameters Computed by Using Results of the Original Data of Exploration with Those from Additional Mining Exposures

Floor	Number of Holes		CaO %		Ms		Ms.Ma	
	E	M	E	M	E	M	E	M
1.	12	94	41.63	41.40	3.37	3.24	8.34	8.32
2.	8	74	41.58	41.43	3.67	3.50	9.37	9.14

E = exploration data   M = mining data

of view, we may conclude that the described methodological approach to the reserve estimation takes into consideration most of the variables. The data, as gathered and evaluated, represent valuable dynamic information to be used for a long period of perspective mining and production activity. Technical progress or changes of economic parameters may need corrections in mining plans or perhaps even in the program of the deposit utilization, but in all situations the results of various combinations of megablocks may be a valuable tool for the most serious of decision problems.

Needless to say that already the first computation results serve as a basis for any design work in mining development. In the future, it is hoped to introduce more automation in the decision-making process. The unification of geological, chemical and mining concepts seems to be most significant in the general approach presented. Practical applications of the method have been made since 1966 in about fifteen cases of deposits of limestone in Bohemia.

As mentioned previously, much work is to be done on improving the method. A critical approach is essential and thus each new practical problem to be solved can lead to improvements in the theory.

Although this paper deals only with limestones and other cement raw materials, it is evident that the method is adaptable for other sedimentary deposits. The abbreviated computing part of the methodology with systems of micro-, macro- and megablocks has been used successfully in the study of building stones and of industrial minerals, where only a dynamic evaluation of the quantities of reserves under various mining and other conditions was demanded, and not of qualitative factors. The final results were readily obtained even without using computers, and decision-making was possible utilizing all facets of information.

ACKNOWLEDGMENTS

The author acknowledges the comments and critical remarks of D. F. Merriam and D. G. Krige on the prepared manuscript and the help of P. H. Heckel in constructing *Figure 3*.

REFERENCES

Chayes, F., "On Correlation Between Variables of Constant Sum", *Jour. Geophys. Research*, Vol. 65, pp. 4185-4193, 1960.

Nemec, V., "Disadvantages of the Polygonal Method for Computations of Mineral Raw Materials Reserves", Short course and symposium on computers and computer applications in mining and exploration, Univ. of Arizona, Tucson, Vol. 1, 1965.

Nemec, V., "Kombinovana metoda vypoctu zasob nerostnych surovin" (Combined method of computing ore reserves — in Czech), *Geologicky pruzkum*, No. 9, 1966.

Nemec, V., "A Contribution to the Application of Isolines: moving averages as regionalized variables", *XXIII Internat. Geol. Congress*, Prague, Vol. 13, pp. 175-180, 1968.

Nemec, V., "Law of Regular Structural Pattern", in preparation.

Pana, M. T., "The Simulation Approach to Open-Pit Design", Short course and symposium on computers and comp. applic. in mining and exploration, Univ. of Arizona, Tucson, 1965.

Serra, J., "Buts et réalisation de l'analyseur de textures", *Revue de l'industrie minérale*, Sept. 1967.

Vistelius, A. B., and Sarmanov, O. V., "On the Correlation Between Percentage Values: major component correlation in ferromagnesium micas", *Jour. Geology*, Vol. 69, pp. 145-153, 1961.

Weiss, A., "Mathematical Techniques and Computer Applications as Incorporated in Exploration, Development and Mining Systems", Short course and symp. on computers and comp. applic. in mining and exploration, Univ. of Arizona, Tucson, 1965.

Computer programs used for computing moving average data, correlation analysis and all systems of blocks are not published but are available from GEOINDUSTRIA, Prague 7, Komunardu 6, Czechoslovakia.

19

Copyright ©1978 by the Oil and Gas Journal
Reprinted from *Oil and Gas Jour.* **76**(23):156–158 (1978)

FIND RESERVES QUICKLY WITH COMPUTER-GENERATED LOG DATA

W. H. Fertl and W. C. Vercellino

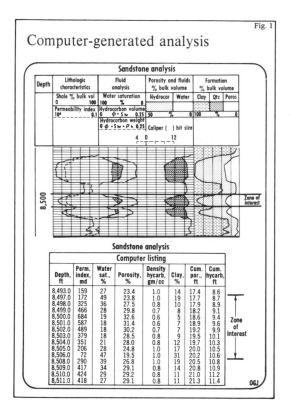

Fig. 1

Computer-generated analysis

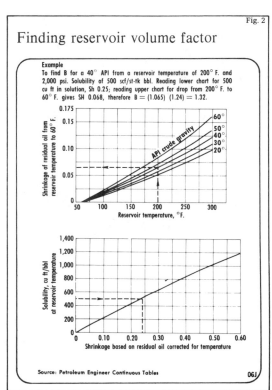

Fig. 2

Finding reservoir volume factor

Fig. 3

Formation volume of bubble-point liquids

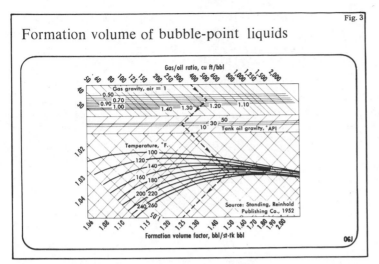

Source: Standing, Reinhold Publishing Co., 1952

06J

Fig. 5

Compressibility factor

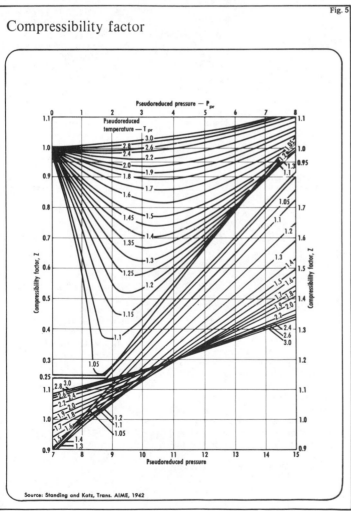

Source: Standing and Katz, Trans. AIME, 1942

COMPUTER-generated well log analyses can provide a quick, complete, and economical look at reservoir potential.

Such computer calculations consider formation variables which cannot easily be evaluated by charts. These computations are generally too tedious for hand calculation.

These analyses convert open hole and cased hole log measurements to formation parameters which can be rapidly and realistically evaluated. The computed results are presented ,in a graphically-coded log format and a complete tabular listing to aid decision making and assist in properly engineering completion work.

An example computer-generated analysis is shown in Fig. 1.

Finding two key parameters. An example illustrates how this tool can be used to determine two important reservoir parameters — average porosity and hydrocarbon saturation — for the pay zone at 8,497 ft to 8,506 ft.

Average porosity is:

$$\phi_{av} = (CPF_B - CPF_T)/(FT_B - FT_T) \qquad (1)$$

where:

CPF_B = cumulative porosity feet listed at bottom of the zone of interest

CPF_T = cumulative porosity feet listed at top of the zone of interest

FT_B = depth at bottom of zone of interest

FT_T = depth at top of zone of interest

For the interval 8,497-8,506 ft:

$$\phi_{av} = (20.2 - 17.7)/(8,506 - 8,497) = 0.278 = 27.8\%$$

If the hydrocarbon volume and hydrocarbon weight are included in the computer-generated data, then the type of hydrocarbon — gas or oil — can be estimated.

For example, in Fig. 1 the solid curve represents the hydrocarbon bulk volume, and the dashed curve is the hydrocarbon weight. Increasing separation of these two curves indicates decreasing hydrocarbon density from

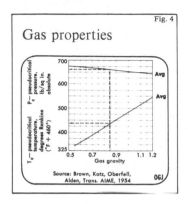

Fig. 4

Gas properties

Source: Brown, Katz, Oberfell, Alden, Trans. AIME, 1954

oil to condensate to dry gas.

Average hydrocarbon saturation is calculated by:

$$SHC_{av} = (CHCF_B - CHCF_T)/$$
$$(FT_B - FT_T) (\phi_{av}) \qquad (2)$$

where:
CHCF$_B$ = cumulative hydrocarbon feet at bottom of the zone of interest
CHCF$_T$ = cumulative hydrocarbon feet at top of the zone of interest

---

The authors . . .

Walter H. Fertl is director of interpretation and field development with Dresser Atlas, Houston. He has engineering degrees from the University of Mining and Metallurgy at Leoben, Austria, and MS and PhD degrees in petroleum engineering from the University of Texas. Fertl has 12 years of worldwide formation evaluation experience dealing with oil-field operations, well-logging research, and teaching. He has authored over 80 technical papers and a book, Abnormal Formation Pressures. He also holds 10 patents. Fertl has long been active in the SPWLA, holding a variety of offices. He is currently vice-president—technology in charge of the technical program for this year's symposium in El Paso.

W. C. Vercellino has been serving as manager of field sales for Dresser Atlas, Houston, since 1975. He joined the company in 1950 and has held a number of positions, including area log analyst, region sales manager, and area sales manager. He was graduated from Texas A&M University in 1948 with a BS in geology. He has published works on perforating and log analysis.

FT$_B$ = depth at bottom of zone of interest
FT$_T$ = depth at top of zone of interest

Then for the interval 8,497-8,506 ft:

$$SHC_{av} = (10.6 - 8.7)/$$
$$(8,506 - 8,497) (0.278) = 0.76 = 76\%$$

Since the average hydrocarbon saturation is 76%, average water saturation is:
$$SW_{av} = 1 - SHC_{av} = 0.24 = 24\%$$

Oil in place. Using these calculated values for porosity and hydrocarbon saturation based on data from the computer-generated log analysis, oil in place can be calculated.

Oil reserves, in stock tank bbl of oil/acre (STB), can be calculated by:

$$STB = 7,758 (\phi_{av}) \times$$
$$(SHC_{av}) (H) (1/B_o) \qquad (3)$$

where:
7,758 = conversion factor (1 acre-ft = 7,758 bbl)
ϕ_{av} = average reservoir porosity from Equation 1
SHC$_{av}$ = average hydrocarbon (oil) saturation from Equation 2
H = net pay, or effective reservoir thickness, from computer-generated data
B$_o$ = reservoir volume factor.

The reservoir volume factor, B$_o$, can be determined in several ways depending on the information available. One method is an empirical rule of thumb and uses only the gas-oil ratio.

$$B_o = 1.05 + (0.05) (GOR)/100 \quad (4)$$

Another method is to use the nomograph in Fig. 2. In this case, it is necessary to know the gas-oil ratio in scf/STB; oil gravity, °API; and formation temperature, °F.

Fig. 3 can also be used to find formation volume factor if the gas-oil ratio, scf/STB; oil gravity, °API; gas gravity (separator gas for example); and formation temperature are known.

All these methods for determining formation volume of oil yield answers that are close to each other. For example, assume the zone of interest is from 8,497-8,506 ft; oil is 30° API; GOR = 350 scf/STB; formation temperature is 200° F.; and gravity of the separator gas is 0.75.

With these data, the first method gives formation volume factor as 1.225 bbl/STB; the second method gives 1.228 bbl/STB; and the answer using the third method is 1.22 bbl/STB.

Using the value of B$_o$ as 1.22 bbl/STB, oil in place reserves can be cal-

culated for the 9-ft pay section from Equation 3.

$$STB = (7,758) (0.278) (0.76) \times$$
$$(9) (1/1.22) = 12,092 \text{ STB/acre}$$

Gas reserves. Gas in place can also be calculated using basic data determined by the computer log analysis.

Reserves, scf/acre, can be calculated by:

$$SCF = (43,560) (\phi_{av}) \times$$
$$(SHC_{av}) (H) (B_g) \qquad (5)$$

where:
43,560 = unit conversion factor, 1 acre-ft = 43,560 ft³
ϕ_{av} = average reservoir porosity as determined from Equation 1
SHC$_{av}$ = average hydrocarbon (gas) saturation in reservoir from Equation 2
H = net pay, i.e., effective reservoir thickness, as determined from data listing
B$_g$ = reservoir volume factor for gas, scf/ft³
$$B_g = 35.35(P_f + 15)/Z(460 + T_f) \quad (6)$$
Z = compressibility factor of gas (dimensionless). Fig. 5 gives the Z-factor as a function of reduced temperature (T$_R$) and reduced pressure (P$_R$).
$$T_R = (460 + T_f)/(T_c) \qquad (7)$$
$$P_R = (15 + P_f)/(P_c) \qquad (8)$$
T$_f$ = formation temperature, °F.
P$_f$ = formation pressure, psi
T$_c$ = critical temperature from Equation 4
P$_c$ = critical pressure from Equation 4

T$_R$ and P$_R$ first have to be determined using Equations 7 and 8 and Fig. 4 before entering Fig. 5 to get Z.

As an example of the gas reserves calculation, assume reservoir pressure is 3,300 psi; formation temperature = 350° F.; gas gravity = 0.85; ϕ_{av} = 27.8%; SHC$_{av}$ = 76%; and net pay thickness = 10 ft.

To find gas in place, the procedure is:

1. Determine T$_c$ = 437 and P$_c$ = 662 from Fig. 4.
2. Calculate T$_R$ = 1.85 and P$_R$ = 5.0 from Equations 7 and 8.
3. Find Z from Fig. 5 = 0.92.
4. Determine B$_g$ = 157.3 scf/cu ft from Equation 6.
5. Then using Equation 5, calculate gas in place as 14.48 × 10⁶ scf/acre.

END

Part III

REVIEW OF THE LITERATURE

EVALUATION AND APPRAISAL

Traditional Methods

One of the most noteworthy early references to valuation of mines was in *Mining Magazine* (Anonymous, 1853). Hoskold's landmark treatise (1877), containing rules, formulas, and examples, marked the beginning of serious treatment of systematic mine valuation. Hoskold's work was supplied with a set of valuation tables calculated on the principle of allowing interest at one practical rate and redeeming the invested capital at another practical rate to a purchaser of any annuity, income, or benefit accruing from a mine. Also supplied were tables of value showing discrepancies in ordinary tables of present values and the errors produced by them. Over the years Hoskold's work has been mentioned in numerous papers (see Whitton, 1918; Riddell, 1940; Weaton, 1973) comparing many methods, both modern and traditional. Eldridge (1949), Pardee (1950), and Rice (1961) dealt exclusively with the merits and disadvantages of the Hoskold methods. Such works as Leith's (1938) stated that all mineral investments should be valued on the basis of the Hoskold Principle. Webster (1936) discussed valuation by Hoskold's methods in relation to sinking funds, deferred interest, compound interest, and unequal dividends.

Channing (1903) compared two cases, showing the extreme limit to which "profit in sight" is almost negligible (as opposed to the other limits of "profit in sight") and showing a mean between the two limits. Finlay (1908) discussed Smyth's formula in the cost of mining in relation to effects of losses, ore homogeneity, and so forth. Brinsmade (1909) derived a formula for quick mine evaluation that emphasized flexibility and absence of compound interest. Dilworth (1910), Langton (1911), Beal (1920a), Church (1926), Kennedy (1926), and Clark (1938) considered tables, general formulas, and treatments for speculative, deferred, and wasting assets and calculating sinking funds. O'Donahue (1910, 1914, 1921) proposed interest tables and methodology for the valuation of mineral property. Morkill (1918) presented a method of valuation in which the risk rate applies only to that portion of the capital that has not yet been returned instead of a risk rate on all of the investor's money during the entire life of the mine. Grimes and Craigue (1928) advocated a three-rate valuation formula. Finlay (1913b, 1914, 1915, 1919, 1932) discussed general principles of mine valuation including future value of mineral properties. Publications on valuation and assessment of a very general nature include those by Munroe (1890), Miller (1913), Rickard (1913), Schmidt (1916), Pallister (1918), Fansett (1918), Redmayne and Stone (1920), Crossman (1921), and Louis (1923). Binckley (1915) explained why appraisal is not valuation. Formulas and examples were given by Guignon (1916) for the valuation of bedded mineral deposits based on one of four possible conditions: the mineral is being used, the mineral is not being used for a year, the mineral is being used and the land has surface value, and the mineral is not being used and the land has surface value. Short papers on valuation of placer deposits were written by Hutton (1921), Sen (1921), and Grant (1922). How valuation was influenced by ore reserves, plant capacity, grading of ores, losses in ore mining, sorting, and timing of plant construction was the subject of Berry's paper (1922). Percy (1933) published valuation tables for mineral rents, royalties, and terminable annuities.

Modern Approaches

Generalities and Comparative Analysis

General texts on principles of property valuation have been written by Grunsky and Grunsky, Jr. (1917), Marston and Agg (1936), Bonbright (1937), and Leith (1938). Important texts that cover the varied phases of mineral property evaluation and mineral economics are classic standard works of Hoover (1933), Baxter and Parks (1933, 1957), Trushkov (1935), McKinstry (1948), Schmutz and Rams (1963), Sobolevskiy (1976), and Campbell et al. (1978a). Banfield and Havard (1975) defined in detail eight terms subject to confusion in mineral valuation. Papers presenting good coverage of various aspects of general mineral-property valuation include those by May (1936), Jerrett (1938), Westervelt (1941), Dismant (1950), Marston et al.

(1953), Swanson (1956), Brons and McCarry (1960), Pogrebitskiy and Ternovoy (1967), Bolotov (1970), Brons and Silbergh (1970), Jones and Pettijohn (1973), Bybochkin (1974), Millan (1975a), Magnou (1975), Gentry and Hrebar (1976), Castle (1977), and Atkins (1977). A selected annotated bibliography assembled by Schenck (1966) serves as a guide to modern mine valuation methods.

Papers by Schmitt (1938) and Liddy (1971) gave consideration to appraisal and evaluation of mines and prospects. *The Mining Congress Journal* (1959–1960) discussed economic evaluation of proposed mining ventures. Hotchkiss and Parks (1936) advised appraisal of a mining business on the basis of present value in order to determine a possible revision of procedure that will increase such present value. Auchmuty (1939) discussed factors influencing mineral land values for assessment purposes. Enhancement and hazard factors can be important in relation to mine valuation (Riddell, 1949). Castle (1958) examined five of the most common sources of errors in estimating and appraising. A compromise approach to evaluation by both statistical and intuitive methods, in addition to maintenance, cost reporting, repairs, and so forth were discussed by Peterson and Eshbach (1962). Schwingle (1962) treated problems of economic valuation of industrial assets. Sundeen (1968) indicated that evaluation of a surface mine should occur at several stages in exploration and development with such factors as geography, legal status of land, geology, mining conditions, ore treatment, and economic analysis taken into consideration. Lewis (1969) compared the "optimum" economic life of four projects on the basis of net present value versus project life. Cost and evaluation of equipment were explained by Childs (1970), and Parkinson and Mular (1972). The more popular mineral valuation techniques being used by the mining industry were summarized by Dran and McCarl (1974). The role of social use value in economic evaluation of mineral deposits was emphasized by Bachmann (1978). Wanielista (1978) dealt with economic justification for the development of ore beds by mine working.

Some of the varied graphic approaches to evaluation include ore valuation curves (Hood, 1920), simple method for depicting present worth (Morris, 1923), estimation of yearly revenues based on ore grade, dilution, price, and concentration ratio (Evans, 1960), graphical shortcut for rate of return determinations (Schoemaker, 1963), and use of sensitivity maps as equal value presentations of the profit function and internal rate of return of the project (Niskanen, 1974).

Several good papers surveyed statistical and computer applications to mineral deposit evaluation (see Soukup, 1963; Bader, 1965; Gibbs, 1966; Parks and Galbraith, 1967; Seda-Reyda, 1971; Godfrey, 1972; Kal'chenko et al. 1973; Herbst, 1973; Guarascio, 1974; and Millan, 1975b). Allsman (1965) developed a program to determine the

243

economic value of mineral deposits with respect to time. Phelps (1968) reviewed generalities and problems of estimation and evaluation in surface mining with emphasis on computers for solutions. Heinemann (1971) studied distribution of common costs in the economic evaluation of blocks of mineral deposits. Bennett and Welborn (1971) applied sensitivity and probabilistic analysis methods in evaluating a potential investment from initial exploration through production, in relation to mineral reserves, capital investments, operating costs, and production rates by means of FORTRAN II. Bosman (1973) explained mine evaluation and production scheduling (MEPS) FORTRAN programs that can be used by mining personnel to solve geological and mining problems including contour maps of assays and geological horizons, trend surface analysis, reserve calculation, scheduling, and so forth. Valuation and production optimization of a metal mine was discussed by Guarascio and Raspa (1974). Sani (1976) surveyed some of the mineral investment models in use and the theoretical foundations of the cost of capital from specific sources of financing. Leach (1975) used mathematical techniques for evaluating vein-type deposits.

Thurlow (1939) commented on the role of geology in mine valuation. McLaughlin (1939) classified ore bodies on the basis of the following valuation. *Plenemensurate:* Ore bodies are capable of being fully evaluated in the early stages of development. *Partimensurate:* Early measurement shows positive ore and discoveries of new material can be anticipated until late in the life of the mine. *Extramensurate:* Definite evaluation of ore bodies is not possible much in advance of actual mining and geologic evidence and inference are more important than engineering practice as a basis of valuation. The role of geology in actual mining was stressed by Cruz (1954) who gave examples in the Philippines that illustrated correct evaluation and exploitation of mineral deposits. Nichols (1968) interpreted field techniques for economic and geotechnical evaluation of an opencast mine with severe groundwater problems. The geologist's function is fundamental in bridging the gap between raw material wealth and technology by means of economic evaluation and prospect analysis (Oliver, 1971). Salski (1974) pointed out the significance of tectonic studies in deep mines and their importance in evaluation.

Detailed comparisons of advantages and disadvantages of the various appraisal methods were surveyed and analyzed by Uglow (1914). These include the following: standard ad valorem, Finlay ad valorem, Arizona, Colorado, Wyoming, South Carolina, Nevada, British Columbia, Ontario, gross receipts-tonnage tax, and equated income. Other important interpretations of mine valuation systems include articles by Bain (1937) and Boericke (1940) on valuation systems in the Philippines; Pardee (1957) on valuation systems in

Michigan; and Slamet (1977) on valuation systems in Indonesia.

There are many good papers on mine revaluation and changing monetary factors (see Webber, 1921; McCormack, 1924; Boericke, 1947; Dolbear, 1952, 1953). Of particular interest in recent times are articles by Davidson (1975), dealing with investment evaluation under conditions of inflation; Drechsler and Stephenson (1977), illustrating the *hurdle rate* for new investment and several price indices that may be used in adjusting price increases (inflation) in mine valuation procedures; and Dran and McCarl (1977), examining interest rates and their effects on mineral deposit evaluation.

Specific Deposits

Discussions of valuation of metallic mines by many authors range from traditional to modern geostatistical approaches (Rickard, 1915; Probert, 1915; Hamilton, 1923; McKechnie, 1930b; Preston, 1960; and Roufaiel, 1966).

One of the earliest writings on valuation in the United States was done on iron mines in New York and New Jersey (Smock, 1882). Finlay's valuations of iron mines (1911, 1913a, 1913c) in many respects were very innovative and controversial, inspiring the commentaries of Brinsmade (1914) and White (1914a, 1914b). Gross (1965) had a good section on factors to be considered in evaluating iron deposits. Michelson et al. (1970) developed three computerized mathematical models based on quality, cost, and economic availability to estimate taconite concentrates in the Mesabi Range, Minnesota. Economic feasibility studies were conducted by Zeidler (1969) on the Wadi Fatima iron ore in Saudi Arabia.

Graton (1926) discussed factors for determining copper mine values. Computer contouring was used by Surkan et al. (1964) for evaluation and analysis of mines with emphasis on economic evaluation of copper ores. Kupferschiefer mining economics and its relation to evaluation and planning were discussed by Freudenberg et al. (1972). An integrated computer application system for mine evaluation (MES) was offered by Ichisugi et al. (1974) with porphyry copper mines as examples. Henriques and Mackenzie (1977) described an economic analysis framework for the operation and evaluation of potential copper mill locations in Chile.

Two noteworthy papers by Hellman (1897–1898) and Carter (ca. 1901) described valuation of gold mines in the Rand district, South Africa. Calculations for average mining width, average gold content in vein, gold content in respect to dilution and gross dollar value per ton of ore mined were given by Chico (1961) in solving the problem of valuation of a gold vein. Johnson and Bennett (1968) developed an engineering and economic model of a hypothetical open-pit gold mine for evaluation. Ageton et al. (1969) epitomized all factors bearing on the economic evaluation and establishment of a gold mine in the Golden Sunlight area, Whitehall mining district, Jefferson County, Montana. A simplified model was derived by Joughin (1973) for the financial return from new South African gold mining projects involving planning, control, and valuation.

Chapter 21 in Nininger's book (1954) dealt with the evaluation of discoveries, samplings, reserves, uranium advice, and so forth. Brewer (1957) summarized uranium evaluation. Evaluation and exploitation of the Nisa, Portugal uranium deposit was described by Vicente ('1966). Economic evaluation of borehole mining of low-grade uranium deposits was discussed by Chase et al. (1978).

Comprehensive mine valuation was reviewed by Hoover (1909) for copper, gold, lead, silver, tin, and zinc; Harrison (1954) for alluvial deposits, with emphasis on tin; Verner and Shurtz (1966) for nickel-copper; Hewlett (1967) for copper, uranium, and silver; Salimbayev et al. (1968) for the Sokolov metals mine, USSR; and Chandler (1970) for lead and zinc. Financial analysis, cash flow, operating costs, tax planning, and geology were topics in the valuation of Weisner et al. (1980) of potash occurrences within the nuclear waste isolation pilot plant in the Carlsbad, New Mexico area.

There are many good publications on general oil-property valuation including those by Requa (1912, 1918) Lombardi (1915), Johnson (1915–1916, 1916), Cappeau (1916), Lewis and Beal (1918), Arnold (1919), Beal (1919b), Oliver (1920, 1921), Beal and Lewis (1921), Morris (1922–1923), Panyity (1926), Ingham (1929), Mills (1936), Gabriel (1937), Hall (1939), Whitehead (1957), Arps (1958), Wooddy and Capshaw (1960), and Manefield (1974). More detailed treatments can be found in books by McLaughlin (1921), Paine (1942), Campbell (1959), Society of Petroleum Engineers (1962), and the very excellent texts of Campbell (1973) and Hughes (1978).

Beal (1920b) classified undeveloped oil land for purposes of valuation as *proved, probable, possible,* and *commercially unproductive.* Darnell (1924) discussed valuation of oil properties for all purposes, and also classified various valuations depending on purposes and considered estimates of reserves, discounts, and so forth (Darnell, 1925). Arnold (1920) explained oil geology in relation to valuation. Hager (1921) and Harnett (1970) showed how valuation and buying and operating oil properties could lead to profits, and they pointed out some of the pitfalls to avoid. Stephenson and Grettum (1930) considered methods of valuation of flood oil properties. Hypothetical oil property was valuated as an example of a time saving method for group appraisals (Larkey and Bright, 1925). The *time-to-pay-out* method has advantages over the *per-barrel-value* method of valuation of oil properties (Moyer, 1923). Evaluation of foreign-producing oil properties requires consideration of many factors (Herald, 1933). Hedberg (1937) used the index of refraction of petroleum as a key to the specific gravity in evaluation. The presentation of oil-property appraisal data was detailed by Bradley (1953). Arps (1960) considered economic factors in land transactions. Wilson and Pearson (1962) showed how to determine the market value of secondary recovery projects. An economic analysis of oil production investments and recoverable reserves was given by Hall (1963). Soloman (1970) discussed the relation of book yield to true yield. Koval'chuk and Predtechenskaya (1975) outlined criteria for the evaluation of the commercial importance of an explored oil field.

Valuation and appraisals of oil and gas lands were discussed in the works of Johnson (1922), Wood (1922), Johnson and Ruedemann (1924–1925), South Texas Geological Society (1939), Jones (1942a), Donoghue (1942), Todd (1950), Meyer (1958), Houchin (1958), and Carter (1967). Good comprehensive consideration was given to all phases of oil and gas evaluation in books by Brown (1924), Vance (1959), and Campbell et al. (1978c). Some of the varied factors beyond engineering evaluation that affect fair market values include finding costs, refinery needs, geographical location, tax position of purchaser, rate of return sought, management philosophy, incentive to investors, and bid sales (Sherrod, 1962).

Wyer (1917) noted principles and legal basis of natural gas leasehold valuation and natural gas rights. Shaw (1919), Stephenson (1933), and Davis and Stephenson (1953) explained the principles of natural gas land valuation. Rawlins and Schellhardt (1936) appraised gas properties in the Michigan "Stray" sandstone horizon of central Michigan. Development and evaluation of gas condensate reservoirs were discussed by Goodson (1960).

An early paper by Blakemore (1895) was concerned with the potential economic limit to the output of a coal mine. Garcia (1925) stressed the importance of the examining engineer in financing coal mine property. Comprehensive coverage of all aspects of coal property evaluation may be found in the text by Hesse (1930). Coal evaluation was the subject of a symposium in 1974 at Calgary, Alberta, published by the Research Council of Alberta (1977).

Classic papers on valuation of coal land by Chance (1904, 1913) were analyzed and discussed in detail by Ashley et al. (1913). There were other important classics in valuation of coal lands published around this time (Ashley, 1910; Fisher, 1910; Rogers and Lesher, 1914). The last two papers are concerned primarily with physical valuation of the thickness and depth relationships in coal beds. Appraisal of coal property by such methods as value of reserves and royalties was stressed by Chance (1927). Dilworth's (1928) paper emphasized various coal land appraisal modes and examples. Other early papers on coal property evaluation include those by Norris (1915), Fohl (1915), Coulthard (1916), and Parsons and Hall (1917). Fish (1969) surveyed accepted operating methods for condemnation proceedings and factors to be considered, and evaluated the approach toward sales, mergers, and leasing of coal land. A paper by Jirasek (1970) was on the possibility of economic evaluation by computers of long range studies of open pit lignite mines in north Bohemia especially the Sokolovo district. A paper of importance by Gorrell et al. (1972) was on monetary evaluation of coal properties. Bajwa (1978) programmed a core logging technique for standardizing descriptions and simplifying computerization of data in coal property planning and evaluation. Svenson (1979) discussed efficient coal exploration and evaluation practice in Australia.

Valuation of building and industrial stones, rocks, sands, clays, and so forth of the Saar region were detailed by Graupner (1939) with emphasis on changes in economic value with time. Roberts et al. (1966) reported on economic valuation considerations in the urban threat to the local sand and gravel industry in the Santa Clara River Valley, California. Some representative quarry evaluation publications discussed sandstone quarries in the Contamana region, Ucayali Province, Peru (Cossio, 1966), decorative building stone (Grigorovich, 1968), and the determining of major parameters of a quarry (Arsent'yev et al., 1976). Krasil'shchikov (1967) offered photographic documentation and excavations as a means of evaluation especially of phlogopite deposits.

GENERAL FINANCE AND ACCOUNTING

Finance

There are many publications describing general mining finance and its problems. Some recommended readings are those by Moreau (1906), Anonymous (1915), Dilworth (1922), Elsing (1932, 1936), Fitz Gerald (1938), Boericke and Bailey (1959), Tyler (1959), Rodgers (1965), Baker (1965), Cummings (1965), Neubauer (1971), Douglass (1971), Zambo (1971, 1972), and Lindley et al. (1976). Krumlauf (1960) and Burgin (1965) discussed some aspects of financing small mines including financing their exploration whereas Fernald (1928) considered methods of financing large mining operations. Fielden (1964) provided an analysis of fund flow and its application to financial statements of mining companies and Rau (1980) examined the role of commercial banks in financing international mining ventures, emphasizing project loans, production loans, and production payments as well as completion guarantees, management-finalized sales contracts, reserve calculations, and so forth. Lane (1911) compared stock value with mine value and Pickering (1917) gave an engineering analysis of a mining share. Included in the many good papers on essentials of selling, developing, and financing a prospect into a mine are papers by Willis (1917–1918) and Wright (1936). Zambo (1970) treated the role of interest in the choice of production capacity of mining works. Crandall et al. (1959) considered the cost of acquiring and operating mineral properties. The solution by financial analysis to many common exploration and development problems can locate favorable areas (O'Brian, 1969). More equity can be retained in financing a project if a stage-by-stage financing is obtained to meet each phase of exploration and development, and so forth (Frohling and McGeorge, 1975). Troly (1979) described an accounting method that related exploration expenses to profit expectation and optimization in use of an annual exploration budget. Optimum ore reserves and plant size can be determined by a method of incremental financial analysis (Halls et al., 1969). Terchroew et al. (1965) has written a sophisticated analysis of the theoretical basis of present worth and investment return calculations.

Programming and modeling was done by Lyon (1975) using multiple cutoff rates for capital investment, and by Rudenno and Thomas (1977) for flexibility and direct discounted cash flow (DCF) analysis in mining feasibility studies. Other computer studies included a systematic method for evaluating mineral deposit cash flow analysis and varying cost criteria of producing iron ore pellets from magnetic taconite (Michelson and Polta, 1969); a financial analysis designed to incorporate all aspects of a mining project including mining rate, grades, reserves, changes in metal price and operating costs, taxes, and so forth (Trafton and Sheinkin, 1969); a FORTRAN IV program for calculating capital and operating costs (Johnson and Peters, 1969); and an investigation of capital investment and operating cost estimation in open pit mines (Jarpa, 1977).

General texts on financial analysis of mines have been written by Rickard et al. (1905), Finlay (1909), Skinner and Plate (1915), Marriott (1925), Caudwell (1929), Truscott (1947), and Pryor (1958). Mining accounting is considered in detail in the texts of Lawn (1897), Davis (1909), Wallace (1909), Charlton (1913), Holmes (1920), McGarraugh (1920), McGrath (1921), and Willcox (1938, 1949). Good general mine accounting works are by Pickering (1919), Goodner (1922), Lamb (1950), Institute of Internal Auditors (1950), Hawkins (1952), and Peloubet (1959).

Valuation of inventories of metal mining companies was examined in 1936 (Anonymous). A preliminary economic analysis by Pfleider and Scofield (1967) featured assumed data from hypothetical underground taconite mining. Comprehensive financial analysis of molybdenum deposits explored in the Henderson Project, Colorado was done by Peiker and Forsythe (1969).

Waller (1956) and Porter (1965) presented textbook treatment of petroleum accounting practices. Topics covered on these practices include correct determination of drilling campaign as an aid in allowing financial gain during depressed periods (Cutler and Clute, 1921), payout status (Humphries, 1927), ultimate recovery and its relation to ultimate profits (Shaw, 1933), evaluation table with limiting elements (Morris, 1937), general oil accounting (Hand, 1940), specialized phases of accounting practice (Strain, 1947), valuation of producing properties for loan purposes (Terry and Hill, 1953), aggregation of oil and gas properties (Wittman, 1955), purchase of properties by use of a production payment (Hardwicke, 1955), financing oil and gas transactions (Dunlap, 1960), oil and gas property aquisition (Wilson, 1961), use of engineering reports in financial analysis for producing oil and gas companies (Dodson, 1965), oil and gas loans and reserve estimations (Vance, 1968), principles of processing new applications for oil, gas, and mining loans in commercial banks (Denver United States Bank, 1969), optimization of capital expenditures in petroleum investments (Campbell, 1970), organization and planning of petroleum exploration as a factor in the optimal distribution of capital investment (Milovidov et. al., 1972), financial analysis of proposed 100,000 barrel-per-day oil shale plant (Katell et al., 1974), economic index for increased recovery from stripper wells (Pederson, 1974), and financing oil and gas ventures (Mosberg, 1979). Warren (1956), Dodson (1958, 1960), Ireton (1960), McElvaney (1961), Sonosky (1961), and Dodson (1967) have authored good papers on oil and gas loans and financing.

Lisle (1900) summarized colliery sinking or redemption funds. Capacity calculations, investment allocation, and long-range production scheduling in German coal mines were programmed by Wilke (1973) as a model with an objective to maximize the total discounted cash flow over the planning period. Financing the acquisition of a going coal mine was the topic of Epstein's paper (1975). Finance in relation to evaluating newly discovered coal deposits and ventures was considered by Hammes (1976).

Stalker (1937) and Howington (1939) discussed cost and management in quarry accounting.

Legal Aspects

Monographs written on the general aspects of depletion and depreciation in mineral valuation are by Mathewson (1910), Leake (1920), Lentz (1960), Winfrey and Hempstead (1952), and the Internal Revenue Service (annual). Shorter publications on depletion and depreciation taxation include those by Armitage (1919, 1922), Kircaldie (1923), Finlay (1931), Peloubet (1937), Anonymous (1950), Austin (1952), and Standard Oil Company of New Jersey (1958b). Anderson et al. (1977) traced the history of United States federal tax policy in relation to percentage depletion of minerals.

Taxation of wasting assets in relation to mine accounting and valuation was emphasized by Saliers (1922), Fernald (1922), Reis (1923), Fernald et al. (1939), Burtchett and Hicks (1948), Benson (1951), Lourie (1951), Fagerberg (1952), Holmes (1954), Goulette (1955), White and Brainerd (1956), and Matthews (1969).

Books concerned with oil and gas tax law and depletion/depreciation have been written by the United States Treasury Department (1919), Arnold et al. (1920), Fiske (1958), Blaise (1959), and Lichtblau and Spriggs (1959). Some of the many shorter works dealing with oil and gas depletion and depreciation are by Wyer (1914), Henry (1916), Caudill (1921a, 1921b), Boedeker (1932), Galvin (1943), Pitcher (1947), Kinard (1950), Randolph (1950), Flagg (1951), Plank (1951), Williams (1951), Seale (1951), Fiske (1952), Belin (1952), Baker (1952), McGowen (1952), Adamiam (1953), Cohn (1953), Johnson (1954), Freeman (1955), Bird (1955), Weller and Lipscomb (1955), Miller (1956), Stanley (1957), Nickerson (1958), Standard Oil Company of New Jersey (1958a) Anonymous (1959), and Mid-Continent Oil and Gas Association (1968).

Young (1916) and American Mining Congress (1921a, 1921b, 1922, 1923, 1924, 1927) are authors of major books on mine taxation. Some important shorter contributions to mine and mineral taxation are by Steele (1914), Allen and Arnold (1920), Gibson (1920), Reed (1920), Leith (1921), Montgomery (1923a), Kurtz (1923), Goodner (1923), Williams (1952), Alexander and Grant (1953), Borden (1959), Byrne (1965), Davitt (1974), Fiekowski and Kaufman (1976), and Brown (1980).

Kurtz (1920), Anonymous (1920a), Fisher (1921), Fernald (1923), Graton (1923), the Joint Economic Committee (1964), the University of Arizona (1969), Maxfield (1973), and Martin (1977) are authors of important works on issues of United States taxation of mines. Some brief examples of the many contributions to foreign mine taxation literature are those by Shakespeare (1965) on Canada, Hodgson and Beard (1966) on Canada, Jeal (1956) on South Africa, and Deichmann (1976) on South Africa.

Aspects of state mining taxation and valuation are treated in the following publications by Allen (1915) on Michigan taxation, Ingalls (1922) on taxation of mines by states, Finlay (1922) on appraisal of New Mexico properties, Roberts (1944) on taxation of metallic deposits by states, the Arizona Department of Mineral Resources (1951, 1955a, 1955b, 1956, 1958a, 1958b, 1958c, 1960a, 1960b, 1962, 1964) on Arizona mine taxation and yearly assessed valuations of mining property, the California State Board of Equalization (1965) on appraisal of mining property in California, Martin (1972) on synopsis of state mine taxation and revenues, and Laing (1977) on effects of state taxation on the mining industry in the Rocky Mountain States.

Appraisal and taxation accounting are discussed by Baxter (1915) and Montague (1948) for iron mining, Fox (1934) and Thomas (1954) for gold mining, Dalby (1955), Stoddard (1955), Roe (1955), and Krakover (1955) for uranium mining, Pardee (1952) for copper and iron mining, and the United States Treasury Department (1922) for copper and silver mining.

Some of the numerous contributions to appraisal taxation accounting of oil and gas include those by Beal (1920d), Cox (1921), Montgomery (1923b), Caudill (1925), Stanley (1946), Terry (1946), Tulane University, (1952), Day (1953), Flagg (1956), Breeding and Herzfeld (1958), Powell (1960), Bullion (1962), Boswell (1967), Freling (1968), Dutton (1970), Goodson (1970), and Houghton (1978). Taxation of oil and gas income is the subject of books by Hayes (1866), Tucker (1923), Anderson and Co. (1939), Miller (1948), Breeding and Burton (1954), Frank et al. (1956), and McMurray College (1957-1958). State oil and gas taxation of property is considered by Beal (1920c), Beal et al. (1920), and Anonymous (1920a) for California; Hartman (1929) for Los Angeles County, California; Tracy (1948) for Illinois; and Weinaug (1964) for Kansas. Short papers on assessing and appraising coal property for taxation are by Griffith (1913), Chance (1914), Norris (1914), and Reeves (1950). McCarthy (1951) epitomized tax accounting for quarries.

Morrison (1878), Mills and Willingham (1926), Summers (1927), Costigan (1912), Glassmire (1938), Sullivan (1955), Kuntz (1962-1980), Hemingway (1971), and Williams and Meyers (1975) have authored good textbooks on mineral and petroleum law with some cases on valuation and related topics. Other treatises on mineral and petroleum law of the United States were written by Miller (1907), Shamel (1906), Ricketts (1911), Lindley (1914), and MacDonnell (1976). The Rocky Mountain Mineral Law Foundation (1955) publishes annually a proceedings volume with an abundance of papers on various phases of mineral law. Mining and petroleum laws of the world were summarized by Ely (1960). Shamel (1907), Mathews (1936), Ely and Wheatly (1959), and Brereton (1936) are authors of shorter contributions to mining and petroleum law. Deussen (1940) discussed royalties in relation to the petroleum industry. The textbook of Van Meurs (1971) emphasizes petroleum economics and offshore mining legislation in relation to geological evaluation.

Expert witnessing in court was discussed by Leith (1920) for mining litigation, Powers (1967) for eminent domain cases involving mineral deposits and fair market value, Walker (1968) for log analysis, and the American Institute of Professional Geologists (1974) for evaluation testimony. Hamilton (1933) reviewed legal case histories concerned with valuation of gas leases and fair market value. A classic example of regional mineral resource appraisal and valuation resulted from the litigation involving the state of Utah versus the United States of America over the state-owned lands to be covered by Lake Powell in the Glen Canyon Region of Utah (Cohenour et al. 1963). Mineral policy and laws affecting leasing, evaluation, and so forth were detailed by Cruz (1969) for the Philippines, Hershey (1972) for Tennessee, Leon (1973) for Peru, and Herbst (1974) for Minnesota. Barrows and Webendorfer (1976) reviewed the influence of community impacts and the acceptance of mining operations on evaluation, legislation, and so forth. Mosberg (1979) reviewed securities law considerations. Zoning and land-use planning can be major problems in mineral property valuation as are so well described by Ahern (1964), Stearn (1966), and the Wisconsin Department of Natural Resources (1976). Bell (1975) analyzed the effects of radiological and waste management legislative controls on uranium production costs with emphasis on Eldorado Nuclear Ltd's "Beaverlodge operation" in Saskatchewan. A symposium on uranium law was presented by the Rocky Mountain Law Review (1955).

GENERAL SAMPLING, EXAMINING, AND ESTIMATING

Rickard's (1903) paper was one of the first good considerations of general sampling in relation to estimation of ore in a mine. Other general publications on sampling and estimating ore deposits include an anonymous paper (1925) and those by Lewis (1931), Frick and Dausch (1932), Jackson and Knaebel (1932), Bakhvalov (1937a), Jewett (1956), the California Department of Natural Resources (1960), Zenkov (1963), Basden (1965), Cornish (1966), Gorzhevskiy (1968), Gy (1971), Hodgson (1971), Kindl (1973), and Vasilev (1973). Classic texts on mining and exploration geology are by Stretch (1904), Park (1907), Forrester (1946), Zeschke (1964), and the masterpiece by Peters (1978). Comprehensive treatment of placer sampling and valuation occurs in the texts by Herzig (1914) on mine sampling and valuing, Thorne (1926) on testing and estimating alluvials for gold, platinum, diamonds, and tin, Grosjean (1953) on economic variables, functions, and errors inherent in detrital and placer prospecting with emphasis on estimation of placers, based on Pearson curves, and Wells (1973) on placer examination. In the evaluation and economic potential of pegmatitic mineral bodies, Rowe (1953) deals with sampling and grain counts, and Pavlishin and Vovk (1970) deal with K/Rb ratios. Development drilling and bulk sampling practices directed toward grade, volume, and shape of previously located ore zones are detailed by Waterman and Hazen (1968). In their work iron, copper, coal, uranium, bauxite, beach-sand, and molybdenum mines are exemplified. Northern Miner Press Ltd. (1968) compiled a series of articles that explained elements of prospecting, diamond drilling, sampling, mine workings, evaluation, and so forth. Mills (1972) outlined such problems as those involved in planning for risk reduction, reserve estimation, and modeling by the modern senior mining geologist. The subject of a paper by Astaf'yeva et al. (1978) was economic evaluation of deposits at different stages of their exploration and development.

Some important works on examination and surveying of mines and prospects are authored by Kirby (1894), Park (1905), Gunther (1912), Janin (1913), Haddock (1926), Joralemon (1928), and Schmitt (1929). Lintern's (1872) classic text on mine surveying and valuation gave particular attention to iron and coal properties. An interesting paper on the theory and philosophy of geological valuation and examination of mines in relation to time and empirical methods was authored by Clark (1935). Varvill (1935) was concerned with the many pitfalls in the examination of abandoned mines. Reid and Huston (1945) discussed the preparation of reports and the practical examination of mineral prospects. Gaydin (1974) used geotechnical methods for surveying deposits during their exploitation.

Geostatistical factors in mine sampling are treated in many good papers such as Sichel's (1951–1952), which proposes an estimator based on the mathematics of the theory of probability and on R. F. Fischer's Theory of Maximum Likelyhood. Hazen and Berkenkotter

(1962) discussed an experimental mine-sampling project designed for statistical analysis. Hazen (1962) utilized statistical analysis to plan sampling programs. Koch and Link (1964) treated accuracy of diamond drill-hole data in estimating metal content and tonnage of ore bodies. Hewlett (1964c) presented methods developed for simulation of populations of assays that were representative of a deposit and provided the added data for solution of specific evaluation problems. Bandemer (1966) was concerned with the number of drill holes needed in evaluation and exploration projects. Hazen and Meyer (1966b) explored by computer simulation the relation between the degree of correlation present and the difference between weighted and unweighted average grade in relation to assay values and unequal sample interval lengths. Shurtz (1966a) used Fourier approximation with trigonometric functions in the problem of estimating the precision of systematic samples from a mineral deposit. Linear discriminant analysis employing the Mahalanobis d^2 method provided a means of discerning interrelationships between mutivariate data and was applied effectively to assay data from the Frisco mine, San Francisco del Oro, Chihuahua, Mexico (Link and Koch, 1967). Hazen (1967c) applied statistical techniques to mine sampling, evaluation, grade, and tonnage problems with numerous examples. Koch and Link (1971) advocated the coefficient of variation c (equal to the ratio of standard deviation to mean) as a guide to the amount of sampling that should be done in an ore deposit. A plot of means and c values calculated from 50,057 observations for amounts of various substances in 484 ore deposits established an empirical relationship between c and the mean and results in probable values of c for various substances. A discussion of the problems of the application of operations research to ore search, ore evaluation, investment decisions, and so forth was presented by Blais and Berry (1971). Pierre (1974) reviewed statistical methods including modeling for sampling broken ores. Geostatistical factors in mine sampling and design were well covered by Royle et al. (1974).

The text by Koch and Link (1970–1971) on statistical analysis of geological data is outstanding for its comprehensive treatment of sampling. The standard general text on geomathematics by Agterberg (1974a) contains a wealth of information of interest to geologists and engineers engaged in sampling, exploration, and so forth.

The subject of general assaying methodology and practice is too broad a field to be treated in a book of this type; however, a few references of special interest are included here. Gilbert (1948) described a new assay slide rule to aid in computing complex ore values. Hazen (1967b) used statistical methods in assigning an area of influence for an assay obtained in mine sampling. Salting of mines and its problems in assaying were noted in detail by McDermott (1894–1895), Rickard (1941), and Bassett (1974).

EXPLORATION AND PROSPECTING

Risk Analysis and Decision-Making

General risk analysis is a very extensive field and is out of the scope of this book. Nevertheless, some noteworthy publications are listed here for the sake of continuity for the reader. The treatise by Matheron (1962–1963) is an excellent source for exploration and estimation strategies. Lewis (1946), Morse (1964), and Hertz (1964) treated general risk analysis in capital investment in mining ventures. The philosophy and meaning of risk venture in exploration are different from evaluation (Farrell, 1939). Ways of appraising chances in conformance with probability principles and graphs are suggested to show tonnages, grades, and the corresponding judged expectancies by Fitzhugh (1947).

Concerning exploratory ventures, profitability was discussed by Arps (1961) whereas Brant (1968) was concerned with preevaluation of the possible profitability. Clark (1963) recommended pay out, the ratio of capital expenditure to annual cash return (a quantity numerically but not physically equal to a factor by which cash returns may be discounted to equal required investment) as a gage of the merits of a proposed mining venture. Dickinson (1965) surveyed general exploration finance and ore-reserve assessment. Neter and Wasserman (1966) provided an understanding of statistical methods as tools for decision making in business and economics. The text on the capital budgeting decision by Bierman and Smidt (1966) advocated the *present-value method* instead of such investment decision criteria as *cash payback* and *return on investment* for gaging economic worth. Economic analysis of mining ventures was summarized by Jones (1968). Griffiths (1968) reviewed the decision-making methods in exploration and evaluation of mineral properties. Roginskiy (1968) determined the economic efficiency of geologic exploration. Peters (1969) pointed out that the largest financial losses often are not in exploration, sampling, and prospecting, but in the abandonment of an unprofitable mining operation after the commitment of capital on the basis of incomplete geologic investigation.

Changing conditions in mining practice, processing technology, marketing opportunities, and concepts of an ore body also represent real risks. Bergman and McLean (1971) stressed the importance of risk analysis for evaluating capital investment proposals. David (1972) noted the importance of grade-tonnage curves and statistical models as decision tools in ore-reserve estimation. Five case histories of feasibility studies show what mine owners, developers, and loan officers must know when evaluating proposed mining ventures (Lewis and Bhappu, 1975). Parker and Switzer (1975) and Royle (1978) interpreted ore reserve statements on a probabilistic basis. Shestakov (1976) reviewed scientific foundations of choices and economic analysis of ore deposit mining systems. Procedure for feasibility determination in conjunction with small subsurface mines was outlined by Gentry and Hrebar (1977). The role of weighted average cost of capital was featured by Sani (1977) in evaluating a mining venture. The textbook by Stermole (1977) is an excellent comprehensive integration of evaluation procedure and investment decision methods. The paper by Schwab and Drechsler (1978) provides a comparative analysis of the average cost approach with the use of net present value as an investment criterion.

Some important concepts in statistical methods applied to risk and decision include stochastically variable treatment of all geologic decisions that could have economic implications when integrated into the search effort (Brown, 1961). Chacon et al. (1970) statistically demonstrated that mining production follows cycles that are aperiodic but regular, and that the maximum period can be predicted for use in indicating at which moment work should be started in poor mines suffering from spasmodic productions. Geostatistics forecasted the value of information resulting from drilling a hole in a known ore body for use in exploration (Rendu, 1971), while probabilistic techniques were applied by Imai and Itho (1971) in a priori distribution and simulation approaches involving the existence of an ore body and in the determination of effective spacing and number of drill holes to accomodate required precision in grade and tonnage estimations. A probabilistic method of ranking underground exploration proposals by comparing vein intersections of a known nature with those of known ore veins was investigated by De St. Jorre and Whitman (1972). Trend analysis was employed in mining exploration and exploitation by Merriam (1972), and formulation of exploration strategies was based on profit and corporate risk criteria (Mackenzie, 1973), Denisov (1974) estimated mining investment effectiveness and the limiting parameters in the definition of mineral reserves and resources, and economic effectiveness was determined by Stammberger (1974) in geologic exploration.

There are several good works on computer, statistical modeling, and risk and decision theory. Exploration was discussed by Hazen and

Meyer (1966a) and Kol'berg (1969) covered statistical evaluation of geological exploration for planning of mining with relation to the information theory. Discounted cash flow, utility theory, taxation, changes in government policy, and general finance were researched by Brown (1970). Optimizing mine production and cutoff grade was explained by Noren (1971). Risk analysis in an iterative approach to the optimal design of open pits may be found in the writings of Leigh· and Blake (1971). Woodtli (1971) wrote on laboratory estimation of prospecting progress, success, and cost in an unexplored region by simulation, and the correlated randomness problem in ore body evaluation was explored by Borgman (1972). Klinge (1972) explained the application of computers to decision making in valuation, planning, grade calculations, and preliminary viability studies in mines. Roberts (1972) used an estimator with an expression of uncertainty in individual parameters and the inclusion of multivariate return on investment measure. An economical program providing for rapid financial evaluation of the economic feasibility of mineral exploration ventures was discussed by Beasley and Niedermayer (1973). Reconnaissance and description of ore reserves was covered by Vitorovic et al. (1975), and practical operating of computers and a management information system was established by Handelsman et al. (1975). An economic model to optimize investment in delineation drilling for tonnage and grade estimation data was set up by Bilodeau and Mackenzie (1977). Geostatistical methods in the analysis and prediction of economic deposits were supplied by Tulcanaza Navarro (1977). Dallin (1977) talked about the development of an investment system, and Boehmke (1979) explored a simple system suitable for FORTRAN IV programming for the prediction of "probable" ore reserves. After comparison with actual results, revisions were supplied.

Bennett et al. (1970) employed sensitivity and probabilistic analyses using FORTRAN II in financial evaluation of mineral deposits (in which uncertainty may exist in the input) with a gold deposit as an example. Ruzicka (1976a) detailed a uranium resources evaluation model for use in strategy and decision making in exploration used by the Uranium Resources Evaluation Section of the Geological Survey of Canada. Agterberg (1969b) applied trend-surface analysis and frequency distribution to interpreting copper values. Hewlett (1964a) presented computer and mathematical methods for the construction of statistical probability models for simulating metallic deposits by Monte Carlo techniques. Zemlyanov and Olonov (1970) applied a relative probability formula for quantitative appraisal of information value of prospecting indications in the western part of the Zeravshano-Gissar antimony-mercury belt in Central Tadjikistan. Bugayets et al. (1970) applied decision algorithms based on discrete analysis methods to solve problems in evaluation of rare-metal pegmatites in Kazakhstan territory, USSR. Beasley and Pfleider (1973) utilized profitability sensitivity analysis in determining venture profitability in a case study of low grade copper-nickel ores in the United States. Vidyarthi and Kala (1975) applied probability models and sequential decision making to the evaluation of the Balaria zinc-lead deposit, Zawar, India. In order to evaluate the limits of economic feasibility of an integrated bauxite-alumina-aluminum plant in western Colombia, Velasquez (1973) used two basic investment models based on cost range estimates. Von Wahl (1973) surveyed criteria and methods to control and calculate risks and chances of decisions under uncertain conditions for the economic valuation of nonferrous ore deposits. Lampietti and Marcus (1974) used a computer model to predict acceptable risk for commercial mining of manganese nodules in the deep ocean. Mathematical methods were applied by Ehrisman et al. (1975) to the exploration and reconnaissance of rutile placer deposits in Sierra Leone.

Probability theory was employed by Pirson (1941) in oil exploration ventures. Hardin (1959) reported on how to evaluate wildcat prospects. Wansbrough (1960) discussed discounted cash flow rate of return and gave an approach to evaluation of oil production capital investment risks. Hulsey (1962) philosophically interpreted risks behind the trading of oil properties and emphasized that property values are determined in the marketplace, not in the computer room. Walstrom et al. (1967) showed how to evaluate uncertainty in engineering calculations. In order to evaluate the risk-return ratio of an oil exploration program, Schwade (1967) provided a check list for numerical evaluation, which organized data into 3 trapping factors that are further subdivided into 198 geologic elements. A simple method for indicating probability distribution of present value of oil was used by Davidson and Cooper (1976). Hill (1975) reviewed resources, risk, and rewards in petroleum exploration. Ickes (1936), Watkins (1959), Rummerfield and Morrisey (1964a), Benelli (1967), and Harbaugh and Prelat (1973) reported on statistical and computer interpretation of prospect evaluation. In relation to computer and modeling applications in oil exploration decisions, Grayson (1960, 1962a, 1964) presented a good general outline of economic evaluations, Smith (1970) constructed 4 probabilistic models based on trend, prospect, development, and production phases utilizing the Monte Carlo method, and Bradley and Kaufman (1973) treated reward and uncertainty in exploration programs. Probability and decision analysis in the petroleum industry are well covered in 3 recent texts by McCray (1975),

Newendorp (1975), and Harbaugh et al. (1977). King (1974) examined Bayesian decision theory in petroleum exploration. Moreover, Grayson (1962b) and Newendorp (1972) recommended Bayesian analysis in updating risk estimates as new information becomes available. Upadhyay et al. (1977) used a modified risk analysis computer program, a probabilistic evaluation, and sensitivity analyses in evaluating the economics of oil shale mining in the central portion of the Piceance Creek Basin, Colorado. Roebuck's (1979) textbook on economic analysis of petroleum ventures is a good epitome of this broad subject.

Papers on oil and gas prospect evaluation, risk, and decision were written by Carter and Whitaker (1961), Kaufman (1963), Gotautas (1963), Hardin (1966), Hardin and Mygdal (1968), Campbell and Schuh (1970), Fuda (1971), Wintermute (1971), Mints (1972), and Vassoevich et al. (1979). Boyd and Miller (1967) advised better investment decisions and planning of gas-producing operations by means of a gas sales simulation computer program. Strategies and competitive bidding for oil and gas in high risk situations were covered by Arps (1970) and Capen et al. (1972).

The development of the lignite district of the Rhine is given as an example by Gaertner (1970) of synthesis of the decisions of miners and exploration. In solving problems with cutoff grades and ratios, equipment capabilities, and economics, Falkie and Porter (1973) developed an economic decision-making model for daily use in multiple seam mining of bituminous coal.

Mendes and Melcher (1975) used a simulation model for production planning at the Jacupiranga phosphate mine, Brazil. Engineering geology maps were used by Sverzhevskiy (1975) for evaluating and predicting conditions of exploitation of the Donets Basin (USSR) coal deposits. Hajdasiński (1977) investigated the problem of value equivalence for models of continuous and discrete representation of value changes in time exemplified by the construction of an underground coal mine.

Aspects Other than Risk Analysis and Decision Making

Basic prospecting, and the locating and valuating of mines was surveyed by Stretch (1904) and McKechnie (1930a). Under-runs and over-runs, by the application of certain factors adapted from prospecting in frozen ground to adjacent thawed ground, were controlled by Doheny (1941) in placer valuation in Alaska. Hammer (1944) estimated ore masses by gravity prospecting. Blondel (1951) discussed criteria for evaluating distribution of ore deposits. Bose (1959) called attention to the role of geology in exploration, evaluation, and development of concealed mineral deposits. Anderson (1959) gave a general treatment of prospecting in Washington. Matveev and Nikiforov (1961) emphasized the proper mixture of exploration and evaluation of deposits, and used examples in recommending detailed preliminary geologic exploration and evaluation before commercial development be started. Geochemical analysis of stalactite deposits found in old mine areas such as Cripple Creek, Colorado was used by Baxter and Poet (1964) in an innovative aid to mine evalution and rehabilitation. Cordwell (1965) presented a good account of theoretical and technical aspects of rapid appraisal of prospects and mines. Geologic and economic evaluation of useful deposits during exploration were the subjects of Kazhdan and Kobakhidze's writings (1967a, 1967b, 1969). Karpushin et al. (1968) used economic geology in evaluating geophysical methods in surveys. Kreiter (1968) published an excellent text on geological prospecting and exploration whereas Dutt (1970) advocated a standardization of prospecting techniques for mineral exploration. Noteworthy advance deposit estimates in underground mining were determined by geologic-economic principles (Bachmann and Bintig, 1978).

Works dealing with regional resources development and potential are too numerous , theoretical, and specialized to be considered here in much detail. The following representative publications are mentioned as examples: Blondel (1956), Allais (1957), Harris (1966, 1967, 1969b), Sokolovskiy (1966), Miroshnikov and Prokhorov (1967), Pasternak (1969), Bugayets (1969), Manhenke (1969), Harris and Euresty (1969), Harris (1970), De Geoffroy and Wu (1970), Sinclair and Woodsworth (1970), Zezulka (1971), Karpova (1971), Koch and Link (1970–1971), Griffiths and Singer (1971), De Geoffroy and Wignall (1971), Harris et al. (1971), Domitrovic (1972), Olson (1972), Rudkevich (1972), Harris (1973), Kaganovich (1974), Agterberg et al. (1975), Cameron (1975), Agterberg (1975), Gansauge (1975), Banks and Franciscotti (1976), Vegh (1976), Phillips (1977), Dagbert and David (1977), Chung (1977), Menzie et al. (1977), Watson (1977), Kingston et al. (1978), Hansen et al. (1978), Griffiths (1978), Hanisch et al. (1978), Missan et al. (1978), Sinclair (1979), and Barnes (1980).

Mine management and planning were detailed by Chappell (1962), Whitford (1970), and Journel (1979). A probabilistic mathematical approach was employed by Marshall (1964) in the determination by modeling of optimum drilling pattern in locating and evaluating an ore body. Link et al. (1964, 1966) fitted surfaces to assay and other data by regression analysis. A basic criterion for exploration is the magnitude of errors in reserve calculations (Ivanov and Kuznetsova, 1968). Heath and Kalcov (1971) graphed valuation methods for rapid analysis of tax law effects on profits in proposed exploration programs. Sinclair (1974) graphed probability of ore tonnages in mining camps as a guide to exploration. Other works on computers and geostatistics in prospection and exploration include Dotson (1961), Ruskin (1967), Bugayets (1968), Hodgson (1971), Makarov et al. (1973), Sergeev (1973), Kazhdan et al. (1974), Latyshova and D'Yakonova (1974), Mach (1975), Steckley (1978), and Kužvart and Böhmer (1978).

Hoover's (1904b) paper considered physical and geological valuation of gold mines. Nordale (1947) recommended the open-hole method over the cased-hole drilling method in the valuation of gold dredging ground in the subarctic region across Alaska and Yukon. Sichel (1947) investigated various bias errors that are caused by cutting samples and contribute to the discrepancy between grade and ore determined by mine sampling and mill heads with special reference to narrow gold reefs. Sichel and Rowland (1961) reviewed recent advances in mine sampling and valuation practice in the South African gold fields. In the valuation of large, gold-bearing placers, Daily (1962) described prospective drilling, recovery estimates, the Radford Factor, valuation in frozen ground, and a reliable procedure for estimating the amount of gold in a soil. Koch (1969) pointed out the importance of computer application in analysis of trace element data, particularly gold assays. A statistical estimation study of the joint distribution of variables was made for effective pay limits on the basis of peripheral values in stoping through a block of ore in the highly variable Kolar gold fields of India (Sarma, 1970).

Publications on computer and statistical applications in exploration and evaluation in the South African gold fields include significance of a limited number of borehole results in exploration (Krige, 1962), conceptual geological models in the Witwatersrand Basin (Pretorius, 1966), geological determination of the pattern of gold distribution in a reef (Knowles, 1966), conceptual geological models in exploration in prediction of gold content (Whitten, 1966a), and quantitative models in the economic evaluation of rock units with trend-surface analysis, illustrated by the Virginia mine in South Africa (Whitten, 1966b).

Dodd (1966) used quantitative logging and interpretation systems to evaluate uranium deposits. Bostick et al. (1968) provided resource evaluation and geologic data processing systems for prediction, location, and volume of sedimentary host rocks of uranium ore. Bogushevskiy et al. (1969) tested a method of uranium exploration· typifying and evaluating deposits on the basis of mechanical characterization of their ores. Exploration and methodology of reserve calculations in the Zirovski Vrh uranium deposits of Ljubljana, Yugoslavia were summarized by Omaljev (1974). Curry (1976) evaluated uranium resources in the Powder River Basin, Wyoming. Sandefur and Grant (1976) evaluated a roll-front uranium deposit in the Shirley Basin in Wyoming and closely studied spaced drilling densities geostatistically. Ruzicka (1976b, 1978) evaluated several important uranium-bearing regions in Canada. Stol and Drever (1977) evaluated techniques for assessing the uranium potential of lake-covered areas by underwater radiometry. Mortimer (1980) analyzed resource economics of uranium.

Products of oxidation and leaching of sulphides at outcrops can be used as indicators of the value of porphyry copper deposits (Zenin, 1938). Nordeng et al. (1964) applied trend analysis in the White Pine copper deposit of Michigan as a test of the possible utility of multiple regression in ore search. Schillinger (1964) applied induced polarization techniques successfully in the footwall of the Osceola amygdaloid lode of the Michigan native copper district. The Whalesback copper deposit in Newfoundland was exemplified by Agterberg (1969a) in a detailed statistical analysis involving interpolation of areally distributed data. A preliminary evaluation of ore potential by Cameron and Baragar (1971) in the Yellowknife and Coppermine River Groups of Northwest Territories, Canada, was based on frequency distributions of copper from geochemical sampling and analysis. Pryor et al. (1972) described sampling and evaluation practice in a low-grade copper open-pit operation at Cerro, Colorado, Rio Tinto, Spain, and compared results of core drilling with initial production results. Distribution laws for copper content, thickness of copper mineralization reflected by exploratory boreholes, and main parameters in economic evaluation can be used for studying the optimization based on mathematical models (Murgu and Sandu, 1972). Wimpfen and Bennett (1975) reported on regional and national copper resource appraisal.

Prospecting, estimating, sampling, and valuing of lead-zinc ore bodies in the Mississippi Valley were discussed by Boericke (1923) whereas Netzeband (1929) wrote on the tri-state district of Kansas, Oklahoma, and Missouri. The text by Amiraslanov (1957) is the standard for comprehensive treatment of prospecting, exploration, and evaluation of world lead-zinc deposits. Statistical applications in mineral exploration and ore valuation of the Zawar lead-zinc deposits in Rajasthan were described in 1974 (Anonymous). Evaluation of the Black Angel lead-zinc mine, Marmorilik, Greenland, was done by Masters (1975) in exploration and mining under permafrost conditions. Wellmer and Podufal (1977) made a statistical model for the exploration of the lead-zinc bearing vein deposits of the Ramsbeck mine of the Sauerland area of West Germany.

Moberg (1962) treated new developments in the exploration and

263

investigation of iron ore properties. Complexity of form was determined by thickness variability, configuration, and continuity of mineralization. This was done by using the Kallistov variation factor, *area method,* and outline modulus of Zenkov and Lemenov in the exploration and appraisal of the complex nature of the iron ore deposits in Krivoy Rog Basin of the USSR (Kazak, 1965). Magnetitic ore-tonnage estimates from an aerial electromagnetic survey were useful in illustrating that the economic potential of a known magnetite deposit had not been fully appreciated (Fraser, 1973). Kantor (1977) used mineralogical criteria in the economic evaluation of magnetite occurrences with examples from Siberia, the Turgaj, and other regions of the USSR.

Chernyshev (1967) reported on evaluation of manganese ore deposits during exploration and detailed prospecting. A gravity survey of the manganese deposit at Woodie Woodie, Western Australia ($D =$ 4.0 g/cm^3) defined the subsurface configuration, and the reserve estimates closely corresponded to those later proven by drilling (Nowak, 1970).

In the Cape Mountain district of Seward Peninsula, Alaska, detrital cover sampling of tin lodes was undertaken to assess its effectiveness in exploration of areas of permanently frozen ground (Mulligan, 1966). Lampietti and Marcus (1971) used computer simulation in aiding offshore sampling for tin in which a deposit is to be evaluated with a limited number of drill holes. Kuscevic et al. (1974) compared development and corresponding stope samples obtained during subsequent mining operations at South Crofty tin mine in Cornwall, England, and found that the sampling interval could be doubled with no significant loss of accuracy; the use of the estimator gave improved results when a small sample contained a random high.

Evaluation of beryllium deposits during prospecting and exploration was the subject of the excellent text by Beus (1962). Beryl-bearing pegmatites of the Copper Mountain mining district, Fremont County, Wyoming, were evaluated by Kopp (1965) with emphasis on geological principles and comparison with commercially successful pegmatites.

Smirnov (1960) reviewed geological bases for the exploration and prospecting of metallic ore deposits, emphasizing methods of estimating. Examination of the *mineral belt concept* in recent years has led to a better understanding of deposition and an extending of reserves in the Coeur d'Alene district of Idaho (Farmin and Crosby, 1962). Semenyuk et al. (1964) authored an important paper on the reliability of such prospected ore reserves of molybdenum, tungsten, tin, and gold as compared with mining reserves in Transbaykal' vein deposits. Harris (1969a) presented a probabilistic regional appraisal of base and precious metal resources in Alaska with expected residual values based on probability distribution for each cell that resulted from combining probabilities computed by two separate discriminant and classification analyses. A paper by Savosin (1971) emphasized accuracy of evaluation of thickness, content, and quantity of metals in the exploratory stage. In estimation by quantification of lateritic-type structures, Matheron's theory of universal kriging was utilized (Serra, 1971). Resources of copper, zinc, lead, gold, and uranium were inferred from "reasonably assured" reserves and previous production by a log-binomial, equal probability model known as the IRIS-diagram (Brink, 1972).

Some recent publications on exploration reserve assessment and evaluation considered the following metals and ores: gold and rare polymetal deposits (Kuz'min, 1972), copper and zinc (Agterberg et al., 1972), nickel, copper, zinc, lead, and silver (Agterberg, 1974a), lead (Bozdar and Kitchenham, 1972), nickel-bearing layered trap intrusions of the northern Siberian Platform (Vasil'yev et al., 1973), copper, iron pyrites, and chromites (Pantazis, 1978), bauxite and other ores (Matveev and Nikiforov, 1961; Chapman and Evans, 1978), barite (Wharton, 1972), and global lateritic nickel resources (De Vletter, 1978).

There is an abundance of publications on the subject of exploration and prospecting for oil and gas. The following are only a few examples: Huntley and Huntley (1921), Dunbar and Gabriel (1936), Hügel (1942), Mallory (1963), Rummerfield and Morrisey (1964b), Martinez (1966), Tixier and Curtis (1967), Shamkova and Saubanova (1969), McDaniel (1970), Condon (1971), Megill (1971), McCrossan et al. (1972), Armstrong and Heemstra (1973), Demina (1975), and Nalivkin et al. (1976). There are many noteworthy publications on oil and gas formation evaluation utilizing log and core analysis such as Koepf and Granberry (1961), Wilshusen (1962), Lynch (1962), and the excellent text by Pirson (1963).

A method for evaluating drillable oil and gas prospects was explained by Mabra (1956). Callaway (1959) explained evaluation of waterflood prospects. Hill et al. (1961) reduced oil-finding cost by use of hydrodynamic evaluation. Data on predicted oil and gas reserves were employed for prospecting and exploration planning (Aksenov et al., 1973). As a basic factor for national allocation of exploration activity Yatselenko (1973) employed comparative estimation of recoverable, potential reserves. McNaughton and Garb (1975) investigated finding and evaluation of oil and gas in fractured reservoir rocks. A synopsis of procedure for petroleum reserves (models) in the *Accelerated National Oil and Gas Resource Evaluation* (ANOGRE) was given by Mallory (1975). Kontorovich and Demin (1979), using a mathematical model, derived a method of assessing the amount and distribution of oil and gas reserves in large oil and gas basins. Relationships of gasses dissolved in ground water were discussed in papers by London et al. (1964) on the principles of evaluating gas prospects and by Kortsenshteyn (1975) on the estimates of oil and gas reserves.

The prominent text by Pierce and Kennedy (1960) on mine examination, reports, and valuation contained much detailed information on the examination of bedded sedimentary coal deposits. Collinson and Elliott (1960) summarized data on exploratory borings and proving coal reserves in the Nottinghamshire and north Derbyshire coalfields, England. Carmichael (1969) applied simple air-photo techniques in regional coal-reserve evaluation to pare exploration and mining costs. Bogdanov et al. (1974) recommended methodology for the estimation of potential coal and lignite resources. Good general papers on modeling and computer evaluation of coal fields are by Pester and Planert (1969), Wilson (1976), and McQuillin (1977a).

Messel (1947) explained examination and valuation of chrysotile asbestos deposits in massive serpentine and Shurtz (1966b) compared methods of statistical analysis in evaluation of asbestos deposits. Comprehensive chrysotile asbestos evaluation presents many unique problems; coring and logging are done because ore and gangue have the same composition, and assays are useless (Dean and Mann, 1968).

Prospecting, surveying, and evaluation are the main subjects of the treatise on salt deposits of the USSR by Ivanov (1953). Well logs can be used to locate and evaluate deposits of various commercially important minerals if the mineral of interest, such as coal or oil shale, represents a significant fraction of the formation bulk volume and has properties measurable by logs (Tixier and Alger, 1970). Applin (1972) explained the details of development sampling procedures for ore-reserve valuation of alluvial diamond deposits in Ghana and Sierra Leone where a variety of sampling techniques are necessary to suit a wide range of diamond sizes and field conditions. Rodionov and Ronenson (1972) assessed mica deposits in prospecting and exploring. Kolm (1974) applied ERTS MSS imagery to mapping and economic evaluation of sand dunes in Wyoming. Estimation of the optimal exploration grid density for mica deposits was done by Chesnokov (1975). Jorgensen (1978) reviewed some factors involved in prospecting for gypsum deposits.

RESERVES, GRADE CONTROL, AND PLANNING

Generalities

Classification, definition, and nomenclature problems of ore reserves were reviewed in papers by Hiessleitner (1937), Lasky (1945), Vogel (1954), Jones (1954–1955), King (1955), Blondel and Lasky

(1955, 1956), Kogan (1961), Cox (1968), Blondel and Lasky (1970), Bogushevskiy et al. (1971), Benko (1971), Engalychev and Mukhin (1973), and Kiselev (1974). Petrascheck (1951) recognized the following categories of possible reserves: *indicated* (from outcrop data or geophysical prospecting and geological knowledge), *inferred* (from knowledge of the specific geological situation), and *existing* (from general geological analogies and information).

The following publications contain good treatments of traditional, conventional, and general applications of ore reserve, grade control, and planning practice methodology: Hoover (1904a), Knox (1908), Probert and Earling (1911), Garrison (1911), Magnus (1916), Harding (1921, 1923), Truscott (1930), Juravsky (1931), Fennell (1939), Keil (1942), Read (1943), Wolf (1958), Smirnov et al. (1960), Pozharitsky (1963), King (1965), Didyk and Tulcanaza Navarro (1970), Nemec (1973), Vasilev (1974), Kiselev (1974), Govett and Govett (1974), Paione (1975), Carpenter (1976), Earll et al. (1976), Burdo (1977), and Petrov (1977).

One of the first to use probability theory on a mining problem was Watermeyer (1919) who tried to minimize the discrepancy between the grade of ore calculated at the mill head and the grade of ore calculated by underground sampling. Zenkov (1937) attempted to solve estimation and sampling problems with coefficient of variability of properties of an ore body. Bray (1943) recognized the importance of rock structures such as bedding, lamination, cleavage planes, joints, and grain in the economics of mining and quarrying operations whereas Barbosa (1946) stressed the bearing of structure on the calculation of reserves. Volodomonov (1944) treated methods of estimating vein deposit reserves. Criteria involved in the calculation of known, possible, and probable reserves were summarized by Sorrentino (1948). Norton and Page (1957) proved that the grade of a pegmatite deposit can be determined satisfactorily by measuring grains on representative exposures, by production records, or by assays and reserves of an unmined pegmatite by projections down the structure.

A new method of lineation-schistosity fabric analysis for use in predicting thickness variations in ore bodies was introduced by Sugiyama (1955, 1958). Yui (1959) studied the sources of error in estimating ore reserves in fully developed ore blocks. Popoff's (1966) comprehensive review analyzed conventional methods of computing reserves, such as analogy, geologic blocks, cross-section, isolines, and polygons, and pointed out that accuracy or results usually depend more on geologic interpretation and assumptions than on method. Illustrations of difficulties and misconceptions that affect estimation

267

of reserves were examined by Lovering (1969). Tapi and Satyanarayana (1973) used specific gravity in ore-reserve estimation. The text by Thomas (1973) on mining and exploration has a good chapter on determination of ore reserves. In a text by Kogan (1974) there is comprehensive consideration of all phases of reserve calculation and geological-industrial evaluation of ore deposits. Economics, reserves, and resources in relation to mineral supply as a stock were reviewed comprehensively by Brooks (1976).

Fundamental works by Matheron (1962–1963, 1965, 1969a) established a definitive foundation for exploration and estimation geostatistics. The more recent texts by David (1977) and Journel and Huijbregts (1978) give superb coverage to the whole field of mining geostatistics, the former stressing the practical aspects of mathematical solution to ore-reserve estimation problems and the latter emphasizing more theoretical aspects of mining. Solovev (1939) applied methods of variation statistics to prospecting and calculation of reserves. Application of the lognormal distribution to ore deposits goes back at least to Razumovsky (1940). The following publications deal with various topics related to computing of ore reserves and grade control: Hazen (1956), Hewlett (1961a), Petrov (1965), Hall (1966), Sohnge (1966), Kuz'min (1966), Kuck (1966), Prokof'yev (1967), David and Blais (1968), Davids (1968), Blais and Carlier (1968), Kaas (1968), Perets (1969), Urasin and Shamanskii (1970), Vlasov et al. (1970), Kosygin et al. (1970), Sergijko et al. (1972), Prazsky (1972), Vasilev (1972), Yumatov et al. (1973), Ambrus (1973), David and Blais (1973), Wilke and Von der Linden (1974), Azcarate (1975), Yershov (1975), Brooker (1976), Lallement (1976), Journel (1976b), Switzer and Parker (1976), Quinlan and Crosby (1977), Perišić, et al. (1977), Lallement (1977), Echeverria and Salas (1978), and Royle (1979a, 1979b). Some noteworthy papers on various phases of grade calculation are by Weaver (1964), Hazen and Gladfelter (1964), Koch and Link (1967), Rao (1968), Leach (1970), Taylor (1972), Guretskiy and Asadulin (1974), Daud (1974), and Huijbregts (1976). Various treatments of mine planning, reserves, and finance were presented by Hartman and Varma (1966), Zorileanu (1970), Stucke (1970), Hargreaves (1972), Marino and Slama (1973), Coyle (1973), David et al. (1974), Dowd (1975), Gignac (1975), Khokhryakov (1975), Dowd and David (1976), Clark (1976), Maréchal (1976b), and Johnson (1978). Shalmina et al. (1974) recommended methods for estimation of multicomponent mineral deposits. A rolling mean technique of trend analysis was employed to discover trends in mineral values over a deposit, which was sampled extensively by Williamson and Thomas (1974). Pirow et al. (1966), Krige (1966a), Phillips (1968), and Godfrey (1970) reported on applications of trend surface analysis. Various methods of kriging were considered by Matheron (1969*b*). Huijbregts and Matheron (1971) researched universal kriging, which is an optimal method for estimating and contouring in trend surface analysis. Brooker (1975) examined optimal block estimation whereas Maréchal (1976*a*) wrote on linear least-square approximation of conditional expectation observed in a two-dimensional simulated ore body. Matheron (1976*a*) discussed disjunctive kriging, a new procedure for nonlinear estimation, as a simple substitute for conditional expectation. Damay (1976) explained local evaluation and David (1976) and Brooker (1979) authored papers on general practice and examples.

Hewlett (1961*a*) gave descriptions and comparisons of polygonal, triangular, cross sectional, and statistical methods for evaluating mineral reserves. Nemec (1965) explained the disadvantages of the polygonal method; on the other hand, Hewlett (1962, 1963*a*, 1963*b*) computed reserves and grades by means of statistical, polygonal, and triangular methods. Munro (1966) reviewed statistical techniques in ore valuation. Various methods of calculating ore reserves and inherent problems in the use of each method were compared by Hazen (1967*a*, 1968). In an examination of major computer methods available for determining ore reserves, O'Brian and Weiss (1968) introduced the mineralization inventory as an intermediate stage in ore-reserve estimation. Koch (1969) reviewed computer application of various techniques used in evaluation of grade, reserves, classification, delineation of deposits, and so forth. Reference was made to a number of practical examples of estimation of mineral reserves with computers by mining companies in Canada, especially IBM application program 1130-CX-IIX, numerical surface techniques, and contour map plotting (Sassano, 1970). Side-by-side comparison with manual methods displayed conclusive proof that computer techniques are valuable in ore-reserve calculations (Denny and Batcha, 1970). The distribution of an ore is expressed as a complex function modeling the chemical composition and having the spatial characteristics that change systematically within ore fields, forming the basis of a comparative analysis of different deposit types (Bogushevskiy et al., 1971). Kraft and Whitford (1972) compared computer and realized grades. Knudsen et al. (1975) did a comparative study of the geostatistical ore-reserve estimation method in contrast to conventional methods. Rutledge (1976) gave a critique and comparison of conventional methods of ore reserve and block estimation versus geostatistical techniques.

Lane (1964) used a model for mining, concentrating, and refining in choosing optimum cutoff grade. Excellent reviews of the use of mathematical models in ore evaluation were written by Agterberg (1967) and Newton and Royle (1973). Various topics dealing with modeling of ore deposits, grades, and planning were mentioned by Nemec (1974, 1975), Perry et al. (1974), Journel (1974), Zharov and Trofimov (1975), and, Matheron (1976b). Hiatt (1969) explained a program that established a statistical reliability envelope for model specification, performed mathematical operations necessary for model construction, and presented a visual display of resultant models in mineralization pattern prediction and ore evaluation. After a short survey of different types of predictive models developed to date, such as the deterministic, independent random, and correlated random, Rendu (1970) discussed in detail the geostatistical or correlated random model. This model is characterized by the assumption that ore grades are values taken by regionalized variables generated by a random process with the property that any two variables are correlated. Rendu researched the correlated random model's use in conjunction with the kriging estimator and kriging error, which enable the estimation of the average grade of any block of ore with a minimum of error.

Polygonal, triangular, cross-sectional, or other geometric configurations are used to facilitate calculations, which relate to the tonnage much more than they relate to the grade of ore. These geometric models are based on the assumption that the area of influence of any assay in any one drill hole is a function of the distance to any other adjacent drill hole without any consideration to the mineralization that actually exists between the drill holes. Applications of classical-statistical techniques to ore grade, sampling, and other problems can have many advantages but they are not a simple and complete answer to all sampling problems (Hazen, 1972). Geostatistics and models were formulated by Phillips and Edwards (1976) into a method of determining various metal prices as a function of ore grade. Astafyev (1977) based a simulation ore-reserves control model on reliability characteristic of open-pit equipment. Becker's use of the binomial model was modified according to the standard deviation of the distribution, corresponding with the maximum possible assay value for k for the chemically pure metal, which enabled the estimation of quality of ore reserve by means of a statistical model and sequential decision process (Mukherjee and Mukherjee, 1977). A very excellent explanation of the theory behind the semivariogram and how to fit a simplistic model to an experimental semivariogram was written by Clark (1979).

One of the most significant early works on sampling and estimating ore reserves was by Charleton (1881) who delineated ore bodies with graphic methods. Melvill (1929) employed graphic methods of showing variations in values of ore bodies. Altshuler (1938) devised an innovative triangular diagram from which all possible combinations of cutoff grade values might be read for any three specific metals. A simple graphic method of statistical estimation using step curves and the analysis of the different facies of mineralization of a mineral deposit was proposed by Baty (1948). Gavich (1967) solved problems of contouring mineralization and concentration involved in estimating reserves by modeling methods. García (1969) used longitudinal sections in volume calculations. Both digital plotting and cathode-ray tube methods of computer graphics were applied to such problems as ore grades, tonnages, and reserve estimates by Kaas (1969). Use of automatic computers permits an objective approach to map making and rapid calculation of reserves by the isolines method explored by Hradek (1972). Reserve calculations by macrostereologic methods were done by Csitneki (1973) with data obtained from vertical geological sections through the center of the ore bodies. Savinskiy et al. (1975) made use of mathematical and automatic data processing methods in increasing the accuracy of ore-reserve calculations and in construction of isopleth maps. Dowd (1975) presented an interactive graphic method for the estimation of ore reserves using kriging, which involves finding the Minimum Variance Unbiased Linear Estimator. Clark (1979) reviewed the theoretical foundations of geostatistics and the practical methods of constructing a semivariogram.

Metallic Deposits

Development, sampling, estimating, and valuation were reviewed by Horwood and Park (1908) for gold mines and by Fansett (1917) for gold placers. A differentiated method for developing an output program from a gold placer was worked out by Bakhvalov (1937b). Swanson's (1945) topic was probability in estimating the grade of gold deposits. Calculations were given by Chico (1961) for average mining width, average gold content in respect to dilution, and gross dollar value per ton of ore mined, in solving the problem of valuation of a gold vein. Matheron (1967) studied the best methods of estimating gold content during exploration and concluded that kriging procedure is valid for stationary or intrinsic random functions whereas the polynomial interpolation procedure should be applied only in specific cases. Lognormal distribution of the gold sample assay values in conjunction with a grade control chart developed on the basis of this distribution was applied by De Gast (1968) to problems in the grade control of on-stream broken ore. Krige and Munro (1968) reviewed problems in spatial variation in gold content of ore and some conceptual and practical implications of the use of valuation surfaces for gold ore-reserve estimation. Prokof'yev (1968) recommended flexible cutoff and substitution depending on the grade of ore, number of assays, and so forth as a remedy for the problem of extraordinarily high values ("bonanzas") of gold in assays. Salgado (1968) evaluated gold mine samples for reserve estimation by digital computer. In the study of mathematical models, sample sets, and ore-reserve estimation by Royle and Newton (1972), kriging produced the optimum estimates in such bodies as hydrothermal gold and tin veins, which may be difficult to value and may have local anisotropies and nugget effects. The reliability of ore-grade determination from the geological documentation of gold mines was discussed by Lobach and Moshkin (1974).

Statistical studies on ore valuation in the Kolar gold fields of India were done by Sarma (1968), involving gold values that follow a two-parameter lognormal distribution pattern. Qureshy et al. (1969) wrote on practical applications of statistical gold value distribution, and Verma (1972) wrote on correcting the discrepancy between samples taken from peripheral and central portions of blocks by utilizing normal regression and lognormal regression of internal and total block values.

From the study of 235,000 channel samples at the Hollinger Consolidated Gold Mines in Ontario, using percentage frequency and assay value as weighting factors, Jones (1943) found close agreement between sampling grade and mill heads. Ore-reserve estimation and grade control at the following Canadian gold sites were reviewed by Butler (1968), Hollinger mine, Ontario; Paransevicius (1968), Pamour mine, Porcupine district, Ontario; Weeks (1968), Bralorne gold mine, British Columbia; Rancourt and Evans (1968), East Malartic Mines Ltd., Quebec; Krause (1968), Campbell Chibougamau, Quebec; and Dadson and Emery (1968), Giant Yellowknife mine, Northwest Territories. Hester (1970) compared reserve evaluation from spaced churn-drilling samples with actual values recovered in the study of Klondike, Yukon Territory placer gold deposits.

In their statistical evaluation of two methods to predict grade in underdeveloped parts of the Getchell mine in Humboldt County, Nevada, Koch and Link (1970) concluded that deposits with gold particles of submicron size are likely to be less expensive to sample and evaluate than deposits with coarse-grained gold.

Topics concerning the South African gold fields are to be found in publications by Leggett and Hatch (1904) on estimate of production and life of the Main Reef Series; Krige (1951–1952) on application of lognormal curve to frequency distribution of the Witwatersrand gold ore values; Cousins (1956) on value distribution of economic minerals in the Witwatersrand gold reefs and their relation to a placer origin; Krige (1960) on departure of ore value distribution from the lognormal model; Krige (1964a, 1964b) on reviews of recent developments in statistical ore valuation; Koch and Link (1966) on statistical interpretations of gold distribution at City Deep mine; Shaw (1966) on computer evaluation practice at Western Areas Gold Mining Co. Ltd; Storrar (1966) on ore valuation practice in the Gold Fields Group; and Krige (1966b) on ore value trend surfaces based on a weighted moving average. Krige et al. (1969) used contour surfaces as predictive models for gold values. Krige (1970a) grouped mathematical statistical techniques used in ore valuation in South African gold mines into three categories. First is the correlation, regression, and weighted-moving average trend-surface analysis for ore reserve calculations. Second is the quality control for underground sampling in relation to a specialized distribution model and for routine assaying based on the lognormal with variance negatively correlated with gold values. Third is the small random sampling for borehole valuations based on three-parameter categories. Krige (1970b, 1976) reviewed the theory and application of statistical and computer techniques in evaluation of gold prospects and reserves in South Africa.

Vickers (1961) explained marginal analysis and its application in finding cutoff grade of copper. A method that combines data of an isopach map and a contoured grade/thickness map was introduced by Gilmour (1964) with a copper deposit as an example for calculating ore reserves in tabular ore bodies where there is positive correlation between thickness and grade. Dadson (1968) pointed out the importance of specific gravity in copper-reserve determinations. Clark and Garnett (1974–1975) identified multiple mineralization phases by statistical methods in the estimation of copper reserves. Krige and Rendu (1975) worked on the fitting of contour surfaces to hanging and foot wall data for an irregular copper deposit. From a study of 267 porphyry, strata-bound, and massive sulfide-copper deposits, Singer et al. (1975) concluded that geologic factors influencing tonnage and grade in a specific deposit were probably distinct, there was no significant correlation between tonnage and grade for strata-bound and porphyry deposits and frequency distributions of tonnages and grades were approximately lognormal, enabling the feasibility of predicting the probability of various tonnage-grade classes and testing correlations between variables. They also concluded that significant negative correlation was found between tonnage and grade for the massive sulfide subset and between tonnage and grade for the mixture of deposit types in the whole sample. Porphyry copper deposit tonnage and grade-estimation relationships were treated in detail by Parsons (1933), Lasky (1950), and David et al. (1977).

Uchida et al. (1970) applied two-dimensional trend-surface analysis for ore-reserve estimation at Lorax copper mine, British Columbia. The case study by Vallée et al. (1977) compared the actual production figures for the copper ore bodies of the Société Minière Louvem, Val d'Or, Québec with the preproduction estimates and subsequent geostatistical estimation using kriging in order to avoid overestimation of grade. Raymond (1979) examined ore-estimation problems in an erratically disseminated copper ore body at Newmont Mines Ltd, Similkameen Division mine in British Columbia. Open-pit copper production in British Columbia was modeled by Bradley (1980) with emphasis on planning effects of different price levels on output and stocks of reserves. Ore-reserve estimation and grade-control were explained for Canadian copper sites by Weeks (1968) on the Quemont mine, Quebec; Ruttan (1968) on the Sherritt Gordon's Lynn Lake operation, Manitoba; Kermeen (1968) on the Phoenix mine, British Columbia; Parrish (1968) on the Copper Rand Mines Division, Chibougamau, Quebec; Ewanchuk (1968) on the Bethlehem Copper Corporation, Ltd, British Columbia; and Cairns (1968, 1969) on the Coronation mine, Sasketchewan. A comparison was made by Heim (1968) between ore estimates made during the operations at North Coldstream Mines Ltd in Ontario and the actual production. Agterberg (1968a) evaluated the Whalesback copper mine in Newfoundland by the technique of trend-surface analysis. Piloski (1968) commented on Agterberg's observations at Whalesback and implied a tendency toward overestimation when using trend analysis; Agterberg's (1968b) reply gave reasons for this and offered alternate approaches.

Sampling and estimation practice with copper in Arizona were given by Prouty and Green (1925) on the Copper Creek Branch, Phelps Dodge Corporation, Bisbee district; Ziesmer (1926) on the Sacramento Hill, Bisbee district; Dickson (1925) on the Warren district; Joralemon (1925) on the Ajo district; and Williamson and Mueller (1977) on the Cyrus Pima Mining Company, Pima County. It is interesting to note that statistical studies on the Cyrus Pima Mining Company showed relationships that could be utilized in subdividing ore blocks in a manner similar to the way they are actually mined. This new method of evaluation yielded results that were superior to kriging methods at Cyrus Pima mine. Hewlett (1964b) described an attempt to fit assay data from a copper deposit in Arizona to a three-dimensional polynomial surface as a means of expressing assay points in terms of their mine coordinates by an equation.

Ore-reserve estimation and ore-dilution control practice at the

White Pine copper mine in Michigan were noted by Ensign and Patrick (1968). Computer techniques were used for mine planning and evaluating mineralized blocks at Kennecott Copper Corporation in Bingham, Utah by Carlson et al. (1966). Blackwell (1971, 1973) used models and programs for ore-reserve estimation, optimum pit planning, and economic investigations in the Bougainville Copper Project in the Soloman Islands. Peltola (1962) reviewed methods used in estimating ore reserves at the Outokumpu Copper mine in Finland.

Taylor (1966a, 1966b) discussed ore valuation and underground metal accounting at the Kirila Bomwe South ore body of Bancroft Mines and other copper mines of Zambia. Computer techniques in ore-reserve calculation and grading in copper mines of Zambia are discussed in papers by Mackenzie (1966), Pronk Van Hoogeveen et al. (1973), Spinks and Nicholls (1973), Page and Jaczkowski (1977), and Maxwell (1978).

A geological classification of ore deposits was proposed by Momdzhi and Pastushenko (1965) for use in conjunction with experience and theoretical inferences in prognostic estimates of iron ore reserves. Dagenais et al. (1968) explained ore-reserve estimation and grade-control practice of the Iron Ore Company of Canada at the Knob Lake Division mines in Quebec and the Carol Division mines in Labrador. Zodrow and Harris (1967) made an ore-grading model for the Smallwood mine near Labrador City, Newfoundland, where accurate prediction of a highly variable magnetite content required more than simple-weighted averaging of block values. Bases for the valuation of American iron ore reserves was the subject of Eckel's (1912) paper. Electric computation of the Eagle Mountain iron ore reserves in California was done by Davis (1962). Ore reserves were estimated in the iron districts of Minnesota and Michigan by Zapffe (1925), Wolff et al. (1925), Wolff (1925), Wasson et al. (1962), and Weaton (1965). Bubenicek and Haas (1969) dealt with new geostatistical models for ore-reserve calculations of the minette-type Lorraine iron ores. In Kumaraswamy, India, iron ores were estimated by Mishra (1972). Geological and metallurgical data were used in grade, reserve, and overburden calculations by electronic data processing in production planning at the Bong Range iron deposits, Liberia (Tegtmeier, 1972). Geostatistics were applied by Maréchal (1972) to estimation problems in iron mines.

The successive steps in ore-reserve estimation, exemplified by uranium, were outlined by Patterson (1959). Wright (1959) gave an experimental comparison of uranium ore estimates as opposed to production engineering estimates. Grundy and Meehan (1963), Scott (1963), and Patterson et al. (1964) described computer programs for uranium ore-reserve estimation and analysis. Novik-Kachan (1970) gave a new method of uranium-reserve estimation for underground leaching in permeable sedimentary rocks. A FORTRAN program for routine interpretation of γ-loggings to be used in estimation of uraniferous deposits was described by Coulomb et al. (1970). Lucero (1971) and Klinge (1971) applied geostatistical criteria to uranium ore-reserve estimations and gave several examples. Cokriging and "classical" geostatistical methods were adopted by Guarascio (1976) for improving estimations in a stratiform uranium deposit in Novazza, Central Alps, Italy. Guarascio and Turchi (1977) summarized geostatistical methodologies for exploration data management and evaluation techniques for uranium mining projects. Ellis (1979) employed a FORTRAN program for mine operations and emphasized the derivation of minimum economical specifications for exploration of sandstone-type uranium deposits. Uranium ore estimation and grade-control practice in Canadian mines were explained by Peebles and Ward (1968) on Eldorado's Beaverlodge vein-type deposit in Saskatchewan, Hart and Sprague (1968) on Elliot Lake Camp in Ontario, and Winckworth (1968) on Rio Algom, Elliot Lake area in Ontario. Koch et al. (1964) presented a statistical analysis of assays of samples from the Mi Vida uranium mine, Big Indian district, San Juan County in Utah. The samples were used in grade estimates and predictions on the direction of best mineralization beyond the sampled area. Keefer and Borgman (1979) examined spatially correlated geologic phenomena with respect to their geometrical influence by means of kriging in an underground uranium mining region of Wyoming.

Topics related to grade control and estimation of reserves of manganese ores were considered in several publications. Holloway (1918) covered generalities whereas Hazen (1958) researched statistical analysis and other methods of computing reserves at Maggie Canyon, Mohave County in Arizona. Frazer (1977) discussed marine nodules and O'Leary (1979) studied the Scully mine in Canada.

Computer methods were used by Peltola et al. (1971) in ore evaluation at the Vuonos nickel deposit in Finland. De Gast and James (1972) used retrogression analysis of ore blocks in valuation of nickel deposits. At Western Mining Corp. Ltd's Kambalda nickel operations in Western Australia, Gee and Reichman (1973) and White and Gee (1977) applied computer methods to the evaluation of nickel reserves.

Some important publications on tin reserves and grades are by Lukman (1969) on the alluvial cassiterite on the Plateau tinfield of Nigeria by Hosking (1969) on the computer applications and limitations in dealing with hypothermal deposits, by Evrard and Schaar (1969) on the comparison of various methods of reserve calculation, by Gocht (1973) on the variations of the cutoff grade, by Panou (1974) on the Bukena tin deposit, Katanga, and by Taylor (1979) on the generalities concerning tin reserves and grades.

Reserves, recovery, and gradation of magnesite ores were estimated by computer at Gabbs, Nevada (Shurtz, 1959) and at Chalk Hills of Salem, Tamil Nadu in India (Shrivastava, 1977).

Ore-reserve estimation by means of data processing was done at Sullivan lead-zinc mine, Kimberley, British Columbia (Freeze, 1961). David (1969) examined the notion of *extension variance* and its application to the grade estimation of stratiform lead-zinc deposits. Universal kriging and the concept of regionalized variables were utilized in the ore-reserve estimation at the Tara Exploration and Development Company, Ltd, Navan lead-zinc deposit, Ireland (Dagbert and David, 1976).

Collins (1904) discussed the relative distribution of gold and silver values in the ores of Gilpin County, Colorado. In the study of sampling and estimating Cordilleran lead-silver limestone replacement deposits, Prescott (1925) warned against the dangers of overestimation and underestimation. Gold-and platinum-reserve estimation were considered in detail by Leviatov (1937). Sandier's (1962) text on valuation of metalliferous deposits has an excellent chapter on estimation, tonnage, and grade calculation. Hatch (1966) dealt with factors affecting ore-grade values in production of copper, lead, zinc, gold, and silver concentrates. Studies of gold and uranium distribution patterns in the Klerksdorp gold field, South Africa, by means of a three-parameter lognormal frequency distribution were made by Krige (1966c). Assays for gold, silver, lead, copper, and zinc were interpreted in ore-estimation studies by computer at Frisco mine, San Francisco del Oro, Chihuahua, Mexico (Koch and Link, 1967).

Sampling problems at the Questa molybdenite mine, Sangre de Cristo Mountains, New Mexico were solved with longhole and rotary drilling techniques (Hymas, 1968). Evaluation of significance of the correlation coefficients was done by regressive analysis when a statistical asymmetry was found in the distribution of economic parameters of a rare metals ore body, metasomatically disseminated in hydrothermally altered rocks (Vikent'yev et al., 1968). Vasilev (1968) proposed an expedient method of analysis and calculating reserves of disseminated gold, silver, and cadmium accessory minerals of the Madzarovo lead-zinc-copper complex. Random kriging was applied to estimation of tonnage and grade problems in bauxite lenses (Maréchal and Serra, 1970). Estimation and quantification by geostatistics and computers of the lateritic nickel and cobalt deposits of New Caledonia were done by Serra (1971) and Journel and Huijbregts

(1973). Ore reserves of polymetallic deposits can be treated as a function of economically exploitable minimal concentration (Simakov and Baybusinov, 1973). Guarascio and Deraisme (1974) did a geostatistical evaluation of a zinc deposit in the eastern Alps of Italy. An excellent epitome of mathematical aspects of the grade-tonnage distribution of metals was given by Musgrove (1976).

Ore-reserve estimation and grade-control procedures were presented for metallic deposits on the following Canadian sites: McIntyre mine for copper and gold (Parfitt, 1968); Falconbridge ore bodies at Sudbury, Ontario for copper-iron sulphide (Potapoff, 1968); Lake Dufault Mines, West Norbeck operation at Noranda, Quebec for copper, zinc, silver, and gold (Purdie 1968); Anglo-Rouyn, Saskatchewan for copper-gold (Skopos and Lawton 1968); Buchans ore bodies, Newfoundland for copper, lead, and zinc (Swanson 1968); Con, Sullivan, Pine Point, and H. B. mines, British Columbia for gold, lead, and zinc (Swanson and Irvine, 1968); Brunswick ore bodies, Bathurst district, New Brunswick for massive lead, zinc, and copper sulfides, with pyrite (Welwood 1968); Geco Division of Noranda Mines, Ontario for massive sulfides and zinc (Brooks and Bray 1968); and Whitehorse copper mine, Yukon for copper-magnetite skarn (Sinclair and Deraisme, 1976).

Nonmetallic Deposits

Campbell et al. (1978*b*) presented a comprehensive energy source and system appraisal of coal, oil, gas, and geothermal reserves.

The first published estimate of natural gas reserves was by Coste (1911) who used the volumetric method. Theory and problems of gas-reserve estimation were treated by Versluys (1928) and Parsons (1928). Biddison's (1935) article is representative of the many good publications on gas reserves calculations with detailed explanations of formulas fitting various cases and situations. Stephenson (1968) did a comparative study of the several methods of estimation of natural gas reserves. Keplinger (1977*a*) showed that gas-reserve calculation review accounted for differences in engineering estimates. Ruedemann and Gardescu (1922) examined the relationship of production to closed pressure in gas-reserve estimation and Johnson and Morgan (1926) researched the equal pound loss method. Bell (1931) discussed the pressure-volume, and porosity or saturation methods whereas Bartle's (1933) writings included examination of the production decline curve. Heithecker (1937) was noted for his work on the pressure-volume and White (1940) wrote on back-pressure open-flow test data and curves in gas-reserve estimation. Browkaw (1941) explained the correction for temperature and deviation from Boyle's Law.

The California Railroad Commission (1942) studied the effects of pressure drop and volume withdrawal on gas-reserve estimation. Elfrink et al. (1949) wrote on compressability correlation and its application to estimates of gas in place and Young (1949) was concerned with pressure-volume and pressure-decline. Volumetric and decline curve were explored by Gruy and Crichton (1947), and Brownscombe and Collins (1949) examined reference pressure and production history. Davis's (1951) article was about reservoir pore space and volume of gas in pore space. Katz et al. (1952) analyzed sample grading in gas estimation and Houpeurt and Lacroix (1952) wrote on the derivation of equations for calculating reserves and volume of the reservoir, both under hydrostatic pressure and not under hydrostatic pressure. Cornell (1952) gave a quick, accurate method for gas-reserve estimation involving only one pressure measurement. Cornell (1953) discussed the expected deliverability method using extrapolation of back-pressure tests. Davis and Meltzer's (1953) writings concerned predicting availability of gas based on average reservoir performance (decline curve) and the *Oil and Gas Journal* (1953*a*, 1953*b*) contained information on the pressure-volume and porosity area methods. Mensch (1959) supplied gas density data whereas Schoemaker (1957) wrote about a graphic short-cut. Nyun (1965) listed three procedures for obtaining bottom hole shut-in and flowing pressure of a natural gas well. Watson (1966) discussed volumetric and production pressure decline or material balance methods. Chierici et al. (1967) explored uncertainty in reserve evaluation from past history in water drive gas reservoirs. Mayer-Gürr (1969) proposed coordination of data on gas. Koranyi (1970) gave a comparison of methods used in different countries for classification and estimation of natural gas reserves and resources. Hammerlindl (1972) wrote on shut-in bottom-hole pressures versus cumulative production (P/Z plot). Diot and Delpit (1971) investigated computer determination of gas reserves in a water drive reservoir. Richard's (1973) writings on gas considered a three-dimensional model that predicts new discoveries and subsequent appreciation of existing discoveries. Huppler's (1974) concerns included scheduling gas field production for maximum profit. Buyalov and Golovastov (1974) wrote on estimation of total yield of gas deposits and Rossen (1975) discussed the regression approach to estimating gas-in-place. Bukreeva et al. (1975) was interested in gas-reserve estimation under experimental and industrial exploitation conditions.

The following authors have written on various topics of general interest on oil reserves and estimation: Washburne (1915), Pack (1917), Collingwood (1921), Veatch (1922), De Golyer (1922), Shaw (1935), Judson et al. (1935), Albertson (1936), Deussen (1936), Bilibin

(1937), Pirson (1944), Lees (1950), Lahee (1955), Arps (1956), Eggleston (1962), Polster (1964), Napolskii (1964), Cole (1969), Oblitas (1969), Drake (1974), Smith (1975), Coste (1977), and Campbell et al. (1978b, 1978c). There are so many items related to oil-reserve estimation and evaluation, such as waterflood performance, reservoir behavior and modeling, and steam drive recovery, that space does not permit us to include all of them here. Publications dealing with calculation of undiscovered or purely theoretical reserves, such as those by Ishayev and Rudkevich (1978) and Riesz (1978), can be given only brief attention in the review.

Listed here are good examples from the abundant literature on the decline curve type applications of oil-reserve estimation: Huntley (1913), Beal (1919a), Johnson (1919), Cutler (1920), Alvey (1920), Lewis and Cutler (1920), Knapp (1921), Johnson (1921), Culter (1924), Roeser (1925), Larkey (1925), Van Ostrand (1926), Johnson and Bollens (1927), Herold (1930), Owen (1931), Ritchie (1931), Weeks (1932), Pirson (1935), Gross (1938), Cutler and Johnson (1940), Jones (1942b), Lefkovits and Mathews (1958), Arps (1970), Higgins and Lechtenberg (1971), and Slider (1976). Cutler (1921) offered the rate of production curve as an auxiliary to the production decline curve, with time as the abscissa and barrels of production as the ordinate. Another auxiliary curve to the production decline curve was advocated by Beal and Nolan (1921). Production rates can be predicted with rate-time decline curves (Ramey and Guerrero, 1970).

Short papers of historical interest on the barrel-day method of reserve calculation and valuation are by Johnson and Foster (1920), Alvey and Foster (1921), and Ruedemann (1922b).

Some aspects of the various volumetric methods of oil-reserve estimation were mentioned by Sage and Lacey (1935) on formation volume and viscosity in the Dominguez field, California; Schilthuis and Hurst (1935) on variations in reservoir pressure in the East Texas oil field; Buckley (1939) on pressure dependence on time and production rate; Hügel et al. (1943) on calculation of oil volume of a deposit in a geologic structure; Nowak (1953) on survey of factors in volumetric analysis and its uses; Jeffries (1967) on well data system (logging); Buyalov and Zakharov (1970) on estimating prognostic reserves; Borisenko and Soson (1973) on a volume method; and Meyer (1978) on computer-resource estimation.

Brace (1934) and Bilibin (1941) wrote important publications comparing and contrasting the decline curve and volumetric methods and discussing their associated problems.

Fralich (1931), Panyity (1931), Horner (1936), Clough (1936), Horn (1942), Lewis (1957), Smith et al. (1963), Preston and Van Scoyoc (1964), and Kotyakhov et al. (1968) published on core analysis in oil-reserve estimates.

In a study evaluating functions containing indeterminate variables in calculation of recoverable oil reserves and the determination of water saturation in a reservoir rock, Walstrom (1964) used induction-electric and sonic logs. Jeffries and Reichl (1968) utilized wireline logs for computer evaluation of oil reserves. Automatic data processing was employed by Gaymard et al. (1968) for petroleum-reservoir evaluation in the Middle East and by Bauerschmidt et al. (1974) for an energy reserves (petroleum) and resources submodel. Hirakawa et al. (1971) presented a probabilistic method for oil-reserve estimation. Sultanov et al. (1973) used computers for assessing oil supplies in major deposits. Final oil recovery of Azerbaijan pools was statistically evaluated by Abasov et al. (1974). Abasov et al. (1975) and Aziz and Settari (1979) supplied examples of modeling for estimation of oil reserves.

Studies concerned with sundry methods and techniques in the estimation of oil reserves have been done by Ruedemann (1922a) who wrote about graphical methods that compare valuations in relation to fixed dates and variable dates. Moore (1933) discussed the questionable value of open flow test in gauging oil reserves and Morris (1933) researched the probable recovery estimate made on the basis of uniform potentials. Use of Coleman's formula for prorated limestone fields under volumetric control, and use of rise in the oil-water contact together with changes in the extent of the gas cap for prorated limestone fields under hydraulic control was explained by Gregory (1935). Schilthuis (1935) studied relationships between reservoir energy and oil production and Gardescu (1940) studied relationships of average percentage of water to cumulative oil production. Evinger and Muskat (1941) wrote on calculation of theoretical productivity factor. Ginter's (1941) interests included the amount of source bed required to furnish the Oklahoma City oil pool whereas Charnuii (1945) discussed available oil field reserves. Dykstra and Parsons (1950) covered prediction of oil recovery by water flood. Chilingar (1959) was concerned with determining approximate oil reserves in fractured rocks by average height of fractures. Roberts (1959) invented a permeability block method of calculating a water drive recovery factor and Maksimov (1960) explored relationships between extracted quantities of oil and water in the final stage of exploitation of an oil field under conditions of displacement of oil by water. Arrington (1960) wrote about comparing oil-finding costs with the amount of oil found. Stevens and Thodos (1961) explained estimation from production data and Dobrynin (1968, 1969a, 1969b) researched use of elastic characteristics of the stratum as a means of calculating oil reserves. Thomas and Hellums (1972) discussed a nonlinear automatic history matching technique for reservoir-simulation models. Zhdanov et al. (1974) explained integrated evaluation of geologic nonhomogeneity in analysis of oil yield and final recovery ratio and Keighin (1975) analyzed resource appraisal of oil shale in the Green River Formation, Piceance Creek Basin, Colorado.

Many writers were involved with two or more approaches to estimation of oil reserves such as Hager (1916) with production curves and barrel-day; Lewis and Beal (1918) with appraisal and future production methods; Cutler (1926) with average production decline curve, average ultimate production per acre, and productive acreages; Bilibin (1936) with curves, saturation, thickness, and so forth; Ginter (1937) with total reserves in place, ultimate recovery, and reservoir void space methods; Lahee (1941) with "probable-area" and *proved area* methods; Earlougher and Guerrero (1961) with analysis and comparison of five methods of predicting water flood reserves and performance; Sarkisyan (1961) with production from a unit, productivity decrease curve, and volume methods; Keplinger (1967) with case histories of actual performance compared with appraised prognostication; and Mayer-Gürr (1976) with volumetric, expectation curves, dynamic methods (cumulative production), and so forth.

Lewis (1917), Deussen et al. (1934), Pow et al. (1963), Granberry et al. (1968), and Keplinger (1977b) discussed oil-reserve estimation problems in sandy reservoirs. Gebauer and Gornig (1967), Haas and Mollier (1974), and Borisenko and Soson (1975) authored papers evaluating errors and reliability of oil-reserve estimates.

Classification of oil- and gas-reserve estimates was given by Sproule (1955) with emphasis on western Canada and by Abramovich (1962) with emphasis on general usage and application in prospective areas of folded oil and gas regions. A summary study that defines, compares, and classifies various types of methodology in oil, gas, and gas liquids is by Lovejoy and Homan (1965). Subjects dealing with problems in oil-and gas-reserve estimations are covered by Schilthuis (1938) on connate water in oil and gas sands; Kaye (1939) on recovery of condensate from distillate wells; Dodge et al. (1941) on comparative review of methods using decline curve, volumetric, and new modifications; Carter (1944) on applications of reserve estimates of oil, gas, and condensate; Lahee (1950) on plea for uniform terminology; Webber (1961) on estimation of discovered oil and gas reserves; Kochetov (1962) on problem of simplifying techniques; Zhdanov and Gutman (1969) on estimate of initial parameters in calculation of oil and gas reserves; and Buyalov and Vagerov (1978) on possible variants of prospective hydrocarbon-reserve evaluation. Ryan (1966), McKelvey (1966), and Moore (1966) considered the limitation of statistical methods for predicting oil and gas reserves. Sanders (1936), Huckaba and Meyer (1958), Zhdanov et al. (1973), Farkash (1976), and Dakhnov (1979) reported on errors and fallacies in oil- and gas-reserves estimation. Works on the relation of porosity and permeability to oil and gas reserves are plentiful; space limitations prevent most of these from being listed here.

The formation volume method was used by Coleman et al. (1930) in relation to quantitative effect of gas-oil ratios on decline of average rock pressure, and Schilthuis (1936) in relation to active oil and reserve energy. Huntington (1936) also used this method for material balance in reserve estimation. Material balance formulas for reserve estimation were used by Van Everdingen et al. (1953), Hawkins (1955), Tracy (1955), Hurst (1958), Rose (1960), and Havlena and Odeh (1970). Prediction of depletion performance was employed as a mode of reserve evaluation by Muscat and Taylor (1946), Jacoby and Berry (1958), Reudelhuber and Hinds (1958), Wahl et al. (1959), and Jacoby et al. (1959). Topics based on a wide range of methods can be found in works by Katz (1935) on properties of the oil-gas mixture and oil and gas production data; Sage and Lacey (1937) on formation volume, reservoir, gas-oil ratio and relation of formation volume to gas-oil ratio; Cook et al. (1951) on special considerations in predicting reservoir performance of highly volatile hydrocarbon reserves; Katz (1953) on development of techniques; Zhdanov (1960) on oil reserves with water drive, with dissolved gas drive, and with a gas cap; Dakhnov (1963) on geophysical techniques; Buyalov et al. (1963) on prognostic calculations; Mikolayevskiy and Aliyev (1968) on estimation of the oil and gas capacity of shaly reservoirs; Green et al. (1971) on two new modeling procedures for reserve estimates based on automatic history matching; Xhacka (1972) on quantitative evaluation of oil and gas reserves; Franks (1973) on estimation of oil and gas reserves by a method that allows volumes to be defined on a quantitative probability scale; Koval'chuk and Predtechenskaya (1973) on derivation of an intregal version for the volumetric method of calculating oil and gas reserves; Onoprienko et al. (1975) on delimitation of various reserve groups in the volumetric method of oil- and gas-reserve estimation; Anderson (1975) on core analysis; Borisenko et al. (1976) on complex of average statistical sections for use in oil and gas exploitation and reserve calculation; and Markov et al. (1978) on construction of oil-saturated thickness by computer. Such problems as estimation by layers of the reserves of an oil field and estimation by grid blocks of the reserves of a gas field are approached with geostatistics and kriging for working out a satisfactory subsurface model (Haas and Jousselin, 1976). Graphics and mapping, as well as paleogeography, facies data, and simulation, were used by Ulyanov (1954) and Borisenko (1974) in oil-and gas-reserve estimations.

There is an abundance of general descriptive publications on local and regional coal reserve estimates of which these sample topics are only a few of many excellent ones: Berryhill (1955) on Pittsburgh (No. 8) bed in Belmont County, Ohio; Smith and Berggren (1963) on strippable coal reserves of Illinois; Sara (1963) on bituminous coal in the Liverpool and Paparoa areas, Greymouth coalfield, New Zealand; Skipsey (1970) on role of the geologist in modern coal mining with special reference to the East Midlands coalfield of Britain; Robertson (1971) on evaluation of coal reserves of Missouri; Anonymous (1973) on coal-reserve evaluation; and Van Rensburg (1975) on confidence limits, differences in economic exploitability, and dynamic nature of coal reserves as a leading indicator with the South African coal industry as an example. Carmichael's (1968) innovative procedures for rapid estimation of Fort Union coal reserves in the northcentral United States depended on aerial photographs supplemented by instrumental leveling, barometric determination of elevations, and drilling. There are many important aspects of statistical and computer utilization in coal-reserve estimation discussed by Pundary (1966) and Koch and Gomez (1966) on statistical techniques; Lautsch (1970) on determination of economically viable coal reserves in Ruhr Carboniferous rocks depending on functions of factors of geology in mining, rock mechanics, and surface damage; Deist et al. (1971) on computer-assisted evaluation of coal reserves from trend surfaces assembled from assay data; Simic et al. (1972) on computer calculation of mass volumes in open pit mining with special reference to the Kosovo coal basin; Haycocks et al. (1973) on a computer program to aid in the assessment of tonnages of a proposed mining property based on exploratory drill hole information; McGiddy and Whitfield (1974) on coal mine planning with a computer system; Shumin (1974) on economic limit to shale content in estimation of reserves in coal seams; Hoeflinger and Bengal (1975) on compilation of estimates of coal reserves in Illinois by automatic data processing; Deist et al. (1975) on coal resources of South Africa; Venter (1976) on calculation of reserves of coal for the plains region of Alberta by the polygon method; McQuillin (1977b) on a computer system for geologic evaluation and reserve estimation of a surface mine, Witbank coalfield, South Africa; Smith (1978) on computer evaluation, classification, and management of coal reserves; and Gillette (1978) on coal reserves as a computer evaluation and mine planning tool.

Statistical and computer techniques were applied by Hallam and Hawthorne (1966) and Sichel (1973) to valuation of diamond deposits. Phillips (1971) applied these techniques to estimating the grade of diamond deposits with poisson probability distribution, and Sharp (1976) to a lognormal distribution of alluvial diamonds with an economic cutoff. Oosterveld (1973) used statistical and computer techniques in ore-reserve estimation and depletion planning for an ancient beach diamond deposit by Consolidated Diamond Mines of South West Africa Ltd.

Statistical and computer analysis of phosphate deposits was done by Berkenkotter (1964) in evaluating beds in Florida, Hazen (1964) in the Phosphoria Formation of Idaho, Journel (1976a) in mine scheduling and grade control in the Togo deposit of west Africa, and Powell (1974) in evaluating phosphate resources in southeastern Idaho.

Weis (1978) estimated reserves of amorphous graphite in Sonora, Mexico. Shaw (1929), Royle and Hosgit (1974, 1975), Herdendorf and Braidech (1970), and Chester (1980) treated reserve estimation of stone, sand, and gravel deposits. Debyser et al. (1974) were involved with evaluation of mineral gravel reserves from a dredging network off the coast of Brittany in the southern part of the Western (English) Channel.

REFERENCES

Abasov, M. T. et al., 1974, Statisticheskaya otsenka konechnoy nefteotdachi po zalezham Azerbaydzhana, *Geologiya Nefti i Gaza* **2:**21-24.

Abasov, M. T. et al., 1975, Modeling of Oil Output of Rocks for the Estimation of Exploitable Oil Reserves, *Akad. Nauk Azerbaydzhan. SSR Izv. Ser. Nauk o Zemle* **1:**9-14.

Abramovich, M. V., 1962, Estimate of Reserves of Prospective Areas in Folded Oil and Gas Regions, *Petroleum Geology* **4:**315-318.

Adamiam, G. H., 1953, Oil Industry and the Tax Depletion Allowance, *Monthly Dig. Tax Art.* **3:**35-46.

Ageton, R. W. et al., 1969, A Systems Approach to Recovering Gold Resources in Jefferson County, Montana, *U. S. Bur. Mines Rept. Inv.* **7305:**1-16.

Agterberg, F. P., 1967, Mathematical Models in Ore Evaluation, *Canadian Operational Research Soc.* **5**(3):144-158.

Agterberg, F. P., 1968a, Application of Trend Analysis in the Evaluation of the Whalesback Mine, Newfoundland, *Canadian Inst. Mining and Metallurgy Spec. Vol.* **9:**77-88.

Agterberg, F. P., 1968b, Further Comments on the "Application of Trend Analysis in the Evaluation of the Whalesback Mine, Newfoundland," *Canadian Inst. Mining and Metallurgy Spec. Vol.* **9:**95-96.

Agterberg, F. P., 1969a, Interpolation of Areally Distributed Data, *Colorado School Mines Quart.* **64**(3):217-237.

Agterberg, F. P., 1969b, Trend Analysis and Frequency Distribution of Whalesback Copper Values, in *Proceedings of a Symposium on Decision-Making in Mineral Exploration II,* A. M. Kelley and A. J. Sinclair, eds., Extension Department of the University of British Columbia, Vancouver, pp. 177-196.

Agterberg, F. P., 1974a, Computer-Based Statistical Analysis of Geological Data: Past and Current Activities, *Canada Geol. Survey Paper* **74-60:**15-18.

Agterberg, F. P., 1974b, Geomathematics, Elsevier, Amsterdam, 596p.

Agterberg, F. P., 1975, Statistical Models for Regional Occurrence of Mineral Deposits, in *Thirteenth International Symposium on the Application of Computers and Mathematics for Decision Making in Mineral Industries,* G. Dorstewitz, ed., Verlag Glueckauf GmbH, Essen, pp. C.I.1-C.I.15.

Agterberg, F. P. et al., 1972, Geomathematical Evaluation of Copper and Zinc Potential of the Abitibi Area, Ontario and Quebec, *Canada Geol. Survey Paper* **71-41:**1-55.

Agterberg, F. P., C. F. Chung, and S. R. Divi, 1975, Preliminary Statistical Analysis of Project Appalachia Data, *Canada Geol. Survey Paper* **75-1C:**133-140.

Ahern, V. P., 1964, *Land Use Planning,* National Sand and Gravel Association, Washington, D. C., 30p.

Aksenov, A. A., Yu. M. Vasil'yev, and Yu. M. L'vovskiy, 1973, Ispol'zovaniye dannykh po prognoznym zapasam nefti i gaza dlya perspektivnogo planirovaniya poiskovorazvedochnykh rabot, *Geologiya Nefti i Gaza* **6:**1-4.

Albertson, M., 1936, Estimation of Developed Petroleum Reserves, *Am. Inst. Mining and Metall. Engineers Trans.* **118:**13-17.

Alexander, D. C., and J. P. Grant, 1953, Mine Development and Exploration Expenditures, *Tax Law Rev.* **8:**401-424.

Allais, M., 1957, Method of Appraising Economic Prospects of Mining Exploration Over Large Territories, *Management Sci.* **3:**285-347.

Allen, R. C., 1915, Valuation of Mines for Taxation Purposes, in *Michigan State Conference on Taxation Proceedings,* pp. 14-24.

Allen, R. C., and R. Arnold, 1920, Principles of Mine Taxation, *Am. Inst. Mining and Metall. Engineers Trans.* **61:**649-661.

Allsman, P. T., 1965, Developing a Program to Determine the Economic Value of Mineral Deposits with Respect to Time, in *Short Course and Symposium on Computer and Computer Applications in Mining and Exploration,* vol. 1H, J. C. Dotson and W. C. Peters, eds., College of Mines, University of Arizona, Tuscon, 27p.

Altshuler, H. I., 1938, Determining Cut-Off Grades with Triangular Coordinates, *Eng. and Mining Jour.* **139**(10):35-38.

Alvey, G. H., 1920, Decline Curve Prediction from First Day and First Thirty Days, *Am. Assoc. Petroleum Geologists Bull.* **4:**209-216.

Alvey, G. H., and A. W. Foster, 1921, Barrel-Day Values, *Am. Inst. Mining and Metall. Engineers Trans.* **65:**412-417.

Ambrus, W. J., 1973, Calculation of Reserves in an Open-Fissured Mine, *Minerales* **28**(121):3-10.

American Institute of Professional Geologists, 1974, *Suggested Practices and Guides,* Denver.

American Mining Congress, 1921a, *Proceedings of the 23rd Annual Convention, Denver, Colorado, 1920,* Washington, D.C., 843p.

American Mining Congress, 1921b, *Proceedings of the 24th Annual Convention, Washington, D.C., 1921,* Washington, D.C., 863p.

American Mining Congress, 1922, *Proceedings of the 25th Annual Convention, Conference on Mine Taxation, Cleveland, Ohio, 1922,* Washington, D.C., 174p.

American Mining Congress, 1923, *Proceedings of the 26th Annual Convention, Conference on Mine Taxation, Milwaukee, Wisconsin, 1923,* Washington, D.C., 172p.

American Mining Congress, 1924, *Proceedings of the 27th Annual Convention, Conference on Mine Taxation, Sacramento, California, 1924,* Washington, D.C., 149p.

American Mining Congress, 1927, *Proceedings of the 29th Annual.Convention,*

Conference on Mine Taxation, Washington, D.C., 1926, Washington, D.C., 100p.

Amiraslanov, A. A., 1957, *Principal Types of Lead and Zinc Deposits (Methods of Prospecting, Exploration, and Evaluation of Lead-Zinc Deposits),* Gosgeoltekhizdat, Moscow, 212p.

Anderson, A. A., and Co., 1939, *Oil and Gas Federal Income Tax Manual,* Arthur Anderson & Co., Houston, 30p. (See also subsequent editions).

Anderson, D. L., 1959, Prospecting in Washington, *Washington Div. Mines and Geology Inf. Circ. 31,* pp. 1–26.

Anderson, G., 1975, *Coring and Core Analysis Handbook,* Petroleum Publishing Company, Tulsa, 200p.

Anderson, R. C., A. S. Miller, and R. P. Spiegelman, 1977, United States Federal Tax Policy, *Resources Policy* **3:**165–178.

Anonymous, 1853, To Determine the Value of a Mine, *Mining Mag.* **1:**607–611.

Anonymous, 1915, Value of Mining Property: A Discussion of the Relations Between Northern Capital and South American Mines, *Pan American Scientific Congress Proc.* **8:**987–992.

Anonymous, 1920a, Mine Valuation for Federal Taxation, *Mining and Metallurgy,* Jan., sec. 1, pp. 24–28.

Anonymous, 1920b, Valuation of California Oil Properties for Federal Taxation, *Mining and Metallurgy,* Feb.–Mar., sec. 1, pp. 14–17, 37–39.

Anonymous, 1925, Sampling and Estimating Ore Deposits, *Am. Inst. Mining and Metall. Engineers Trans.* **72:**591–605.

Anonymous, 1936, Valuation of Inventories of Metal Mining Companies, *Jour. Accountancy* pp. 153–155.

Anonymous, 1950, What You Should Know About Depletion Allowances, *Eng. and Mining Jour.* **151**(3):60–63.

Anonymous, 1959, The Case for 27½% Depletion, *Oil and Gas Jour.* Feb. 23, pp. 99–114.

Anonymous, 1973, Coal Reserve Evaluation, *Ann. Res. Rev. Chamber Mines South Africa* **11:**95–99.

Anonymous, 1974, Statistical Applications in Mineral Exploration and Ore Valuation of Zawar Lead-Zinc Deposits, Rajasthan, *Hyderabad, India Natl. Geophys. Res. Inst. Ann. Rept.,* pp. 114–116.

Applin, C. E. S., 1972, Sampling of Alluvial Diamond Deposits of West Africa, *Inst. Mining and Metallurgy Trans.* **81**(785):A62–A77.

Arizona Department of Mineral Resources, 1951, *Six Years of Mine Property Taxation in Arizona, 1945–1950,* 5p.

Arizona Department of Mineral Resources, 1955a, *Mining Taxes in Arizona— Six-year Totals and Averages,* 9p.

Arizona Department of Mineral Resources, 1955a, *Final Valuation of All Properties Assessed to Producing Mining Companies for Year 1955,* 2p.

Arizona Department of Mineral Resources, 1956, *Assessed Valuations of Each Class of Property in Arizona, 1951–1955,* 3p.

Arizona Department of Mineral Resources, 1958a, *Arizona Mine Tax Laws— Past and Present* 3p.

Arizona Department of Mineral Resources, 1958b, *Mine Taxation in Arizona,* 10p.

Arizona Department of Mineral Resources, 1958c, *Assessed Valuation of All Arizona Properties of Producing Mining Companies—July 1958,* 2p.

Arizona Department of Mineral Resources, 1960a, *Final Valuation of All*

Properties Assessed to Producing Mining Companies for Year 1960—by Counties, 2p.

Arizona Department of Mineral Resources, 1960b, Mine Taxation in Arizona, 1956-1960, 2p.

Arizona Department of Mineral Resources, 1962, Mine Taxation in Arizona, 1959-1962, 11p.

Arizona Department of Mineral Resources, 1964, Mine Taxation in Arizona, Fiscal Years 1961-1964; Statistical Analysis for Year 1964, 11p.

Armitage, P., 1919, Depletion of Mineral Deposits and the Federal Income Tax, Eng. and Mining Jour., April 26, pp. 750-752.

Armitage, P., 1922, Distributions from the Depletion Reserve Under the Federal Tax Law, in Proceedings of the 25th Annual Convention, Conference on Mine Taxation, Cleveland, Ohio, 1922, Washington, D.C., pp. 61-100.

Armstrong, F. E., and R. J. Heemstra, 1973, Radiation Halos and Hydrocarbon Reservoirs: A Review, in U. S. Bur. Mines Inf. Circ. 8579, Washington, D.C., 5p.

Arnold, R., 1919, Problems of Oil Lease Valuation, Am. Assoc. Petroleum Geologists Bull. **3:**389-406.

Arnold, R., 1920, Oil Geology in Relation to Valuation, Geol. Soc. America Bull. **31:**433-440.

Arnold, R. et al., 1920, Manual for the Oil and Gas Industry Under the Revenue Act of 1918, Wiley New York, 190p. (See also revised edition, 1921, 245p.)

Arps, J. J., 1945, Analysis of Decline Curves, Am. Inst. Mining and Metall. Engineers Trans. **160:**451-470.

Arps, J. J., 1956, Estimation of Primary Oil Reserves, Jour. Petroleum Technology **8**(8):182-191.

Arps, J. J., 1958, Profitability of Capital Expenditures for Development Drilling and Producing Property Appraisal, Jour. Petroleum Technology **10**(July):13-20.

Arps, J. J., 1960, Economic Factors to be Considered in Land Transactions, Natl. Inst. Petroleum Landmen Proc. **1:**43-51.

Arps, J. J., 1961, The Profitability of Exploratory Ventures, in Economics of Petroleum Exploration, Development, and Property Evaluation, International Oil and Gas Education Center, Dallas, pp. 153-173.

Arps, J. J., 1970, A Strategy for Sealed Bidding, Am. Inst. Mining and Metall. Engineers Petroleum Trans. Reprint Ser. No. 3, pp. 220-226.

Arrington, J. R., 1960, Predicting the Size of Crude Reserves is Key to Evaluating Exploration Programs—And Here's a Practical Way to Evaluate Reserves, Oil and Gas Jour. **58**(Feb. 29):130-132, 134.

Arsent'yev, A. I. et al., 1976, Opredeleniye glavnnykh parametrov kar'yera, Izd. Nedra, Moscow, U.S.S.R., 213p.

Ashley, G. H., 1910, The Value of Coal Land, U.S. Geol. Survey Bull. **424:**5-47.

Ashley, G. H., H. N. Eavenson, and R. B. Brinsmade, 1913, Discussion of Paper by H. M. Chance, "Valuation of Coal Land" in Am. Inst. Mining and Metall. Engineers Bull. **79**(July):1315-1341, **83**(Nov.):2693-1701.

Astafyev, Y., 1977, Simulation Ore Reserves Control Model Based on Reliability Characteristics of Open-Pit Equipment, in Application of Computer Methods in the Mineral Industry, R. V. Ramani ed., American Institute of Mining and Metallurgical Engineers, New York, pp. 626-639.

Astaf'yeva, M. P. et al., 1978, Osobennosti ekonomicheskoy otsenki mestorozhdeniy v razlichnyye periody razvedki i osvoyeniya, *Vyssh. Ucheb. Zavedeniy Izv., Geologiya i Razved.* **4:**144–152.

Atkins, W. F., 1977, Fair Market Value of a Mining Property, *Canadian Inst. Mining and Metallurgy Bull.* **70**(785):113–115.

Auchmuty, R. L., 1939, Factors Influencing Mineral Land Values for Assessment Purposes, *Mining and Metallurgy* **20**(393):412–415.

Austin, W. L., 1952, Percentage Depletion; Its Background and Legislative History, *Univ. Kansas City Law Rev.* **20**(1):22–30.

Azcarate, J. E., 1975, La valoración geomatematica de mineralizaciones filonianas, in *II Congreso Ibero-Americano de geologia economica: La geológia en el desarrollo de los pueblos,* vol. 3, Dept. Arg. Sci. Publ., Buenos Aires, pp. 315–331.

Aziz, K., and A. Settari, 1979, *Petroleum Reservoir Simulation,* Appl. Sci. Publ., London, 449p.

Bachmann, H., 1978, Die Rolle des gesellschaftlichen Gebrauchswerts bei der ökonomischen Bewertung von Lagerstätten, *Zeitschr: Geol. Wiss.* **6:**855–868.

Bachmann, H., and K. H. Bintig, 1978, Geologischökonomische Grundsätze zur Festlegung des notwendigen Vorlaufes an Lagerstättenvorräten, *Tiefbaubetrieben des Bergbaus, Neue Bergbautechnic* **8:**686–690.

Bader, J. W., 1965, *Mineral Property Evaluation with Computers,* J. C. Dotson and W. C. Peters, eds., Short Course and Symposium on Computer and Computer Applications in Mining and Exploration, vol. 3DD, Col. Mines, University of Arizona, Tucson, 17p.

Bain, H. F., 1937, Method for Mine Valuation in the Philippines, *Mining and Metallurgy* **18:**255–256.

Bajwa, L. Y., 1978, Computerized Methods for Coal-Property Planning and Evaluation, in *Colorado Geological Survey Resources Series 4,* pp. 153–163.

Baker, R. P., 1952, Nature of Depletable Income, *Tax Law Rev.* **7:**267–299.

Baker, W. D., 1965, Analysis of a Mining Project, *Mines Mag.* **55**(3):15–18.

Bakhvalov, A. P., 1937a, Materials on the Method of Calculating the Metal Content in a Drill Hole, *All-Union Gold & Platinum Prosp. Trust, Tr.,* issue 4, pp. 107–129 (In Russian.)

Bakhvalov, A. P., 1937b, A Differentiated Method for Working Out the Program of Output from a Gold Placer, *All-Union Gold & Platinum Prosp. Trust, Tr.,* issue 4, pp. 130–146. (In Russian.)

Bandemer, H., 1966, Die für eine vorgegebene Genauigkeit notwendige Bohrlochanzahl bei quantitativer Auswertung der Ergebnisse, *Zeitschr. Angew. Geologie* **12:**122–127.

Banfield, A. F., and J. F. Havard, 1975, Let's Define Our Terms in Mineral Valuation, *Mining Eng.* **27**(July):74–78.

Banks, C. E., and B. C. Franciscotti, 1976, Resource Appraisal and Preliminary Planning for Surface Mining of Oil Shale, Piceance Creek Basin, Colorado, *Colorado School Mines Quart.* **71**(4):257–285.

Barbosa, A. F., 1946, *Estrutura, pesquisa e reserva de certos depósitos minerais,* São Paulo, Univ., Escola Politéc., Geol. et Met., B. 4, pp. 5–19.

Barnes, M. P., 1980, *Computer-Assisted Mineral Appraisal and Feasibility,* Society of Mining Engineers, American Institute of Mining and Metallurgical Engineers, New York, 167p.

Barrows, D., and B. Webendorfer, 1976, Community Impacts and Acceptance of Mining Operations, *Geol. Nat. History Survey Spec. Rept. 6,* Wisconsin University, pp. 129–151.

Bartle, G. G., 1933, Application to Methods of Estimating Reserves—Production Decline Curve Method, in *The Geology of the Blue Springs Gas Field, Jackson County, Missouri,* Missouri Bureau of Geology and Mines 57th Biennial Report, pp. 37–51.

Basden, K. S., 1965, Principles of Sampling and Valuation. in *Exploration and Mining Geology,* L. J. Lawrence, ed., 8th Commonwealth Mining and Metallurgical Congress, Australia and New Zealand, vol. 2 Melbourne, pp. 301–316.

Bassett, R. C., 1974, "Salting"—An Age-Old Practice, *Rocks and Minerals* **49:**756–757.

Bassett, R. C., 1975, "Salting"—An Age-Old Practice, *Rocks and Minerals* **50:**62.

Baty, V., 1948, Statistical Estimation of Mineral Deposits, *Mining Mag.* **79:**9–16.

Bauerschmidt, R., R. Denton, and H. H. Maier, 1974, Energy Reserves and Resources Submodel, in *Multilevel Computer Model of World Development System, Int. Inst. Appl. Syst. Analys.* **4:**B691–B771.

Baxter, C. H., 1915, Observations on the Appraisal of the Iron Mines of Michigan, *Eng. and Mining Jour.* **99:**439–440.

Baxter, C. H., and R. D. Parks, 1933, *Mine Examination and Valuation,* Michigan College of Mining and Technology. (See also subsequent editions, 1939, 1952.)

Baxter, C. H., and R. D. Parks, 1957, *Examination and Valuation of Mineral Property* 4th ed., Addison-Wesley, Reading, Mass., 507p.

Baxter, C. H., and S. E. Poet, 1964, Possible Use of Stalactite Deposits in Mine Evaluation, *Mines Mag.* **54**(4):24–25.

Beal, C. H., 1919*a,* The Decline and Ultimate Production of Oil Wells, with notes on the Valuation of Oil Properties (and Bibliography), *U.S. Bur. Mines Bull.* **177:**215.

Beal, C. H., 1919*b,* Essential Factors in Valuation of Oil Properties, *Am. Inst. Mining and Metall. Engineers Bull.* **153:**2219–2227.

Beal, C. H., 1920*a,* Valuation of Speculative, Deferred and Wasting Assets, *Eng. and Contracting,* Jan. 21, pp. 77–80.

Beal, C. H., 1920*b,* The Classification of Undeveloped Oil Land for Purposes of Valuation, *Econ. Geology* **15:**315–327.

Beal, C. H., 1920*c,* Report of Sub-Committee. Valuation of California Oil Property for Federal Taxation, *Mining and Metallurgy,* March, pp. 37–39.

Beal, C. H., 1920*d,* Undeveloped Land Classifications, *Oil and Gas Jour.* **18:**80–83.

Beal, C. H., 1921, Essential Factors in Valuation of Oil Properties, *Am. Inst. Mining and Metall. Engineers Trans.* **65:**344–352.

Beal, C. H., and J. O. Lewis, 1921, Some Principles Governing the Production Oil Wells, *U. S. Bur. Mines Bull.* **194:**1–58.

Beal, C. H., and E. D. Nolan, 1921, Application of Law of Equal Expectations to Oil Production in California, *Am. Inst. Mining and Metall. Engineers Trans.* **65:**335–343.

Beal, C. H. et al., 1920, California Oil Property Valuation, *Oil and Gas Jour.* **18**(Jan. 16):54–55, 58, 59, 60, 62.

Beasley, C. A., and L. B. Niedermayer, 1973, Econeval Program, in *Computer Applications in Underground Mining Systems,* Virginia Polytechnic Institute and State University, Blacksburg, 70p.

Beasley, C. A., and E. P. Pfleider, 1973, Profitability Sensitivity Analysis of a Mining Venture, in *Application of Computer Methods in the Mineral Industry,* D. G. Salamon and F. H. Lancaster, eds., South African Institute of Mining and Metallurgy, Johannesburg, pp. 109–114.

Belin, O. F., 1952, Percentage Depletion, *Illinois Certified Public Accountant,* **15**(Dec.):45–50.

Bell, A. H., 1931, Methods of Estimating Natural Gas Reserves, *Western Soc. Engineers Jour.* **36**(3):168–177.

Bell, D. D., 1975, Effects of Radiological and Waste-Management Legislative Controls on Uranium Production Costs with Specific Reference to the Beaverlodge Operation of Eldorado Nuclear Limited, in *Radon in Uranium Mining,* International Atomic Energy Agency, Vienna, Austria, pp. 37–48.

Benelli, G. C., 1967, Forecasting Profitability of Oil-Exploration Projects, *Am. Assoc. Petroleum Geologists Bull.* **51**:2228–2245.

Benko, F., 1971, Development of Resource and Reserve Classification in Hungary, *Földtani Kutatas* **14**(4):20–34. (In Hungarian.)

Bennett, H. J., and L. E. Welborn, 1971, Application of Sensitivity and Probabilistic Analysis Methods to Mineral Deposit Evaluation, *Am. Inst. Mining and Metall. Engineers Council of Economics Proc.* **6**:165–223.

Bennett, H. J., J. G. Thompson, H. J. Quiring, and J. E. Toland, 1970, Financial Evaluation of Mineral Deposits Using Sensitivity Analysis and Probabilistic Analysis Methods, *U.S. Bur. Mines Inf. Circ. 8495,* 82p.

Benson, K. S., 1951, Depletion in Federal Income Taxation of Mines, *Am. Inst. Mining and Metall. Engineers Trans.* **190**:612–616.

Bergman, K. G., and J. H. McLean, 1971, Risk Analysis for Evaluating Capital Investment Proposals, *Canadian Inst. Mining and Metallurgy Spec. Vol.* **12**:236–239.

Berkenkotter, R. D., 1964, Application of Statistical Analysis in Evaluating Bedded Deposits of Variable Thickness—Florida Phosphate data, *U.S. Bur. Mines Rept. Inv. 6526,* 38p.

Berry, E. S., 1922, Present Value in Its Relation to Ore Reserves, Plant Capacity, and Grade of Ore, *Mining and Metallurgy* **187**:11–16.

Berryhill, H. L., 1955, Coal Reserves of the Pittsburg (No. 8) Bed in Belmont County, Ohio, *U.S. Geol. Survey Circ. 363,* 15p.

Beus, A. A., 1962, *Evaluation of Deposits During Prospecting and Exploratory Work,* Freeman, San Francisco, 171p.

Biddison, P. M., 1935, Estimation of Natural Gas Reserves, in *Geology of Natural Gas,* H. A. Ley, ed., American Association of Petroleum Geologists, Tulsa, Oklahoma, pp. 1035–1052.

Bierman, H., and S. Smidt, 1966, *The Capital Budgeting Decision,* Macmillan, New York, 420p.

Bilibin, V. V., 1936, Methods of Estimating Underground Oil Reserves, *17th International Geological Congress, Leningrad Moscow, 1937.*

Bilibin, V. V., 1937, Calculation of Oil Reserves Underground, *Azerbaidzhan Oil-Sci. Tech. Inst., Baku, USSR,* 192p. (In Russian.)

Bilibin, V. V., 1941, Estimation of Underground Reserves of Petroleum, *Petroleum Engineer* **12**(8):141, **12**(9):72.

Bilodeau, M. L., and B. W. Mackenzie, 1977, The Drilling Investment Decision in Mineral Exploration, in *Application of Computer Methods in the Mineral Industry,* R. V. Ramani, ed., American Institute of Mining and Metallurgical Engineers, pp. 932–949.

Binckley, G. S., 1915, Why Appraisal is Not Valuation, *Engineering Record,* Oct. 23, pp. 515–517.

Bird, B. L., 1955, "Property" for Purposes of Depletion, *Texas Law Rev.* **33:**785–791.

Blackwell, M. R. L., 1971, Some Aspects of the Evaluation and Planning of the Bougainville Copper Project, *Canadian Inst. Mining and Metallurgy Spec. Vol.* **12:**261–269.

Blackwell, M. R. L., 1973, A Model of Bougainville Copper's Panguna Orebody, in *Application of Computer Methods in the Mineral Industry,* M. D. G. Salamon and F. H. Lancaster, eds., South African Institute of Mining and Metallurgy, Johannesburg, pp. 35–40.

Blais, R., and C. Berry, 1971, Operations Research, Or Quantified Common Sense, An Aid to Managers—A Discussion, *Canadian Inst. Mining and Metallurgy Spec. Vol.* **12:**3–18.

Blais, R., and P. A. Carlier, 1968, Applications of Geostatistics in Ore Evaluation, *Canadian Inst. Mining and Metallurgy Spec. Vol.* **9:**41–68.

Blaise, F. J., 1959, *The Case for Percentage Depletion for Oil and Gas Wells,* Pure Oil Company, Chicago, 60p.

Blakemore, W., 1895, Is There an Economic Limit to the Output of a Coal Mine, *Federated Canadian Mining Inst. Trans.* **1:**257.

Blondel, F., 1951, La répartition géographique des gisements minéraux, *Chronique Mines Coloniales* **19**(175):2–6.

Blondel, F., 1956, Les lois statistiques de la répartition géographique des productions minières, *Rev. Ind. Minér.,* num. spéc. 1R, pp. 319–328.

Blondel, F., and S. G. Lasky, 1955, Concepts of Mineral Reserves and Resources, in *Survey of World Iron Resources,* United Nations Department of Economic and Social Affairs, New York, pp. 169–174.

Blondel, F., and S. G. Lasky, 1956, Mineral Reserves and Mineral Resources, *Econ. Geology* **51:**686–697.

Blondel, F., and S. G. Lasky, 1970, *Concepts of Mineral Reserves and Resources,* United Nations Department of Economic and Social Affairs, New York, pp. 53–58.

Boedeker, M. H., 1932, Controlling Depreciation and Depletion Charge by Reducing Asset Values. *Petroleum World,* April, pp. 17–19, 22.

Boehmke, F. C., 1979, Computerized System of Progressive Ore Reserve Assessment and Comparison with Actual Production, in *Proceedings of the 11th Commonwealth Mining and Metallurgical Congress, Hong Kong, 1978,* M. J. Jones, ed., Institution of Mining and Metallurgy, London, pp. 497–505.

Boericke, W. F., 1923, Sampling and Estimating Zinc and Lead Ore Bodies in the Mississippi Valley, *Am. Inst. Mining and Metall. Engineers Trans.* **68:**417–422.

Boericke, W. F., 1940, Mine Valuation in the Philippines, *Canadian Mining Jour.* **61**(10):644–647.

Boericke, W. F., 1947, Valuation Should Recognize Present Interest Rates, *Eng. and Mining Jour.* **148**(Jan.):76–77.

Boericke, W. F., and C. C. Bailey, 1959, Mine Financing, *Economics of the Mineral Industries,* E. H. Robie, ed., American Institute of Mining and Metallurgical Engineers, New York, pp. 239–269.

Bogdanov, V. V. et al., 1974, Estimation of Potential Coal and Lignite Resources: Recommended Methodology, VSEGEI, Leningrad, USSR, 40p.

Bogushevskiy, E. M. et al., 1969, O vozmozhnosti tipizatsii i prognozirovaniya rudnykh mestorozhdeniy na osnove ispol'zovaniya matematicheskikh kharakteristik rud: 2, Eksperimental'naya proverka metoda na detal'no ravedannom uranovom mestorozhdenii, *Akad. Nauk SSSR Izv. Ser. Geol.* **12:**87–98.

Bogushevskiy, E. M. et al., 1971, Classification and Prediction of Ore Deposits by Mathematical Modeling of Geochemical Distributions, *Sbornik Voprosam. Kibernetik.*, Vyp. 44, Tashkent, USSR, pp. 101–112. (In Russian.)

Bolotov, L. A., 1970, Über den Erkundungsgrad von Erzlagerstätten, die für die industrielle Nutzung vorbereitet werden, *Zeitschr. Angew. Geologie* **16**(2):66–69.

Bonbright, J. C., 1937, *Valuation of Property*, McGraw-Hill, New York 1271p. (2nd ed., 1965, Michie, Charlottesville, Virginia.)

Borden, G. S., 1959, Taxation of Mineral Properties, in *Economics of the mineral industries*, E. H. Robie, ed., American Institute of Mining and Metallurgical Engineers, New York, pp. 451–495.

Borgman, L. E., 1972, Correlated Randomness: A Central Problem in Estimating Reliability of Orebody Evaluation, in *Decision-Making IV, Cooperative Decision-Making by the Geologist and Mining Engineer, Feb. 10–12, 1971,* Center for Continuing Education, University of British Columbia, Vancouver, 386p.

Borisenko, Z. G., 1974, Metodika postroyeniya graficheskikh modeley zalezhey pri podschete nefti i gaza, *Geologiya Nefti i Gaza* **11:**42–45.

Borisenko, Z. G., and M. N. Soson, 1973, *Podschet zapasov nefti ob"yemnym metodom*, Izd. Nedra, Moscow, 175p.

Borisenko, Z. G., and M. N. Soson, 1975, Otsenka podtverzhdayemosti zapasov promyshlennykh kategoriy, *Geologiya Nefti i Gaza* **3:**69–72.

Borisenka, Z. G., M. N. Soson, and V. L. Samoylovick, 1976, Kompleks srednestatisticheskikh razrezov pri podschete zapasov i razrabotke zalezhey nefti i gaza, *Geologiya Nefti i Gaza* **4:**41–47.

Bose, M. K., 1959, Geology in Mineral Exploration. *Jour. Mines, Metals and Fuels* **7**(11):17–18.

Bosman, H. C. W., 1973, Mine Evaluation and Production Scheduling (MEPS), *Canadian Inst. Mining and Metallurgy Bull.* **66**(734):92–96.

Bostick, N. H. et al., 1968, Resource Evaluation and Geologic Data Processing Systems for Sedimentary Hosty Rocks of Uranium Ore, *U.S. Atomic Energy Comm. Rept., GJO-115-1*, 382p.

Boswell, J. L., 1967, Tax Consequences on Investment Yardsticks in Oil Property Evaluations, *Jour. Petroleum Technology* **19:**1547–1551.

Boyd, W. E., and J. Miller, 1967, Better investment planning of gas-producing operations. *Jour. Petroleum Technology* **19:**157–162.

Bozdar, L. B., and B. A. Kitchenham, 1972, Statistical Appraisal of the Occurrence of Lead Mines in the Northern Pennines, *Inst. Mining and Metallurgy Trans.* **81:**B183–B188.

Brace, O. L., 1934, Factors Governing Estimation of Recoverable Oil Reserves in Sand Fields, *Am. Assoc. Petroleum Geologists Bull.* **18:**343–357.

Bradley, P. G., 1980, Modelling Mining, *Resources Policy* **6:**44–59.

Bradley, P. G., and G. M. Kaufman, 1973, Reward and Uncertainty in Exploration Programs, *Am. Assoc. Petroleum Geologists Mem.* **19:**638–645.

Bradley, R. I., 1953, How to Present Property Appraisal Data, *World Oil* Jan., pp. 68–72.

Brant, A. A., 1968, The Pre-evaluation of the Possible Profitability of Exploration Prospects, *Mineralium Deposita* **3:**1–17.

Bray, A., 1943, Economic Significance of Rock Structures, *Mine and Quarry Eng.* **8**(4):83–85.

Breeding, C. W., and A. G. Burton, 1954, *Taxation of Oil and Gas Income,* Prentice-Hall, New York, 352p.

Breeding, C. W., and J. R. Herzfeld, 1958, Effects of Taxation on Valuation and Production Engineering, *Jour. Petroleum Technology* **10**(Sept.):21–25.

Brereton, P. M., 1936, Underground Trespass: The Apex Extralateral Rights, *Mines Mag.* **26**(3):11–13.

Brewer, L. J., 1957, Evaluation of uranium deposits, *Mines Mag.* **479:**35–38.

Brink, J. W., 1972, The Prediction of Mineral Resources and Long term Price Trends in the Non-Ferrous Metal Mining Industry, *Internat. Geol. Congress (24th) Proc., Ottawa, Canada,* pp. 3–15.

Brinsmade, R., 1909, Calculation of Mine Values, *Am. Inst. Mining and Metall. Engineers Trans.* **39:**243–249.

Brinsmade, R., 1914, Commentary on J. R. Finlay, "Valuation of Iron Mines" *Am. Inst. Mining and Metall. Engineers Trans.* **45:**282–297, 322–326.

Brons, F., and M. W. McCarry, 1960, Methods for Calculating Profitabilities, *Am. Inst. Mining and Metall. Engineers Petroleum Trans. Reprint Series No. 3,* pp. 129–139.

Brons, F., and M. Silbergh, 1970, The Relation of Earning Power to Other Profitability Criteria, *Am. Inst. Mining and Metall. Engineers Petroleum Trans. Reprint Ser. No. 3,* 2nd ed., pp. 147–153.

Brooker, P. I., 1975, Optimal Block Estimation by Kriging, *Australasian Inst. Mining and Metallurgy Proc. No. 253,* pp. 15–19.

Brooker, P. I., 1976, Block Selection According to Deposit Variability, *Australasian Inst. Mining and Metallurgy Proc. No. 257,* pp. 33–36.

Brooker, P. I., 1979, Kriging. Geostatistics, Part 4, *Eng. and Mining Jour.* **180**(9):148–153.

Brooks, D. B., 1976, Mineral Supply as a Stock, in *Economics of the Mineral Industries,* 3rd ed. W. A. Vogely, ed., American Institute of Mining and Metallurgical Engineers, New York, pp. 127–207.

Brooks, L. S., and R. C. E. Bray, 1968, A study of the Tonnage and Grade Calculations at the Geco Division of Noranda Mines, *Canadian Inst. Mining and Metallurgy Spec. Vol.* **9:**177–182.

Browkaw, A. D., 1941, A Chart to Provide Approximate Correction for Temperature and Deviation from Boyle's law, *Am. Inst. Mining and Metall. Engineers Tech. Pub. 1375,* 3p.

Brown, B. W., 1961, Stochastic Variables of Geologic Search and Decision, *Geol. Soc. America Bull.* **72**(11):1675–1685.

Brown, G. A., 1970, The Evaluation of Risk in Mining Ventures, *Canadian Inst. Mining and Metallurgy Bull.* **63**(702):1165–1171.

Brown, R. D., 1980, Sale of Mining Operations—Basic Tax Considerations, *Canadian Inst. Mining and Metallurgy Bull.* **73**(813):128–130.

Brown, R. W., 1924, *Valuation of Oil and Gas Lands.* McGraw-Hill, New York, 223p.

Brownscombe, E. R., and F. Collins, 1949, Estimation of Reserves and Water Drive from Pressure and Production History, *Am. Inst. Mining and Metall. Engineers Trans.* **186:**92–97.

Bubenicek, L., and A. Haas, 1969, Method of Calculation of the Iron Ore Reserves in the Lorraine Deposit, in *A Decade of Digital Computing in the Mineral Industry,* A. Weiss, ed., American Institute of Mining and Metallurgical Engineers, New York, pp. 179-210.

Buckley, S. E., 1939, *The Pressure Production Practice, 1938,* American Petroleum Instituter, New York, pp. 140-145.

Bugayets, A. N., 1968, Statisticheskiye metody prinyatiya resheniy v zadachakh poiska i otsenki pegmatitov po geokhimicheskim dannym, *Akad. Nauk SSSR Sibirsk. Otdeleniye Geologiya i Geofizika* **12**:65-73.

Bugayets, A. N., 1969, Use of Statistics During the Search for Economic Minerals, in *Sbornik: Geol. i geokhimiya mestorozhd. blagor. met.,* Kazakhstan (Alma-Ata), pp. 131-132. (In Russian.)

Bugayets, A. N., A. P. Matsak, and Yu. A. Sadovskiy, 1970, Primeneniye metodov diskretnogo analiza pri otsenke mestorozhdeniy poleznykh iskopayemykh na territorii Kazakhstana, *Geol. Rud. Mestorozhd.* **12**(6):62-72.

Bukreeva, N. A. et al., 1975, Gas reserve estimation under experimental and industrial exploitation conditions, *Neftegazovaya Geologiya i Geofizika* **2**:33-35. (In Russian.)

Bullion, J. W., 1962, Tax Considerations in Various Oil Transactions, *Jour. Petroleum Technology* **14**(Aug.):825-828.

Burdo, L. P., 1977, Ekonomicheskaya etsenka i obosnovaniye konditsiy dlya podscheta zapasov mestorozhdeniy poleznykh iskopayemykh, *Sovetskaya Geologiya* **5**:44-55.

Burgin, L., 1965, Some Aspects of Financing for Small Mine Operators, *Colorado School Mines Mineral Industries Bull.,* vol. 8, July, 15p.

Burtchett, F. F., and C. M. Hicks, 1948, Depreciation and Wasting Assets, in *Corporation Finance,* pp. 461-473.

Butler, F. C., 1968, Ore Reserve Estimation and Grade Control at the Hollinger Mine, *Canadian Inst. Mining and Metallurgy Spec. Vol.* **9**:203-208.

Buyalov, N. I., and D. S. Golovastov, 1974, Estimation of the Total Yield of Gas Deposits, *Neftegazovaya Geologiya i Geofizika* **10**:28-30. (In Russian.)

Buyalov, N. I., and V. S. Vagerov, 1978, Vozmozhnyye varianty prognoznoy otsenki zapasov uglevodorodov, *Neftegazovaya Geologiya i Geofizika* **4**:21-23.

Buyalov, N. I., and Ye. V. Zakharov, 1970, Use of the Volume Method for Estimating Prognostic Reserves of Oil, *Petroleum Geology* **8**:372-375.

Buyalov, N. I. et al., 1963, Method of Estimating Reserves of Natural Gas and Oil, *Petroleum Geology* **5**(1):11-15.

Bybochkin, A. M., 1974, Geologo-ekonomicheskaya etsenka mestorozhdeniy poleznykh iskopayemykh, *Razved. i Okhrana Nedr* **9**:1-15.

Byrne, W. J., 1965, Paper on Carved-out Production Payments, *Am. Mining Congress Tax Forum,* January, 34p.

Cairns, R. B., 1968, Calculation or Ore Reserves in the Coronation Mine, *Canadian Inst. Mining and Metallurgy Spec. Vol.* **9**:144-146.

Cairns, R. B., 1969, *Calculation of Ore Reserves in the Coronation Mine, Canada Geol. Survey Paper 68-5,* pp. 321-329.

California Department of Natural Resources, 1960, Sampling Mineral Deposits, *California Div. Mines and Geology Mineral Inf. Service* **13**(11):1-7.

California Railroad Commission, 1942, *Estimate of the Natural Gas Reserves* of the State of California as of January 1, 1941, and as of January 1, 1942, California Railroad Commission and Department of Natural Resources, Oil and Gas Division, Case No. 4591, Special Study No. S-334, San Francisco, 29p.

California State Board of Equalization, 1965, *Appraisal of Mining Property,* Assessment Standards Division, Sacramento, 30p.

Callaway, F. H., 1959, Evaluation of Waterflood Prospects, *Jour. Petroleum Technology* **11**(Oct.):11–16.

Cameron, E. M., 1975, Integrated Studies on Mineral Resource Appraisal in the Beechey Lake Belt of the Northern Shield, *Canadian Geol. Survey Paper 75-1, Part A,* pp. 189–192.

Cameron, E. M., and W. R. A. Baragar, 1971, Distribution of Ore Elements in Rocks for Evaluating Ore Potential: Frequency Distribution of Copper in the Coppermine River Group and Yellowknite Group Volcanic Rocks, N. W. T., Canada, *Canadian Inst. Mining and Metallurgy Spec. Vol.* **11**:570–576.

Campbell, J. M., 1959, *Oil Property Evaluation,* Prentice-Hall, Englewood Cliffs, New Jersey, 623p.

Campbell, J. M., 1970, Optimization of Capital Expenditures in Petroleum Investments, *Am. Inst. Mining and Metall. Engineers Petroleum Trans. Reprint Ser. No. 3,* 2nd ed., pp. 140–147.

Campbell, J. M., 1973, *Petroleum Reservoir Property Evaluation,* Campbell Petroleum Series, Norman, Oklahoma, 466p.

Campbell, J. M., J. M. Campbell, Sr., and R. A. Campbell, 1978*a, Economics, Principles and Strategies,* Campbell Petroleum Series: Mineral Property Economics, vol. 1, Norman, Oklahoma, 353p.

Campbell, J. M., J. M. Campbell, Sr., and R. A. Campbell, 1978*b, Energy Sources and Systems,* Campbell Petroleum Series: Mineral Property Economics, vol. 2, Norman, Oklahoma, 213p.

Campbell, J. M., J. M. Campbell, Sr., and R. A. Campbell, 1978*c, Petroleum Property Evaluation,* Campbell Petroleum Series: Mineral Property Economics, vol. 3, Norman, Oklahoma, 303p.

Campbell, W. M., and F. J., Schuh, 1970, Risk Analysis: Over-all Chance of Success Related to Number of Ventures, *Am. Inst. Mining and Metall. Engineers Petroleum Trans. Reprint Ser. No. 3,* pp. 168–175.

Capen, E. C., R. V. Clapp, and W. M. Campbell, 1972, Competitive Bidding in High-risk Situations, *Natl. Inst. Petroleum Landmen Proc.* **13**:171–215.

Cappeau, J. P., 1916, Some Commercial Factors Involved in the Appraisement of Petroleum Properties, in *The American Petroleum Industry,* vol. 1, R. F. Bacon and W. A. Hamor, eds., McGraw-Hill, New York, pp. 368–373.

Carlson, T. R., J. D. Erickson, D. T. O'Brian, and M. T. Pana, 1966, Computer Techniques in Mine Planning, *Mining Eng.* **18**(May):53–56, 80.

Carmichael, V. W., 1968, Procedures for Rapid Estimation of Fort Union Coal Reserves, *U.S. Bur. Mines Inf. Circ. 8376,* pp. 10–18.

Carmichael, V. W., 1969, Simple Air-photo Techniques Pare Exploration and Mining Costs, *Mining Eng.* **21**(Aug.):69–72.

Carpenter, R. H., 1976, A Review of Low-grade Ore Potential, in *Recent Advances in Mining and Processing of Low-grade and Submarginal Mineral Deposit,* United Nations Center of Natural Resources, Energy and Transportation, New York, and Pergamon, New York, pp. 12–23.

Carter, D. V., 1944, Application of Reserve Estimates of Hydrocarbon Fluids (Crude Oil, Gas, and Condensate), *Am. Assoc. Petroleum Geologists Bull.* **28**:630–632.

Carter, D. V., 1967, Engineering Appraisals, *Jour. Petroleum Technology* **19:**193–196.

Carter, F. B., and M. T. Whitaker, 1961, Economic Analysis of Exploratory Project, in *Petroleum Exploration Handbook,* G. B. Moody, ed., McGraw-Hill, New York, pp. 14-1–14-12.

Carter, T. L., ca. 1901, Notes on Valuing a Gold Mine, *Chem. and Metall. Soc. South Africa Jour.* **3:**81.

Castle, J. E., 1958, Economic Evaluation of an Industrial Mineral Project, *Mining Eng.* **10:**675–677.

Castle, J. E., 1977, How to Value a Mineral Property, *Eng. and Mining Jour.* **178**(9):138–141.

Caudill, S. J., 1921a, The Application of Depletion Allowances to Oil Property Taxation, *Am. Assoc. Petroleum Geologists Bull.* **5:**484–489.

Caudill, S. J., 1921b, Depletion and Oil Property Taxes, *Oil and Gas Jour.,* Dec. 9, pp. 92–93.

Caudill, S. J., 1925, Valuing Oil Producing Property for Income Taxation Purposes, *Oil and Gas Jour.* **19:**84–86.

Caudwell, F. W. H., 1929, *A Preface to Mining Investment,* E. Wilson, London, 57p.

Chacon, E. et al., 1970, The Application of Mathematical Statistics to the Exploitation of Poor Mines, *6th Internat. Mining Congress, Madrid, Paper I-A1.*

Chance, H. M., 1904, Appraisal of the Value of Mineral-lands, with Special Reference to Coal-lands, *Am. Inst. Mining and Metall. Engineers Trans.* **35:**347–359, (1905).

Chance, H. M., 1913, Valuation of Coal Land, *Am. Inst. Mining and Metall. Engineers Trans.* **47:**111–146, (1914).

Chance, H. M., 1914, Appraisal of Coal Land for Taxation, *Am. Inst. Mining and Metall. Engineers Bull.* **81:**1461–1466; *Am. Inst. Mining and Metall. Engineers Trans.* **50:**625–639.

Chance, H. M., 1927, Appraisal of Coal-Property Values, *Am. Inst. Mining and Metall. Engineers Trans.* **74:**443–445.

Chandler, J. W., 1970, Exploration, Evaluation, and Development of Lead and Zinc Ore Bodies, in *American Institute of Mining and Metallurgical Engineers World Symposium on Mining and Metallurgy of Lead and Zinc,* vol. 1, D. O. Rausch and B. C. Mariacher, eds., American Institute of Mining and Metallurgical Engineers, New York, pp. 10–74.

Channing, J. P., 1903, Mine Valuation, *Eng. and Mining Jour.* **76:**383–384.

Chapman, J. H., and H. J. Evans, 1978, Wagina Island Bauxite Exploration and Mining Assessment, *Commonwealth Mining and Metall. Congress Proc.,* No. 11, 7p.

Chappell, C. M., 1962, Simulation on a Digital Computer as an Aid to Mine Management in Evaluation of Exploration Results, in *Computer Short Course and Symposium on Mathematical Techniques and Computer Applications in Mining and Exploration,* vol. 1, College of Mines, University of Arizona, Tucson, pp. 1–15.

Charleton, A. G., 1881, A Graphic Method Applied to Delineating Ore Bodies, with Notes on Sampling and Estimating Ore Reserves, *Inst. Mining and Metallurgy Trans.* **9:**203.

Charlton, W. H., 1913, *American Mine Accounting: Methods and Forms Employed by Leading Mining Companies,* McGraw-Hill, New York, 375p.

Charnuii, I. A., 1945, Method for Evaluating Available Oil Field Reserves, *Akad. Nauk SSSR, Classe Sci. Tech., Bull. 11-12,* pp. 15–21.

Chase, C. K., H. J. Winters, and R. B. Bhappu, 1978, Economic Evaluation of In Situ Extraction of Uranium, *In Situ* **2**(1):49–63.

Chernyshev, G. B., 1970, Evaluation of Manganese Ore Deposits During Exploration and Detail Prospecting, in *Manganese Deposits of the Soviet Union,* D. B. Sapozhnikov, ed., Israel Program of Scientific Translations, Jerusalem, pp. 442–445.

Chesnokov, V. N., 1975, Estimation of the Optimal Exploration and Density for Mica Deposits, *Razved. i Okhrana Nedr,* **2:**13–18 (In Russian.)

Chester, D. K., 1980, The Evaluation of Scottish Sand and Gravel Resources, *Scottish Geographical Mag.* **96:**51–62.

Chico, R. J., 1961, Ten Basic Problems from Professor McKinstry (1896–1961), Harvard's 1961 Mineral Valuation Course, *Canadian Mining Jour.* **82**(11):68–69.

Chierici, G. L., G. Pizzi, and G. M. Ciucci, 1967, Water Drive Gas Reservoirs: Uncertainty in Reserves Evaluation from Past History, *Am. Inst. Mining and Metall. Engineers Trans.,* Part 1, **240:**237–244.

Childs, J. F., 1970, Leases: What They Cost and How to Evaluate Them, *Eng. and Mining Jour.* **171**(12):76–78.

Chilingar, G. V., 1959, Approximate Method of Determining Reserves and Average Height of Fractures in Fractured Rocks: An Interim Report, *Compass* **36:**202–205.

Chung, C. F., 1977, An Application of Discriminant Analysis for the Evaluation of Mineral Potential, in *Application of Computer Methods in the Mineral Industry,* R. V. Ramani, ed., American Institute of Mining and Metallurgical Engineers, New York pp. 229–311.

Church, J. A., 1926, Discount Formulas in Valuation of Wasting Assets, *Eng. and Mining Jour. Press* **121**(1):8–12.

Clark, D., 1963, Pay Out! Its Power to Reflect Mine Profitability, *Am. Inst. Mining and Metall. Engineers Trans.* **226:**407–417.

Clark, F. B., 1938, Problem of Deferment in Valuation of Mineral Properties, *Inst. Mining and Metallurgy Trans.,* Part 4, **95:**361–388.

Clark, I., 1976, Some Practical Computational Aspects of Mine Planning, in *Advanced Geostatistics in the Mining Industry,* M. Guarascio, M. David, and C. Huijbregts, eds., Reidel, Dordrecht, Holland, pp. 391–399.

Clark, I., 1979, the Semivariogram, Parts 1–2, *Eng. and Mining Jour.* **180**(7):90–94; **180**(8):92–97.

Clark, I., and R. H. T. Garnett, 1974–1975, Identification of Multiple Mineralization Phases by Statistical Methods, *Inst. Mining and Metallurgy Trans.* **83:**A43–A52 (1974); **84:**A73–A79 (1975).

Clark, N., 1935, Geological Theory in Mine Examinations, *Pan-American Geologist* **63:**33–40.

Clough, K. H., 1936, The Valuation of Oil Bearing Cores, *Oil Weekly* **82**(June 15):67–69.

Cohenour, R. E., A. J. Eardley, and W. P. Hewitt, 1963, Mineral Appraisal and Valuation of Lands of the Glen Canyon Withdrawal Involved in Litigation—State of Utah vs. United States of America, *Utah Geol. and Mineralog. Survey Rept. Inv. No. 4,* 52p.

Cohn, D. L., 1953, *Oil Depletion Allowance,* American Petroleum Institute, New York, 7p.

Cole, F. W., 1969, *Reservoir Engineering Manual,* 2nd ed., Gulf Publishing Company, Houston, 385p.

Coleman, S., H. Wilde, and T. Moore, 1930, Quantitative Effect of Gas-Oil

Ratios on Decline of Average Rock Pressure, *Am. Inst. Mining and Metall. Engineers Trans.* **86:**174–184.

Collingwood, D. M., 1921, Some Structural and Stratigraphic Features Affecting Relative Amounts of Oil Production in Illinois, *Am. Assoc. Petroleum Geologists Bull.* **5:**311–323.

Collins, G., 1904, The Relative Distribution of Gold and Silver Values in the Ores of Gilpin County, Colorado, *Inst. Mining and Metallurgy Trans.* **12:**480–495.

Collinson, P. L., and R. E. Elliott, Proving Coal Reserves in the Nottinghamshire and North Derbyshire Coalfield, *Inst. Mining and Metallurgy Trans.,* Part 9, **119:**537–560.

Condon, M. A., 1971, Evaluation of Large Areas of Sedimentary Basin, *Australian Petroleum Exploration Assoc. Jour.,* Part 1, **11:**46–48.

Cook, A. B., G. B. Spencer, F. P. Bobrowski, 1951, Special Consideration in Predicting Reservoir Performance of Highly Volatile Type Reserves, *Am. Inst. Mining and Metall. Engineers Trans.* **192:**37–46.

Cordwell, K. S., 1965, Theoretical and Technical Aspects of Rapid Appraisal, in *Exploration and Mining Geology,* L. J. Lawrence, ed., Australasian Institute of Mining and Metallurgy, Melbourne, pp. 61–71.

Cornell, D., 1952, How you Can Compute Gas Reserves More Accurately, *World Oil,* Dec., **135**(7):220, 222, 226.

Cornell, D., 1953, New Method Estimates Gas Well Performance, *World Oil* **136**(1):180, 182, 184.

Cornish, E. C., 1966, Sampling Ore Deposits, *Colorado School Mines Mineral Industries Bull.,* Mar., vol. 9, 8p.

Cossio, A., 1966, Estudio de las canteras de Zanja Seca, Maquía y otras en los alrededores de Contamana (provincia de Ucayali, departmento de Loreto), *Com. Carta Geol. Nac. (Peru), Bull., No. 13,* pp. 273–278.

Coste, A., 1977, Incidence du contexte economique sur le volume des reserves recuperables, *8th Congres national du petrole, Pet. Tech., No. 248,* pp. 42–45.

Coste, E., 1911, Reports to the Canadian Western Natural Gas, Light, Heat and Power Company Ltd, Waterlow and Sons, Ltd, London Wall, London, pp. 3–27.

Costigan, G. P., 1912, *Cases on the American Law of Mining,* Bobbs-Merrill, Indianapolis, 819p. (2nd ed., 1929, 644p.)

Coulomb, R. et al., 1970, Computer Calculation of Loggings in Uraniferous Deposits, *Atomic Energy Comm. Rept. CEA-N-1279,* 45p. (In French.)

Coulthard, R. W., 1916, Evaluating Coal Properties in Western Canada, *Canadian Mining Jour.,* Jan. 1, pp. 21–23.

Cousins, C. A., 1956, The Value Distribution of Economic Minerals with Special Reference to the Witwatersrand Gold Reefs, *Geol. Soc. South Africa Trans.* **59:**95–113.

Cox, H. H., 1968, Definition of Ore and Classification of Ore Reserves, *Canadian Inst. Mining and Metallurgy Spec. Vol.* **9:**1–2.

Cox, T., 1921, Application of Taxation Regulations to Oil and Gas Properties, *Am. Inst. Mining and Metall. Trans.* **65:**374–394.

Coyle, R. G., 1973, Dynamic Control of a Mining Enterprise, in *Application of Computer Methods in the Mineral Industry,* M. D. G. Salamon and F. H. Lancaster, eds., South African Institute of Mining and Metallurgy, Johannesburg, pp. 357–360.

Crandall, J. R. et al. 1959, Cost of Acquiring and Operating Mineral Properties.

Part 2. Petroleum and Natural gas, in *Economics of the Mineral Industries,* E. H. Robie, ed., American Institute of Mining and Metallurgical Engineers, New York, pp. 219–238.

Crossman, R. E., 1921, Value of Written Records in Connection with Valuations, *Am. Gas Jour.,* Oct. 25, pp. 343–345.

Cruz, D. P., 1954, Role of Geology in Actual Mining, *Mining Newsletter* **5**(3):10, 12, 24.

Cruz, P. S., 1969, Aspects of the Mining Law Affecting Exploration, Evaluation and Lease of Mineral Land in the Philippines, in *Proceedings of the Seminar on Mining Legislation and Administration, Manila, Philippines,* United Nations Mining and Research Development Series No. 34, pp. 201–206.

Csitneki, F., 1973, A Macrostereologic Ore Reserve Calculation Method Using Data Obtained from Vertical Geological Section Across the Center of the Body, *Mineralia Slovaca (Prague)* **5**:351–356. (In Czech.)

Cummings, T., 1965, Commercial Bank Financing for the Mineral industries. *Mining Eng.* **17**(May):63–65.

Curry, D. L., 1976, Evaluation of Uranium Resources in the Powder River Basin, Wyoming, *Wyoming Geol. Assoc. Guidebook, 28th Ann. Field Conf.,* pp. 235–242.

Cutler, W. W., 1920, A Mathematical Method of Constructing Average Oil-well Production-Decline Curves, *U.S. Bur. Mines Rept. Inv. 2148,* 7p.

Cutler, W. W., 1921, Rate-of-Production Curve, *U.S. Bur. Mines Rept. Inv. 2285,* 6p.

Cutler, W. W., 1924, Estimation of Underground Oil Reserves by Oil-Well Production Curves, *U.S. Bur. Mines Bull. 228,* 114p.

Cutler, W. W., 1926, Predictions of the Future of Oil Pools by Early Wells, *Am. Assoc. Petroleum Geologists Bull.* **10**:747–752.

Cutler, W. W., and W. S. Clute, 1921, Relation of Drilling Campaign to Income from Oil Properties, *U.S. Bur. Mines Rept. Inv. 2270,* 11p.

Cutler, W. W., and H. R. Johnson, 1940, Estimating Recoverable Oil of Curtailed Wells, *Oil Weekly,* May 27, pp. 19–22, 24, 26.

Dadson, A. S., 1968, Ore Estimates and Specific Gravity, *Canadian Inst. Mining and Metallurgy Spec. Vol.* **9**:3–4.

Dadson, A. S., and D. J. Emery, 1968, Ore Estimation and Grade Control at the Giant Yellowknife Mine, *Canadian Inst. Mining and Metallurgy Spec. Vol.* **9**:215–226.

Dagbert, M., and M. David, 1976, Universal Kriging for Ore-Reserve Estimation—Conceptual Background and Application to the Naven Deposit, *Canadian Inst. Mining and Metallurgy Bull.* **69**(Feb.):80–92.

Dagbert, M., and M. David, 1977, Geostatistical Mineral Resources Appraisal, *Internat. Assoc. for Mathematical Geology Jour.* **9**:313–317.

Dagenais, J. E., W. J. Dunn, C. G. Hamilton, D. J. Selleck, and R. Taylor, 1968, Ore Reserve Estimation and Grade Control at the Iron Company of Canada, *Canadian Inst. Mining and Metallurgy Spec. Vol.* **9**:287–301.

Daily, A., 1962, Valuation of Large, Gold-Bearing Placers, *Eng. and Mining Jour.* **163**(7):80–88.

Dakhnov, V. N., 1963, Use of Geophysical Methods for Study of Borehole Sections in Connection with Calculation of Reserves, *Petroleum Geology* **4**(8-B):487–492.

Dakhnov, V. N., 1979, Influence of Shaliness of Traps on the Error of Calculations of Oil and Gas Reserves, *Internat. Geology Rev.* **21**:533–534.

Dalby, W. E., 1955, Tax Problems of Uranium Development on the Colorado Plateau, *Univ. Denver 5th Ann. Tax Inst.*, pp. 128–139.

Dallin, D. E., 1977, Investment Analysis at Texasgulf, in *Application of Computer Methods in the Mineral Industry,* R. V. Ramani, ed., American Institute of Mining and Metallurgical Engineers, New York, pp. 1085–1093.

Damay, J., 1976, Application de la geostatistique au niveau d'un groupe minier, in *Advanced Geostatistics in the Mining Industry,* M. Guarascio, M. David, and C. Huijbregts, eds., Reidel, Dordrecht, Holland, pp. 313–325.

Darnell, J. L., 1924, Valuation of Oil Properties for All Purposes, *Am. Petroleum Inst. Bull.,* Dec. 31, pp. 146–150.

Darnell, J. L., 1925, Valuation of Oil-Producing Properties, *Oil and Gas Jour.,* Jan. 1, pp. 66, 70.

Daud, B. H., 1974, Computerized Geostatistical Estimation of Block Grades, *Int. Symp. Appl. Comput. Math. Miner. Ind., Proc., No. 12* **2:**F1–F29.

David, M., 1969, The Notion of "Extension Variance" and Its Application to the Grade Estimation of Stratiform Deposits, in *A Decade of Digital Computing in the Mineral Industry,* A. Weiss, ed., American Institute of Mining and Metallurgical Engineers, New York, pp. 63–81.

David, M., 1971, Geostatistical Ore Estimation — A Step-by-Step Case Study, *Canadian Inst. Mining and Metallurgy Spec. Vol.* **12:**185–191.

David, M., 1972, Grade-Tonnage Curve: Use and Misuse in Ore Reserve Estimation, *Inst. Mining and Metallurgy Trans.* **81**(788):A129–A132.

David, M., 1976, The Practice of Kriging, in *Advanced Geostatistics in the Mining Industry,* M. Guarascio, M. David, and C. Huijbregts, eds., Reidel, Dordrecht, Holland, pp. 31–48.

David, M., 1977, *Geostatistical Ore Reserve Estimation,* Elsevier, Amsterdam, 364p.

David, M., and R. A. Blais, 1968, Discussion on "Practical Aspects of Computer Methods in Ore Reserve Analysis" by D. T. O'Brian and A. Weiss, *Canadian Inst. Mining and Metallurgy Spec. Vol.* **9:**114–115.

David, M., and R. A. Blais, 1973, Geostatistical Ore Reserve Estimation, in *Application of Computer Methods in the Mineral Industry,* M. D. G. Salamon and F. H. Lancaster, eds., South African Institute of Mining and Metallurgy, Johannesburg, pp. 27–34.

David, M., P. Dowd, and S. Korobov, 1974, Forecasting Departure from Planning in Open Pit Design and Grade Control, *Int. Symp. Appl. Comput. Math. Miner. Ind. Proc. No. 12* **2:**F131–F153.

David, M., M. Dagbert, and J. M. Belisle, 1977, The Practice of Porphyry Copper Deposit Estimation for Grade and Ore-Waste Tonnages Demonstrated by Several Case Studies, *Int. Symp. Appl. Comput. Oper. Res. Miner. Ind., No. 15,* pp. 243–254.

Davids, N. C., 1968, La estadistica aplicada al calculo de reservas de yacimientos, *Jornadas Geologicas Argentinas, III, Actas* **2:**173–182.

Davidson, L. B., 1975, Investment Evaluation Under Conditions of Inflation, *Jour. Petroleum Technology* **27:**1183–1189.

Davidson, L. B., and D. O. Cooper, 1976, A Simple Way of Developing a Probability Distribution of Present Value, *Am. Inst. Mining and Metall. Engineers Trans.* **261:**1069–1078.

Davis, C. E., 1962, Electric Computation of Eagle Mountain Ore Reserves, *Canadian Mining Manual, 1962,* pp. 17, 19, 21–22; *Computer Short Course and Symposium on Mathematical Techniques and Computer Applications in Mining and Exploration,* vol. 2, Colorado Mines, University of Arizona, Tucson, 1965, pp. E2-1–E2-13.

Davis, R. E., 1951, A Method of Estimating Gas Reserves, *Oil and Gas Jour.* **50**(Sept. 27):99–100, 103, 107.

Davis, R. E., and L. H. Meltzer, 1953, A Method of Predicting Availability of Natural Gas, Based on Average Reservoir Performance, *Am. Inst. Mining and Metall. Engineers Trans.* **198**:249–259.

Davis, R. E., and E. A. Stephenson, 1953, Valuation of Natural Gas Property, *Jour. Petroleum Technology* **5**(7):9–13.

Davis, R. E., and J. M. Wege, 1956, Valuation of Gas Reserves, *Jour. Petroleum Technology* **8**(9):18–21.

Davis, W., 1909, *Simple Mine accounting,* 2nd ed., McGraw-Hill, New York, 78p.

Davitt, W. H., 1974, Engineering in Corporate Tax Determination, *Mining Eng* **26**(July):76–80.

Day, E. B., 1953, Income Tax is a Cost in "Paying Out" Oil Properties, *Natl. Assoc. Cost Accountants Bull.* Sect. 1, **34**:1646–1650.

Dean, A. W., and E. L. Mann, 1968, The Evaluation of Chrysotile Asbestos Deposits, *Canadian Inst. Mining and Metallurgy Spec. Vol.* **9**:281–286.

Debyser, J., L. Berthois, and G. A. Auffret, 1974, Étude des incertitudes affectant l'estimation des réserves en granulats exploitables en mer, *Inst. Géologie Bassin d'Aquitaine Bull.* **16**:51–63.

De Gast, A., 1968, Statistics and Mine Grade Control, *Canadian Inst. Mining and Metallurgy Spec. Vol.* **9**:71–76.

De Gast, A., and R. B. James, 1972, Valuation of Ore Blocks by Regression Analysis, An Aid to the Control of Mine Production Grades, *Canadian Inst. Mining and Metallurgy Bull.* **65**(725):52–57.

De Geoffroy, J., and T. K. Wignall, 1971, A Probabilistic Appraisal of Mineral Resources in a Portion of the Grenville Province of the Canadian Shield, *Econ. Geology* **65**:466–479.

De Geoffroy, J., and S. M. Wu, 1970, A Statistical Study of Ore Occurrences in the Greenstone Belts of the Canadian Shield, *Econ. Geology* **65**:496–504.

De Golyer, E. L., 1922, On the Estimating of Petroleum Reserves, *Econ. Geology* **17**:40–45.

Deichmann, H. H., 1976, Taxation of Mining Profits in the Republic of South Africa, in *Mineral Resources of the Republic of South Africa,* C. B. Coetzee, ed., Geological Survey of the Republic of South Africa, Handbook 7, 5th ed., pp. 9–12.

Deist, F. H. et al., 1971, Computer-Assisted Evaluation of Coal Reserves, *Canadian Inst. Mining and Metallurgy Spec. Vol.* **12**:218–222.

Deist, F. H. et al., 1975, An Investigation of the Coal Resources of South Africa, in *13th International Symposium on the Application of Computers and Mathematics for Decision Making in Mineral Industries,* G. Dorstewitz, ed., Verlag Glueckauf GmbH, Essen, pp. G. I. 1–G.I.15.

Demina, A. M., 1975, Primeneniye matematicheskoy statistiki dlya prognoznoy otsenki lokal'nykh struktur Zapadnoy Bashkirii, in *Geologiya i razrabotka neftyanykh mestorozhdeniy vostoka Volgo-Ural'skoy provinstii,* A. G. Aleksin, ed., Izd. Nauka, Moscow, pp. 105–110.

Denisov, S. A., 1974, Estimation of Mining Investment Effectiveness and the Limiting Parameters in the Definition of Mineral Reserves and Resources, *Razved. i Okhrana Nedr* **10**:39–41. (In Russian.)

Denny, J. R., and J. P. Batcha, 1970, Newmont Computers Pay Off in Mine Planning, *Mining Eng.* **22**(Nov.):65–67.

Denver United States Bank, 1969, Fundamentals of Oil and Natural Resources Financing, *Earth Sci. Bull.* **1**(2):7–16.

De St. Jorre, M. G. F., and W. W. Whitman, 1972, A Probabilistic Method of Ranking Underground Exploration Proposals, *Econ. Geology* **67**:789–795.

Deussen, A., 1936, Acre-Foot Yields of Texas Gulf Coast Oil Fields, Petroleum Development and Technology for 1936, American Institute of Mining and Metallurgical Engineers, New York, pp. 53–55.

Deussen, A., 1940, Royalties, in *Elements of the Petroleum Industry,* E. De Golyer, ed., American Institute of Mining and Metallurgical Engineers, New York, pp. 105–115.

Deussen, A. et al., 1934, Discussion of O. L. Brace, "Factors Governing Estimation of Recoverable Oil Reserves in Sand Fields," *Am. Assoc. Petroleum Geologists Bull.* **18**:1078–1083.

De Vletter, D. R., 1978, Criteria and Problems in Estimating Global Lateritic Nickel Resources, *Internat. Assoc. for Mathematical Geology Jour.* **10**(5):533–542.

Dickinson, S. B., 1965, Exploration and Mining Geology, Vol. 2, L. J. Lawrence, ed., Australasian Institute of Mining and Metallurgy, Melbourne, pp. 283–288.

Dickson, R. H., 1925, Sampling and Estimating Orebodies in the Warren District, Arizona, *Am. Inst. Mining and Metall. Engineers Trans.* **72**:621–627.

Didyk, M., and E. Tulcanaza Navarro, 1970, The Determination of Errors in Reserve Calculations, *Minerales* **25**(111):6–20. (In Spanish.)

Dilworth, J. B., 1910, A Method of Calculating Sinking-Funds, *Am. Inst. Mining and Metall. Engineers Trans.* **41**:533–535, 1911.

Dilworth, J. B., 1922, Capitalization of Mine Development, *Am. Inst. Mining and Metall. Engineers Trans.* **66**:715–728.

Dilworth, J. B., 1928, Valuation of Coal Properties, *Am. Inst. Mining and Metall. Engineers Trans.* **76**:215–236.

Diot, S., and M. Delpit, 1971, Determination des reserves d'un gisement de gaz soumis a un water-drive, *Inst. Francais Pétrole Rev.* **26**(3):199–206.

Dismant, C. I., 1950, Mine Valuation, *Colorado School Mines Quart.* **45**(2B):299–323.

Dobrynin, V. M., 1968, Otsenka zapasov nefti v treshchinno-kavernoznykh kollektorakh metodom uprugogo material'nogo balansa, *Geologiya Nefti i Gaza* **12**(5):50–56.

Dobrynin, V. M., 1969a, Calculation of Initial Oil Reserves with a Water Drive Using the Elastic Characteristics of the Stratum, *Petroleum Geology* **8**(1):62–66 (1964).

Dobrynin, V. M., 1969b, Einschätzung der Erdölvorräte in klüftig-kavernösen Speichen mittels der Methode der elastischen Materialbilanz, *Zeitschr. Angew. Geologie* **15**(2):70–74.

Dodd, P. H., 1966, Quantitative Logging and Interpretation Systems to Evaluate Uranium Deposits, in *Prof. Well Log Analysts Soc. Trans. 7th Ann. Log Symposium, May 9–11, 1966,* pp. P1-P21.

Dodge, J., H. Pyle, and E. Trostel, 1941, Estimation by Volumetric Methods of Recoverable Oil and Gas from Sands, *Am. Assoc. Petroleum Geologists Bull.* **25**:1302–1326.

Dodson, C. R., 1958, Facts the Banker Needs in Making Oil Loans, *World Oil,* Feb. 1, pp. 32–34.

Dodson, C. R., 1960, The Petroleum Engineer's Function in Oil and Gas Financing, *Jour. Petroleum Technology* **12**(April):19–22.

Dodson, C. R., 1965, Use of Engineering Reports in the Analysis of Financial Statements for Producing Oil and Gas Companies, *Exploration and Economics of the Petroleum Industry,* vol. 3, International Oil and Gas Education Center, Southwestern Legal Foundation pp. 51-67.

Dodson, C. R., 1967, Application of the Petroleum Engineer's Report to Financing, *Jour. Petroleum Technology* **19:**187-192.

Doheny, L. C., 1941, Placer Valuation in Alaska, *Eng. and Mining Jour.* **142**(12):47-49.

Dolbear, S. H., 1952, Periodic Mine Revaluation, *Mining Cong. Jour.* **38**(12):75-77.

Dolbear, S. H., 1953, Changing Factors in Mine Valuation, *Am. Inst. Mining and Metall. Engineers Trans.* **196:**925-928.

Domitrovic, J., 1972, An Estimate of the Economic Reserves of Minerals in Yugoslavia, *Minerales* **27**(117):12-15. (In Spanish.)

Donoghue, D., 1942, Notes on Appraisals, *Am. Assoc. Petroleum Geologists Bull.* **26:**1283-1289.

Dotson, J. C., 1961, *Short Course on Computers and Computer Applications in the Mineral Industry,* vols. 1 and 2, Colorado Mines, University of Arizona, Tucson.

Douglass, E. J., 1971, How to Make the Most of a Mining Investment, *Mining Eng.* **23**(Oct.):64-67.

Dowd, P., 1975, Mine Planning and Ore Reserve Estimation with the Aid of a Digigraphic Console Display, *Canadian Inst. Mining and Metallurgy Bull.* **68**(754):39-43.

Dowd, P., and M. David, 1976, Planning from Estimates: Sensitivity of Mine Production Schedules to Estimation Methods, in *Advanced Geostatistics in the Mining Industry,* M. Guarascio, M. David, and C. Huijbregts, eds., Reidel, Dordrecht, Holland, pp. 163-183.

Drake, E., 1974, Oil Reserves and Production, in *Energy in the 1980s,* P. Kent, ed., Royal Society of London, London, pp. 47-56. (Reprinted from *Royal Soc. London Philos. Trans.* vol. 276A, issue 1261.)

Dran, J. J., and H. N. McCarl, 1974, A Critical Examination of Mineral Valuation Methods in Current Use, *Mining Eng.* **26**(July):71-75.

Dran, J. J., and H. N. McCarl, 1977, An Examination of Interest Rates and Their Effect on Valuation of Mineral Deposits, *Mining Eng.* **29**(June):44-47.

Dreschler, H. D., and J. B. Stephenson, 1977, The Effect of Inflation on the Evaluation of Mines, *Canadian Inst. Mining and Metallurgy Bull.* **70**(778):76-82.

Dunbar, C. P., and V. G. Gabriel, 1936, Evaluation of Oil Lands During Exploration, *Oil Weekly* **83**(Dec.):49-52.

Dunlap, J. B., 1960, Financing Oil and Gas Transactions, *Natl. Inst. Petroleum Landmen Proc.* **1:**53-68.

Dutt, G. N., 1970, Standardization of Prospecting Techniques for Mineral Exploration, *Indian Minerals* **24:**102-108.

Dutton, G., 1970, *The Effects of the 1969 Tax Reform Act on Petroleum Property Values, Jour. Petroleum Geology* **22:**1475-1479.

Dykstra, H., and R. L. Parsons, 1950, The Prediction of Oil Recovery by Water Flood, in *Secondary Recovery of Oil in the United States,* American Petroleum Institute, New York. pp. 160-174.

Earll, F. N., K. S. Stout, G. G. Griswald, R. I. Smith, F. H. Kelly, D. J. Emblen, W. A. Vine, and D. H. Dahlem, 1976, Handbook for Small Mining Enterprises, *Montana Bur. Mines and Geology Bull.* **99:**1-218.

Earlougher, R. C., and E. T. Guerrero, 1961, Analysis and Comparison of Five Methods to Predict Water Flood Reserves and Performance, *Mines Mag.* **51**(Jan.):11–20.

Echeverria, R. M., and M. A. Salas, 1978, Determinación de la ley de corte optima, *Minerales* **33**(141):13–21.

Eckel, E. C., 1912, American Iron Ore Reserves; Bases for Their Valuation, *Engineering Mag.*, Oct., pp. 7–15.

Eggleston, W. S., 1962, What are Petroleum Reserves?, *Jour. Petroleum Technology* **14**:719–722.

Ehrisman, W., W. Raufuss, and K. Weggen, 1975, Einsatz mathematischer Methoden bei der Exploration und Erfassung von Rutil-Seifen in Sierra Leone, in *13th International Symposium on the Application of Computers and Mathematics for Decision Making in Mineral Industries,* G. Dorstewitz, ed., Verlag Glueckauf GmbH, Essen, pp. I.V.1–I.V.13.

Eldridge, D. H., 1949, Is the Use of the Hoskold Formula Justified?, *Eng. and Mining Jour.* **150**(8):72–74.

Elfrink, E. B., C. R. Sandberg, and T. A. Pollard, 1949, A New Compressibility Correction for Natural Gasses and Its Application to Estimates of Gas-in-Place, *Am. Inst. Mining and Metall. Engineers Trans.* **186**:219–223.

Ellis, T. R., 1979, Estimation of Minimum Specifications for Economically Explorable Sandstone-type Uranium Deposits, *Colorado School Mines Quart.* **74**:1–71.

Elsing, M. J., 1932, Summary of the Cost of Mining, *Eng. and Mining Jour.* **133**:611–613.

Elsing, M. J., 1936, What is the Cost of Mine Development?, *Eng. and Mining Jour.* **137**(5):245–249.

Ely, N., 1960, Summary of Mining and Petroleum Laws of the World, *U.S. Bur. Mines Inf. Circ. 8017*, 215p.

Ely, N., and C. F. Wheatley, 1959, Mineral Titles and Tenure, in *Economics of the Mineral Industries,* American Institute of Mining and Metallurgical Engineers, E. H. Robie, ed., New York, pp. 81–129.

Engalychev, E. H. A., and V. V. Mukhin, 1973, Estimation of Petroleum Deposits According to the New (Soviet) Classification of Resources and Reserves, *Geologiya Nefti i Gaza* **3**:67–72. (In Russian.)

Ensign, C. O., and J. L. Patrick, 1968, Ore Reserve Computation and Ore Dilution Control at the White Pine Mine, White Pine, Michigan, *Canadian Inst. Mining and Metallurgy Spec. Vol.* **9**:308–318.

Epstein, B. S., 1975, Financing the Acquisition of a Going Coal Mine, *Mining Eng.* **27**(Sept.):37–39.

Evans, J. B., 1960, The Evaluation of Mining Properties—A Graphical Approach, *Canadian Inst. Mining and Metallurgy Trans.* **63**:548–554.

Evinger, H. H., and M. Muskat, 1941, Calculation of Theoretical Productivity, *Am. Inst. Mining and Metall. Engineers Trans.* **146**:126–139, 1942.

Evrard, P., and G. Schaar, 1969, Application of Statistical Methods to the Evaluation of Mining Deposits, in *2nd Technical Conference on Tin,* W. Fox, ed., International Tin Council (London), Bangkok, pp. 519–536.

Ewanchuk, H. G., 1968, Grade Control at Bethlehem Copper, *Canadian Inst. Mining and Metallurgy Spec. Vol.* **9**:302–307.

Fagerberg, D., 1952, Tax Aspects of Metal Mine Depreciation and Depletion, *Jour. Accountancy* **93**:224, 226, 228.

Falkie, T., and W. E. Porter, 1973, Economic Surface Mining of Multiple Seams, in *Application of Computer Methods in the Mineral Industry,*

M. D. G. Salamon and F. H. Lancaster, eds., South African Institute of Mining and Metallurgy, Johannesburg, pp. 177-183.

Fansett, G. R., 1917, Sampling and Estimating Gold in a Placer Deposit, *Univ. Arizona Bull. 51,* Tucson.

Fansett, G. R., 1918, Valuation of Prospects, *Univ. Arizona Bull. 68, Economic Series No. 16,* 9p.

Farkash, I., 1976, Opredeleniye pogreshnosti podscheta zapasov nefti i gaza, *Geologiya Nefti i Gaza* **6:**67-71.

Farmin, R., and G. M. Crosby, 1962, Extending Reserves in the Coeur d'Alene District, *Mining Cong. Jour.* **48**(1):23-26.

Farrell, J. H., 1939, Exploration?—Or Is It Merely Mine Evaluation?, *Eng. and Mining Jour.* **140**(7):39-41.

Fennell, J H , 1939, Ore Reserves, *Inst. Mining and Metallurgy Bull. 422,* 52p.

Fernald, H. B., 1922, Accounting for Depletion and Dividends of Mining Companies, in *American Mining Congress: Proceedings of the 25th Annual Convention, Conference on Mine Taxation, Cleveland, Ohio, 1922,* Washington, D.C., pp. 101-122.

Fernald, H. B., 1923, Inventories as Related to Federal Taxation of Mining Companies, in *American Mining Congress: Proceedings of the 26th Annual Convention, Conference on Mine Taxation, Milwaukee, Wisconsin, 1923,* Washington, D.C. pp. 32-58.

Fernald, H. B., 1928, Methods of Financing Large Mine Operations, *Am. Accountant,* Sept., pp. 44-49.

Fernald, H. B., M. E. Peloubet, and L. M. Norton, 1939, Accounting for Nonferrous Metal Mining Properties and Their Depletion, *Jour. Accountancy* **68**(Aug.):105-116.

Fiekowski, S., and A. Kaufman, 1976, Mineral Taxation, in *Economics of the Mineral Industries,* 3rd ed., W. A. Vogely, ed., American Institute of Mining and Metallurgical Engineers, New York, pp. 673-682.

Fielden, A. P., 1964, Fund Flow Analysis and Its Application to Financial Statements of Mining Companies, *Canadian Inst. Mining and Metallurgy Bull.* Nov., pp. 1133-1146.

Finlay, J. R., 1908, The Cost of Mining—General Conditions, *Eng. and Mining Jour.* **75**(16):795-800.

Finlay, J. R., 1909, The Cost of Mining, McGraw-Hill, New York, 415p. (2nd. ed., 1910; 3rd ed., 1920).

Finlay, J. R., 1911, Appraisal of Michigan Mines, Parts I-V, *Eng. and Mining Jour.* **92,** no. 11, Sept. 9, pp. 488-493 (pt. I); no. 12, Sept. 16, pp. 539-542 (pt. II); no. 13, Sept. 23, pp. 591-594 (pt. III); no. 14, Sept. 30, pp. 641-644 (pt. IV); no. 16, Oct. 14, pp. 749-752 (pt. V).

Finlay, J. R., 1913*a,* Factors in the Valuation of Iron Mines, *Am. Inst. Mining and Metall. Engineers Bull.* Mar., pp. 487-502.

Finlay, J. R., 1913*b,* Principles of Mine Valuation, *School Mines Quart.* **34:**87-95.

Finlay, J. R., 1913*c,* Valuation of Iron Mines, *Am. Inst. Mining and Metall. Engineers Trans.* **45:**282-297, 1914.

Finlay, J. R., 1914, *Western Pennsylvania Eng. Soc. Proc.* Mar., pp. 191-220.

Finlay, J. R., 1915, Value of Mining Property, *Pan American Scientific Congress Proc.* **8:**987-992.

Finlay, J. R., 1919, Nature and Uses of Capital in Mining, *Eng. and Mining Jour.* **107**(18):780-786.

Finlay, J. R., 1922, *Report of Appraisal of Mining Properties of New Mexico,* New Mexico Tax Commission, Santa Fe, 154p.

Finlay, J. R., 1931, Percentage of Depletion, *Eng. and Mining Jour.* **131:**180–181.

Finlay, J. R., 1932, The Future Value of Mineral Property, in *Mineral Economics; Lectures Under the Auspices of the Brookings Institution,* F. G. Tyron and Ec. C. Eckel, eds., McGraw-Hill, New York, 311p.

Fish, S. E., 1969, Methods and Considerations in Appraising a Coal Property, *Kentucky Geol. Survey Spec. Pub. No. 18,* ser. 10, pp. 28–31.

Fisher, C. A., 1910, Depth and Minimum Thickness of Beds as Limiting Factors in Valuation of Coal Land, *U.S. Geol. Survey Bull.* **424:**48–75.

Fisher, C. A., 1921, Effect of Revenue Act of 1918 On Methods of Valuation of Oil Lands, in *American Mining Congress: Proceedings of the 23rd Annual Convention, Conference on Mine Taxation, Denver, Colorado, 1920,* pp. 677–683.

Fiske, L. E., 1952, Determination of Depreciation Rates by a Study of Retirements, *Oil and Gas Tax Quart.* **2**(Oct.):1–18.

Fiske, L. E., 1958, *Federal Taxation of Oil and Gas Transactions,* Matthew Bender & Company, Albany.

Fitz Gerald, N. D., 1938, Optimum Rate Working Mineral Deposits, *Mining and Metallurgy* **19**(Sept.):401–403.

Fitzhugh, E. F., 1947, The Appraisal of Ore Expectancies, *Am. Inst. Mining and Metall. Engineers Trans.* **178:**143–149.

Flagg, D. H., 1951, Legal Aspects of Depreciation in Natural Resource Industries, in *Institute on Oil and Gas Law 24th Annual Proceedings,* Southwestern Legal Foundation, Dallas, pp. 424–440.

Flagg, D. H., 1956, *Oil and Gas Taxes Report,* vol. 1, no. 1, Prentice-Hall, Englewood Cliffs, New Jersey.

Fohl, W. E., 1915, Valuation of Coal Lands, *Colliery Engineer,* Sept., pp. 64–66.

Forrester, J. D., 1946, *Principles of Field and Mining Geology,* Wiley, New York, 647p.

Fox, E., 1934, Taxation of a Canadian Gold Mining Company, *Canadian Chartered Accountant,* June, pp. 24–45.

Fralich, C. E., 1931, Application of Core Drilling and Core Analysis to the Recovery of Oil, *Internat. Petroleum Technology,* April, pp. 157–167.

Frank, S. M., C. W. Wellen, and O. Lipscomb, 1956, *Oil and Gas Taxation Cases and Materials,* Prentice-Hall, Englewood Cliffs, New Jersey, 329p.

Franks, G. D., 1973, On the Estimation of Oil and Gas Resources, in *Proceedings of the 4th Symposium on the Development of Petroleum Resources of Asia and the Far East,* United Nations Mining Research Development Series, vol. 3, no. 41, pp. 76–82.

Fraser, D. C., 1973, Magnetite Ore Tonnage Estimates From an Aerial Electromagnetic Survey, *Geoexploration* **11:**97–105.

Frazer, J. Z., 1977, Manganese Nodule Reserves—An Updated Estimate, *Marine Mineralogy* **1**(1–2):103–123.

Freeman, H. A., 1955, Percentage Depletion for Oil—A Policy Issue, *Indiana Law Jour.* **30**(4):399–429.

Freeze, A. C., 1961, Use of Data Processing Machines for Calculating Ore Reserves at the Sullivan Mine, *Mining Eng.* **13:**382–389.

Freling, R. A., 1968, A Current Primer on Oil and Gas Taxation for the Petroleum Landman, *Natl. Inst. Petroleum Landmen Proc.* **9:**207–237.

Freudenberg, J. et al., 1972, Komplexe ökonomische Lagerstättenbewertung im Kupferschieferbergbau und ihre Beziehung zur Prognose und Planung, *Neue Bergbautech,* **2**(11):847–852.

Frick, C., and H. Dausch, 1932, Taschenbuch für metallurgische Probierkunde, Bewertung und Verkäufe von Erzen für Geolgen, Berg-, Hütteningenieure und Prospektoren, Stuttgart, 272p.

Frohling, E. S., and R. M. McGeorge, 1975, How Stepwise Financing Can Turn Your Prospect Into an Operating Mine, *Mining Eng.* **27**(9):30–32.

Fuda, G. F., 1971, The Role of Decision-Making Techniques in Oil and Gas Exploitation and Evaluation, *Canadian Inst. Mining and Metallurgy Spec. Vol.* **12**:130–138

Gabriel, V. G., 1937, Some Tables on Evaluation of Oil Lands, *Oil Weekly,* **87**(Oct. 25):20, 22, 26.

Gaertner, E., 1970, El desarrollo del distrito de las minas de lignito del Rhin, como ejemplo de sintesis de las decisiones de los empresarios y la investigación cientifica, *Internat. Mining Congress (6th) Proc.,* pp. 791–804. (Also issued separately in German as No. III-D.8, 13p.)

Galvin, C. O., 1943, Percentage Depletion of Oil and Gas Wells, *Texas Law Rev.* **21**(4):410–423.

Gansauge, P., 1975, Notwendigkeit und Probleme der Bestimmung eines volkswirtschaftlich zweckmässigen Vorlaufs an Lagerstättenvorräten mineralischer Rohstoffe, *Zeitschr. Angew. Geologie* **21**(1):39–45.

Garcia, A., 1925, Examining Engineer's Work Plays Vital Part in Financing of a Coal Mine Property, *Coal Age* **28**(Aug. 27):275–280.

García, H. H., 1969, Algunas consideraciones sobre el cálculo de volúmenes usando Secciones longitudinales, *Jornadas Geologicas Argentinas Actas* **2**:107–113.

Gardescu, I. I., 1940, Evaluation of Leases Subjected to Proration and Drainage, *Oil Weekly,* April 22, **97**(7):30, 32, 34, 36, 38, 40, 42.

Garrison, F. L., 1911, Decrease of Value in Ore Shoots with Depth, *Canadian Mining Inst. Quart. Jour.* **16**:63–77; *Trans.* **15**:192–209 (1912).

Gavich, I., 1967, Solving Problems of Contouring Mineralization and Concentration When Estimating Reserves by Modelling Methods, *Geol. i Razvedka, No. 1,* pp. 92–101. (In Russian.)

Gaydin, A. M., 1974, Osobennosti razvedki mestorozhdeniy pri otrabotke ikh geotekhnologicheskimi metodami, *Razved. i Okhrana Nedr* **7**:32–36.

Gaymard, R. et al., 1968, *Computer Processed Interpretation,* Middle East, American Institute of Mining and Metallurgical Engineers, Society of Petroleum Engineers, Saudi Arabia Sect., 2nd Reg. Tech. Symp., pp. 121–137.

Gebauer, A., and J. Gornig, 1967, Die Fehlereinschätzung bei der Vorratsberechnung einer Erdöllagerstätte mittels Materialbilanz, *Zeitschr. Angew. Geologie* **13**(11-12):591–594.

Gee, C. E., and J. P. Reichman, 1973, Computer Assisted Ore Reserve Calculations at Western Mining Corporation, Ltd.'s, Kambalda Nickel Operations, in *Western Australia Conference, 1973,* Australasian Inst. Mining and Metallurgy Papers (Computer Systems), pp. 355–362.

Gentry, D. W., and M. J. Hrebar, 1976, Procedure for Determining Economics of Small Underground Mines, *Colorado School of Mines, Mineral Industries Bull.,* vol. 19, no. 1, 18p.

Gentry, D. W., and M. J. Hrebar, 1977, Procedure for Feasibility Determination: Small Subsurface Mines, in *Subsurface Geology: Petroleum, Mining, Construction,* 4th ed., L. W. LeRoy, D. E. LeRoy, and J. W. Raese, eds., Colorado School of Mines, Golden, Colo., pp. 561–576.

Gibbs, T. L., 1966, A Review of the Contributions of the Symposium and Some Comments on the State's Interest in Ore Value Prediction, in *Symposium on Mathematical Statistics and Computer Applications in Ore Evaluation, Mar. 7–8, 1966, Johannesburg,* South African Institute of Mining and Metallurgy, pp. 375–378.

Gibson, T. W., 1920, Principles of Mine Taxation, *Am. Inst. Mining and Metall. Engineers Trans.* **71:**639–648.

Gignac, L., 1975, Computerized Ore Evaluation and Open Pit Design, *Minnesota Univ., Mining Symp., No. 36,* pp. 46–53.

Gilbert, R. E., 1948, New Assay Slide Rule Computes Complex Ore Values, *Eng. and Mining Jour.* **149**(6):95–98.

Gillette, J. M., 1978, Coal Reserves; Computer-Based Evaluation and Mine Planning Tool, *Coal Miner* **3**(3):23–28.

Gilmour, P., 1964, A Method of Calculating Reserves in Tabular Orebodies, *Econ. Geology* **59:**1386–1389.

Ginter, R. L., 1937, Influence of Connate Water on Estimation of Oil Reserves, *Oil and Gas Jour.,* Oct. 7, **36**(21):97–100, 105.

Ginter, R. L., 1941, Exercise on Amount of Source Bed Required to Furnish Oklahoma City Oil Pool, *Am. Assoc. Petroleum Geologists Bull.* **25:**1706–1712.

Glassmire, S. H., 1938, *Law of Oil and Gas Leases and Royalties,* Thomas Law Book Company, 2nd ed., St. Louis, 467p.

Gocht, W., 1973, Veränderungen der Bauwürdigkeitsgrenze in Zinnlager-stätten, *Deutsch. Geol. Gesell. Zeitschr.,* Part 1, **124:**101–109.

Godfrey, K. V., 1970, On the Application of Three-Dimensional Trend Analysis to Ore Reserve Estimation, in *Decision-making in Mineral Exploration III, Computer Assistance in the Management of Exploration Programs,* pt. 3, Ext. Dept., Eng. Prog., University of British Columbia, Vancouver, pp. 50–53.

Godfrey, K. V., 1972, Case study—Catface Deposit: Statistical Techniques for Evaluation, in *Decision-making IV. Cooperative Decision-making by the Geologist and Mining Engineer, Feb. 10–12, 1971,* Cent. Cont. Educ., University of British Columbia, Vancouver.

Goodner, G. E. H., 1922, Some Problems in Mine Accounting, *Eng. and Mining Jour.,* July 22, **114**(4):145–147.

Goodner, G. E. H., 1923, Discover Value, Practical Application of Provisions to Taxation of Mining Ventures, in *American Mining Congress: Proceedings of the 26th Annual Convention, Conference on Mine Taxation, Milwaukee, Wisconsin, 1923,* Washington, D.C., pp. 97–112.

Goodson, J. A., 1970, Taxes and the Petroleum Landmen—A New Look, *Natl. Inst. Petroleum Landmen Proc.* **11:**211–229.

Goodson, W. C., 1960, Development and Evaluation of Gas-condensate Reservoirs, *Petroleum Eng.,* Mar., pp. B-38-B-50.

Gorrell, H. A., C. A. S. Bulmer, and M. J. Brusset, 1972, Monetary Evaluation of Coal Properties, in *Western Canadian Coal (1st) Geological Conference Proceedings, Information Series No. 60,* Research Council of Alberta, pp. 61–71.

Gorzhevskiy, D. I., 1968, Appraisals of Outcrops of Ore Deposits in *Principles of Prospecting and Exploration for Solid Economic Minerals,* vol. 1, V. Kreyter, ed., Nedra Press, Moscow, pp. 222–236. (In Russian.)

Gotautas, V. A., 1963, Quantitative Analysis of Prospect to Determine Whether it is Drillable, *Am. Assoc. Petroleum Geologists Bull.* **47:**1794–1812.

Goulette, F. A., 1955, Depletion of Minerals, *California Soc. Cert. Public Accountants, 6th Ann. Tax Accounting Conf.,* pp. 1–30.

Govett, C. J. S., and M. H. Govett, 1974, The Concept and Measurement of Mineral Reserves and Resources, *Resources Policy* **1**(1):46–55.

Granberry, R. J., R. E. Jenkins, and D. C. Bush, 1968, Grain Density Values of Cores from Some Gulf Coast Formations and Their Importance in Formation Evaluation, in *Transactions of the Society of Professional Well Log Analysts, 9th Annual Logging Symposium,* New Orleans, pp. N1–N19.

Grant, W., 1922, Valuation of Placer Deposits, *Eng. and Mining Jour.,* Feb. 25, **113**:329–331.

Graton, L. C., 1923, Federal Taxation of Mines, *Am. Inst. Mining and Metall. Engineers Trans.* **69**:1185–1281.

Graton, L. C., 1926, *Factors for Determining Copper Mine Values,* brief to Bureau of Internal Revenue Service, Copper Producers Tax Commission.

Graupner, A., 1939, Die nutzbaren Steine und Erden des Saarlandes und ihre Verwertung, *Zeitschr. Praktische Geologie* **47**(5):85–97; **47**(6):106–118.

Grayson, C. J., 1960, *Decisions Under Uncertainty, Drilling Decisions by Oil and Gas Operators,* Div. Res., Graduate School of Business Administration, Harvard University, Boston, pp. 151–165.

Grayson, C. J., 1962a, Decisions Under Uncertainty, in *Symposium on Petroleum Economics and Valuation,* American Institute of Mining and Metallurgical Engineers, Society of Petroleum Engineers, Dallas, pp. 97–101.

Grayson, C. J., 1962b, Bayesian Analysis, A New Approach to Statistical Decision-making, *Jour. Petroleum Technology* **14**:603–607.

Grayson, C. J., 1964, Computer Applications in Oil Exploration Decisions, *Stanford Univ. Pubs. Geol. Sci.* **9**(1):89–101.

Green, D. W., D. F. Merriam, B. Nadan and G. W. Rosenwald, 1971, Some Recent Advances in Methods of Oil and Gas Reserve Estimates, *World Petroleum Congress Proc. No. 8* **3**:177–188.

Gregory, P. P., 1935, Estimation of Petroleum Reserves in Prorated Limestone Fields, *Mining and Metallurgy* **16**(346):421–423; *Oil Weekly,* Oct. 21, **79**:33–34, 36.

Griffith, W., 1913, Assessing and Taxing Coal in the Ground, *Colliery Engineer* **33**:669–670.

Griffiths, J. C., 1968, Keynote Address: An Overview of Decision-making Methods in Exploration and Evaluation of Mineral Properties, in *Proceedings of a Symposium on Decision-making in Mineral Exploration,* Extension Department of the University of British Columbia, Vancouver, Jan. 26, 42p.; *Western Miner* **41**:5–9.

Griffiths, J. C., 1978, Mineral Resource Assessment Using the Unit Regional Value Concept, *Internat. Assoc. for Mathematical Geology Jour.* **10**:441–472.

Griffiths, J. C., and D. A. Singer, 1971, Unit Regional Value of Nonrenewable Natural Resources as a Measure of Potential for Development of Large Regions, *Geol. Soc. Australia Spec. Pub. 3,* pp. 227–238.

Grigorovich, M. B., 1968, Osnovy otsenki mestorozhdeniy oblitsovochnogo kamnya, *Razved. i Okhrana Nedr* **12**:4–8.

Grimes, J. A., and W. H. Craigue, 1928, *Principles of Valuation,* Prentice-Hall, New York.

Grosjean, P. V., 1953, L'évaluation mathématique des gisements détritiques, *Inst. Royal Colonial Belge Sect. Sci. Tech. Mém., Coll. in-8° **8**(3):1–156.

Gross, G. A., 1965, Geology of Iron Deposits in Canada. General Geology and Evaluation of Iron Deposits, *Canada Geol. Survey Econ. Geology Rept.* **1:**1–181.

Gross, H., 1938, Decline Curve Analysis Indicated the Reserves and the Profits, *Oil and Gas Jour.,* Sept. 15, **37:**55, 56, 58, 61.

Grundy, W. C., and R. J. Meehan, 1963, Estimation of Uranium Ore Reserves by Statistical Methods and a Digital Computer, *New Mexico Bur. Mines and Mineral Resources Mem. 15,* pp. 234–243.

Grunsky, C. E., and C. E. Grunsky, Jr., 1917, *Valuation, Depreciation and the Rate-base,* Wiley, New York, 387p. (2nd ed., 1927.)

Gruy, H., and J. Crichton, 1947, A Critical Review of Methods Used in the Estimation of Natural-gas Reserves, *Oil and Gas Jour.,* Oct. 25, **46**(25):88–89, 116.

Guarascio, M., 1974, Valutazione dei giacimenti minerari—l'approccio geostatistico, *Industria Mineraria* **25**(1):3–14; **25**(2):109–117.

Guarascio, M., 1976, Improving the Uranium Deposits Estimations (The Novazza Case), in *Advanced Geostatistics in the Mining Industry,* M. Guarascio, M. David, and C. Huijbregts, eds., Reidel, Dordrecht, Holland, pp. 351–367.

Guarascio, M., and J. Deraisme, 1974, Valutazione geostatistica di un giaciamento di zinco delle Alpi Orientali, *Industria Mineraria* **25:**249–257.

Guarascio, M., and G. Raspa, 1974, Valuation and Production Optimization of a Metal Mine, *Internat. Symp. Appl. Comput. Math. Miner. Ind. Proc. No. 12,* **2:**F50–F64.

Guarascio, M., and A. Turchi, 1977, Exploration Data Management and Evaluation Techniques for Uranium Mining Projects, in *Application of Computer Methods in the Mineral Industry,* R. V. Ramani, ed., American Institute of Mining and Metallurgical Engineers, New York, pp. 451–464.

Guignon, F. A., 1916, Valuation of Bedded Mineral Land, *Eng. and Mining Jour.* **102**(23):969–971.

Gunther, C. G., 1912, *The Examination of Prospects: A Mining Geology,* McGraw-Hill, New York, 222p.

Guretskiy, V. J., and E. E. Asadulin, 1974, Opredeleniy koeffitsiyenta rudonostnosti po razvedochnym dannym, Vyssh. Ucheb. Zavedeniy Izv., *Geologiya i Razved.* **10:**66–69.

Gy, P., 1971, L'echantillonnage des minerais en vrac. Theorie generale, erreurs operatoires, complements, *France Bur. Recherches Géol. et Minières Mém. 67,* 470p.

Haas, A., and C. Jousselin, 1976, Geostatistics in Petroleum Industry, in *Advanced Geostatistics in the Mining Industry,* M. Guarascio, M. David, and C. Huijbregts, eds., Reidel, Dordrecht, Holland, pp. 333–347.

Haas, A., and M. Mollier, 1974, Un aspect du calcul d'erreur sur les reserves en place d'un gisement; l'influence du nombre et de la disposition spatiale des puits, *Fr. Pet. Rev.* **29:**507–527.

Haddock, M. H., 1926, *The Location of Mineral Fields; Modern Procedure in the Investigation of Mineral Areas, Etc.,* Lockwood, London, 302p.

Hager, D., 1916, Valuation of Oil Properties, *Eng. and Mining Jour.* **101:**930–932.

Hager, D., 1921, Elements of Valuation—Buying Oil Properties, in *Oil-Field Practice,* McGraw-Hill, New York, 310p.

Hajdasiński, M., 1977, The Comparative Analysis of the Estimation of the Economic Effectiveness of Mining Development Projects for the Continuous and Discrete Representation of the Value Changes in Time, in *Application of Computer Methods in the Mineral Industry,* R. V.

Ramani, ed., American Institute of Mining and Metallurgical Engineers, New York, pp. 946–983.

Hall, J. C., 1966, The Valuation of Ore Reserves with the Aid of Computer for the Mines of the Union Corporation Group, in *Symposium on Mathematical Statistics and Computer Applications in Ore Evaluation, March 7-8, 1966, Johannesburg,* South African Inst. Mining and Metallurgy, pp. 229–305.

Hall, R. G., 1963, Payout and Profitability in Deep Well Drilling, *Exploration and Economics of the Petroleum Industry* **1:**83–127.

Hall, T. A., 1939, Appraisal of Oil Properties, *Oil Weekly* **93:**38, 40, 42, 44, 46.

Hallam, C. D., and J. B. Hawthorne, 1966, Notes on Valuation of Diamond Mines, DeBeers Conservation Mines Limited, Kimberley (South Africa), in *Symposium on Mathematical Statistics and Computer Applications in Ore Evaluation, Mar. 7-8, 1966, Johannesburg,* South African Inst. Mining and Metallurgy, pp. 327–374.

Halls, J. L., D. P. Bellum, and C. K. Lewis, 1969, Determination of Optimum Ore Reserves and Plant Size by Incremental Financial Analysis, *Inst. Mining and Metallurgy Trans.* **78:**A-20–A-26.

Hamilton, F. C., 1933, Valuation of Gas Leases, *Natural Gas* **14**(Oct):9–10, 23, 38–39.

Hamilton, O. R., 1923, Valuation of Metal Mines, *Mining and Metallurgy* **4:**568–571.

Hammer, S. I., 1944, Estimating Ore Masses in Gravity Prospecting, *Geophysics* **10:**51–62.

Hammerlindl, D. J., 1972, *Predicting Gas Reserves in Abnormally Pressured Reservoirs,* Soc. Pet. Engineers American Institute of Mining and Metallurgical Engineers, Ann. Meet. 1971, preprints of papers, pap. SPE 3479.

Hammes, J. K., 1976, Financial Considerations in Evaluating Newly Discovered Coal Deposits and Ventures, in *Coal Exploration,* W. L. G. Muir ed., Miller Freeman, San Francisco, pp. 586–607.

Hand, A. H., 1940, Oil Accounting, in *Elements of the Petroleum Industry,* E. DeGolyer, ed., American Institute of Mining and Metallurgical Engineers, New York, pp. 406–429.

Handelsman, S. D., R. V. Longe, and R. Willie, 1975, A Practical, Operating, Management Information System, in *13th International Symposium on the Application of Computers and Mathematics for Decision Making in Mineral Industries,* G. Dorstewitz, ed., Verlag Glueckauf GmbH, Essen, pp. G.III.1-G.III.14.

Hanisch, K. H., 1978, Genauigkeitsfragen bei der Vorratsberechnung von Lagerstätten, *Neue Bergbautechnik* **8:**618–620.

Hansen, M., J. M. Botbol, O. R. Eckstrand, G. Gaál, M. Maignan, T. Pantazis, and R. Sindling-Larsen, 1978, Workshop on Deposit Modeling, *Internat. Assoc. for Mathematical Geology Jour.* **10:**519–531.

Harbaugh, J. W., and A. Prelat, 1973, Research in Oil Exploration Decision-making Estimation of Wildcat Well Outcome Probabilities, in *Application of Computer Methods in the Mineral Industry,* M. D. G. Salamon and F. H. Lancaster, eds., South African Institute of Mining and Metallurgy, Johannesburg, pp. 177–183.

Harbaugh, J. W., J. H. Doveton, and J. C. Davis, 1977, *Probability Methods in Oil Exploration,* Wiley, New York, 269p.

Hardin, G. C., 1959, How to Appraise Wildcat Prospects, *World Oil,* Mar., pp. 119-120; Apr., pp. 138-141.

Hardin, G. C., 1966, Economic Evaluation of Exploratory Prospects, *Exploration and Economics of the Petroleum Industry* **4:**43-77.

Hardin, G. C., and K. Mygdal, 1968, Geologic Success and Economic Failure, *Am. Assoc. Petroleum Geologists Bull.* **52:**2079-2091.

Harding, J. E., 1921, Calculation of Ore Tonnage and Grade from Drill-Hole Samples, *Am. Inst. Mining and Metall. Engineers Trans.* **66:**117-126, 1922.

Harding, J. E., 1923, How to Calculate Tonnage and Grade of an Orebody, *Eng. and Mining Jour.* **116:**445-448.

Hardwicke, R. E., 1955, Purchase of Producing Oil or Gas Properties by Use of a Production Payment, *Texas Law Rev.* **33:**848-854.

Hargreaves, D., 1972, Computer Assistance in Planning Open Pits, *Mining Mag.,* Sept., pp. 249, 251.

Harnett, R., 1970, How to Operate an Oilfield: An Outline for the Inexperienced, *Earth Sci. Bull.* **3**(4):13-18.

Harris, D. P., 1966, A Probability Model of Mineral Wealth, *Am. Inst. Mining and Metall. Engineers Trans.* **235:**199-216.

Harris, D. P., 1967, Operations Research and Regional Mineral Exploration, *Am. Inst. Mining and Metall. Engineers Trans.* **238:**450-459.

Harris, D. P., 1969a, Alaska's Base and Precious Metal's Resources—A Probabilistic Regional Appraisal, *Colorado School Mines Quart.* **64**(3):295-328.

Harris, D. P., 1969b, Quantitative Methods, Computers, and Reconnaissance Geology in the Appraisal of Mineral Potential for Discrete Areas, in *A Decade in Digital Computing in the Mineral Industry,* A. Weiss, ed., American Institute of Mining and Metallurgical Engineers, New York, pp. 83-118.

Harris, D. P., 1970, Problems of Mineral Data and the Estimation of Mineral Endowment, *Earth and Mineral Sci.* **39**(7):54-55.

Harris, D. P., 1973, A Subjective Probability Appraisal of Metal Endowment of Northern Sonora, Mexico, *Econ. Geology* **68:**222-242.

Harris, D. P., and D. Euresty, 1969, A Preliminary Model for the Economic Appraisal of Regional Resources and Exploration Based upon Geostatistical Analysis and Computer Simulation, *Colorado School Mines Quart.* **64:**71-98.

Harris, D. P., A. J. Freyman, and G. S. Barry, 1971, A Mineral Resource Appraisal of the Canadian Northwest Using Subjective Probabilities and Geological Opinion, *Canadian Inst. Mining and Metallurgy Spec. Vol.* **12:**100-116.

Harrison, H. L. H., 1954, Examination, Boring and Valuation of Alluvial and Kindred Ore Deposits, 2nd ed., Mining Publishing Ltd, London, 319p.

Hart, R. C., and D. Sprague, 1968, Methods of Calculating Ore Reserves in the Elliot Lake Camp, *Canadian Inst. Mining and Metallurgy Spec. Vol.* **9:**251-260.

Hartman, J. W., 1929, Appraisal of Oil Properties for Assessment Purposes; Los Angeles County Method, *Natl. Tax Assoc. Bull.,* Oct., pp. 18-21.

Hartman, R. J., and G. C. Varma, 1966, A Three Dimensional Optimum Pit Program and a Basis for a Mining Engineering System, in 6th Ann. Internat. Symp. on Computers and Operations Research, vol. 3,

Pennsylvania State University, *Mineral Industries Exper. Station Spec. Pub. 2-65*, pp. 001–0035.

Hatch, A. L., 1966, Factors Affecting Ore Grade Values, *Mining Eng.* **18**(1):72–75.

Havlena, D., and A. S. Odeh, 1970, The Material Balance as an Equation of a Straight Line, *Am. Inst. Mining and Metall. Engineers Petroleum Trans. Reprint Series No. 3*, 2nd ed., pp. 66–70.

Hawkins, L. D., 1952, Provisions for the Replacement of Wasting Assets, *Accountancy* **63**:92–95.

Hawkins, M. F., 1955, Material Balances in Expansion Type Reservoirs Above the Bubble Point, *Jour. Petroleum Technology* **7**(Oct.):49–55.

Haycocks, C., D. P. Vossler, and M. L. Rahrer, 1973, Reserv-Coal Program, in *Computer Applications in Underground Mining Systems,* section 2, Division of Mining Engineering, Virginia Polytechnic Institute and State University, Blacksburg, 84p.

Hayes, S. S., 1866, Petroleum as a Source of National Revenue, *U.S. Revenue Commission Spec. Rept. 7,* Treasury Department, Washington, D.C., 52p.

Hazen, S. W., 1956, Utilizing the Techniques of Statistical Analysis in Computing Reserves and Grade of Ore, *Univ. Missouri School Mines and Metall. Bull. Tech. Series No. 92*, pp. 3–9.

Hazen, S. W., 1958, A Comparative Study of Statistical Analysis and Other Methods of Computing Ore Reserves, *U.S. Bur. Mines Rept. Inv. 5375*, 188p.

Hazen, S. W., 1962, Using Techniques of Statistical Analysis to Plan Sampling Programs, in *Internat. Symposium on Mining Research Proc., Univ. Missouri, 1961*, vol. 1, Pergamon Press, pp. 441–472.

Hazen, S. W., 1964, Statistical Analysis of Some Sample and Assay Data from Bedded Deposits of the Phosphoria Formation in Idaho, *U.S. Bur. Mines Rept. Inv. 6401*, 29p.

Hazen, S. W., 1967a, Ore Reserve Calculations, in *A Symposium on Industrial Mineral Exploration and Development, April 5–7, 1967, Spec. Distribution Pub. 34*, E. E. Angino and R. G. Hardey, eds., University of Kansas, Lawrence, pp. 138–151.

Hazen, S. W., 1967b, Assigning an Area of Influence for an Assay Obtained in Mine Sampling, *U.S. Bur. Mines Rept. Inv. 6955*, 75p.

Hazen, S. W., 1967c, Some Statistical Techniques for Analyzing Mine and Mineral Deposit Sample and Assay Data, *U.S. Bur. Mines Bull. 621*, 223p.

Hazen, S. W., 1968, Ore Reserve Calculations, *Canadian Inst. Mining and Metallurgy Spec. Vol.* **9**:11–32.

Hazen, S. W., 1972, Ore Grade Prediction Models Based on Statistical Methods, in *Decision-making IV, Cooperative Decision-making by the Geologist and Mining Engineer, February 10–12, 1971*, Center of Continuing Education, University of British Columbia, Vancouver, 51p.

Hazen, S. W., and R. Berkenkotter, 1962, An Experimental Mine-Sampling Project Designed for Statistical Analysis, *U.S. Bur. Mines Rept. Inv. 6019*, 111p.

Hazen, S. W., and G. W. Gladfelter, 1964, Using Unequal Sample Interval Lengths and Weighted Averages in Estimating Grade of Ore for Bedded Deposits, *U.S. Bur. Mines Rept. Inv. 6406*, 23p.

Hazen, S. W., and W. L. Meyer, 1966a, Using Probability Models as a Basis for Making Decisions During Mineral Deposit Exploration, *U.S. Bur. Mines Rept. Inv. 6778*, 83p.

Hazen, S. W., and W. L. Meyer, 1966b, Investigation of Correlation Between Assay Values and Unequal Sample Interval Lengths, *U.S. Bur. Mines Rept. Inv. 6867*, 46p.

Heath, K. C. G., and G. Kalcov, 1971, Graphical Valuation Methods for Use in Prospecting and Exploration, *Inst. Mining and Metallurgy Trans.* **80:**A45–A50.

Hedberg, H. D., 1937, Evaluation of Petroleum in Oil Sands by its Index of Refraction, *Am. Assoc. Petroleum Geologists Bull.* **21:**1464–1476.

Heim, R. C., 1968, A Comparison Between Ore Reserve Calculations and Production at North Coldstream Mines Limited, *Canadian Inst. Mining and Metallurgy Spec. Vol.* **9:**172–176.

Heinemann, Z., 1971, Distribution of Common Costs in the Economic Evaluation of Blocks of Mineral Deposits, *Földtani Kutatás* **14**(4):50–52. (In Hungarian.)

Heithecker, R., 1937, Estimate of Natural-Gas Reserves from the Layton, Oolitic and Oswego-Prue Horizons in the Oklahoma City fields, *U.S. Bur. Mines Rept. Inv. 3338*, 35p.

Hellman, F., 1897–1898. Determination of the Present Value of a Mine on the Rand, *Inst. Mining and Metallurgy Trans.* **6:**229.

Hemingway, R., 1971, *The Law of Oil and Gas*, West Publishing Company, St. Paul Minnesota, 486p.

Henriques, L. N., and B. W. Mackenzie, 1977, A Cost-Benefit Analysis of a Custom Mill for Small Copper Mines in Northern Chile, in *Application of Computer Methods in the Mineral Industry*, R. V. Ramani, ed., American Institute of Mining and Metallurgical Engineers, New York, pp. 1035–1057.

Henry, P. W., 1916, Depreciation as Applied to Oil Properties, *Am. Inst. Mining and Metall. Engineers Trans.* **51:**560–570.

Herald, F. A., 1933, Evaluation of Foreign Producing Oil Properties Requires Consideration of Many Factors, *Oil and Gas Jour.*, Jan., **26:**12–13.

Herbst, F., 1973, Aus der Bewertungspraxis von Bergwerken, *Erzmetall.* **26**(9):447–454.

Herbst, R. L., 1974, Minnesota's Mineral Policy, *Univ. Minnesota Mining Symposium No. 35*, pp. 93–97.

Herdendorf, C., and L. Braidech, 1970, A Study of the Sand and Gravel Deposits of the Maumee River Estuary, Ohio, *Michigan Geol. Survey Misc. 1*, pp. 103–116.

Herold, S. C., 1930, Mechanics of a California Production Curve, *Am. Inst. Mining and Metall. Engineers Trans.* **86:**279–292.

Hershey, R. E., 1972, Tennessee's Proposed Rules and Regulations, *Interstate Oil Compact Comm. Comm. Bull.* **14**(2):12–14.

Hertz, D. B., 1964, Risk Analysis in Capital Investment, *Harvard Business Rev.*, Jan.–Feb., pp. 145–156.

Herzig, C. S., 1914, *Mine Sampling and Valuing*, Mining and Scientific Press, San Francisco, 163p.

Hesse, A. W., 1930, *The Principles of Coal Property Valuation*, Wiley, New York, 183p.

Hester, B. W., 1970, Geology and Evaluation of Placer Gold Deposits in the Klondike Area, Yukon Territory, *Inst. Mining and Metallurgy Trans.* **79:**60–67.

Hewlett, R. F., 1961*a*, Calculating Ore Reserves Using a Digital Computer, *Mining Eng.* **13**(1):37–42.

Hewlett, R. F., 1961*b*, Computers Have Application to Mining Problems, *Mining World* **23**(7):32–35.

Hewlett, R. F., 1962, Computing Ore Reserves by the Polygonal Method Using a Medium-sized Digital Computer, *U.S. Bur. Mines Rept. Inv. 5952,* 31p.

Hewlett, R. F., 1963*a*, Computing Ore Reserves by the Triangular Method Using a Medium-sized Digital Computer, *U.S. Bur. Mines Rept. Inv. 6176,* 30p.

Hewlett, R. F., 1963*b*, A Basic Computer Program for Computing Grade and Tonnage of Ore Using Statistical and Polygonal Methods, *U.S. Bur. Mines Rept. Inv. 6292,* 20p.

Hewlett, R. F., 1964*a*, Simulating Mineral Deposits Using Monte Carlo Techniques and Mathematical Models, *U.S. Bur. Mines Rept. Inv. 6493,* 27p.

Hewlett, R. F., 1964*b*, Polynomial Surface Fitting Using Sample Data from an Underground Copper Deposit, *U.S. Bur. Mines Rept. Inv. 6522,* 27p.

Hewlett, R. F., 1964*c*, Application of Simulation in Evaluating Low-grade Mineral Deposits, *U.S. Bur. Mines Rept. Inv. 6501,* 62p.

Hewlett, R. F., 1967, Mineral Deposit Evaluation Using Mathematical Models and a Digital Computer, in *Computer Short Course and Symposium on Mathematical Techniques and Computer Applications in Mining and Exploration, College of Mines,* vol. 1K, University of Arizona, Tucson, pp. K1-1–K1-55.

Hiatt, J. K., 1969, Mineralization Pattern Prediction and Ore Deposit Evaluation, *Colorado School Mines Quart.* **64**(3):99–106.

Hiessleitner, G., 1937, Zur Frage der "wahrscheinlichen" und "möglichen" Erzvorräte in der praktischen Lagerstättenbeurteilung, *Metall. und Erz.* **34**(7):157–164.

Higgins, R. V., and H. J. Lechtenberg, 1971, Production Decline Curves Using Data from California Oilfields, *U.S. Bur. Mines Rept. Inv. 7547,* 28p.

Hill, G. A. et al., 1961, Reducing Oil-Finding Costs by Use of Hydrodynamic Evaluations, in *Economics of Petroleum Exploration, Development and Property Evaluation,* Prentice-Hall, Englewood Cliffs, New Jersey, pp. 38–69.

Hill, M. L., 1975, The 3 R's of Petroleum Exploration: Resources, Risks and Rewards, *Australasian Oil and Gas Rev.* **21**(9):26–30.

Hirakawa, S., H. Isobe, and S. Onoe, 1971, A Probabilistic Method for the Estimation of Oil Reserves, *Jour. Jap. Assoc. Pet. Tech. (Tokyo)* **36**(6):350–356. (In Japanese.)

Hodgson, E. C., and W. V. Beard, 1966, Summary Review of Federal Taxation and Legislation Affecting the Canadian Mineral Industry, *Dept. Mines and Tech. Surv., Mineral Inf. Bull. M. R. 82,* Ottawa, Canada, 30p. (Revises M. R. 42).

Hodgson, W. A., 1971, Statistics in Mineral Exploration, *Univ. Rhodesia Inst. Mining Research Rept. 8,* 78p.

Hoeflinger, J. P., and L. E. Bengal, 1975, Methods Used to Compile Estimates of Coal Reserves, *Illinois Water Survey, State Geol. Survey Coop. Research Rept. 4,* pp. 76–79.

Holloway, G. T., 1918, Valuation of Manganese Ores, *Eng. and Mining Jour.,* June 29, pp. 1163–1165.

Holmes, C. B., 1920, Metal-Mine Accounting, *U.S. Bur. Mines Tech. Pap. 250,* 63p.

Holmes, R. T., 1954, Percentage Depletion for Virginia Industry, *Virginia Accountant,* April, **7:**27–33.

Hood, K. K., 1920, Curves for Ore-Valuation, *Mining and Scientific Press,* Aug. 21, **121:**270–272.

Hoover, H. C., 1904*a,* The Economic Ratio of Treatment Capacity to Ore Reserves, *Eng. and Mining Jour.* **77:**475–476, 632–633, 712–713.

Hoover, H. C., 1904*b,* The Valuation of Gold Mines, *Eng. and Mining Jour.* **77:**801–804.

Hoover, H. C., 1909, *Principles of Mining; Valuation, Organization and Administration: Copper, Gold, Lead, Silver, Tin, Zinc,* McGraw-Hill, New York, 206p.

Hoover, T. J., 1933, *The Economics of Mining,* Stanford University Press, 547p. (2nd ed., 1948.)

Horn, C. R., 1942, Application of Core Analysis in the Estimation of Oil Reserves, *Mines Mag.* **32**(10):525–526.

Horner, W. L., 1936, Core Analysis as a Control to Well Completion, *Oil Weekly,* June 1, **81**(12):31–36.

Horwood, C. B., and M. Park, 1908, Development, Sampling and Ore-Valuation of Gold Mines, *Am. Inst. Mining and Metall. Engineers Trans.* **39:**685–694.

Hosking, J. A., 1969, The Applications and Limitations of Computers in the Evaluation of Hypothermal Tin Deposits, in *A Decade in Digital Computing in the Mineral Industry,* A. Weiss, ed., American Institute of Mining and Metallurgical Engineers, New York, pp. 163–178.

Hoskold, H. D., 1877, *The Engineer's Valuing Assistant,* Longmans, Green & Co., London. (2nd ed., 185p., 1905.)

Hotchkiss, W. O., and R. D. Parks, 1936, Total Profits vs. Present Value in Mining, *U.S. Bur. Mines Tech. Pap. 708,* 9p.

Houchin, J. M., 1958, Management's Use of Petroleum Engineering Evaluations, *Jour. Petroleum Technology,* July, **10:**11–12.

Houghton, J. L., 1978, *Miller's Oil and Gas Federal Income Taxation,* Commerce Clearing House, Inc., Chicago, 687p.

Houpeurt, A., and J. P. Lacroix, 1952, Estimation de reserves des gisements de gas naturel soumis ou non a la poussee de eaux, *Convegno Naz. Metano e Petrolio, 7th, Atti,* **1:**103–116.

Howington, K. D., 1939, How Much Does it Cost? Simple but Comprehensive Accounting System Requires Small Staff, Etc. *Rock Products,* Sept., pp. 23–25, 38.

Hradek, J., 1972, Calculation of Reserves by the Isolines Method Using an Automatic Computer, *Geologicky Pruzkum* **14**(4):103–107. (In Czech.)

Huckaba, W. A., and W. G. Meyer, 1958, *Common Fallacies in Oil and Gas Reserve Estimates,* Meyer & Associates, Dallas.

Hügel, H., 1942, Druck- und Temperaturmessung in Erdölbohrungen, Probenahme und Untersuchung von Öl und Gas von Bohrlochboden und Auswertung der Ergebnisse für die Ausbeutungsplanung, *Öl und Kohle* **38:**919–937.

Hügel, H., M. Stephenson, and M. I. Predoescu, 1943, Eine Methode zur Berechung des Ölvolumens einer Lagerstätte in einer zusammenhängenden tektonischen Einheit, *Öl und Khole* **39:**301–308.

Hughes, R. V., 1978, *Oil Property Valuation,* Krieger, Huntington, New York, 331p.

Huijbregts, C., 1976, Selection and Grad-Tonnage Relationships, in *Advanced Geostatistics in the Mining Industry,* M. Guarascio, M. David, and C. Huijbregts, eds., Reidel, Dordrecht, Holland, pp. 113-135.

Huijbregts, C., and G. Matheron, 1971, Universal Kriging (An Optimal Method for Estimation and Contouring in Trend Surface Analysis), *Canadian Inst. Mining and Metallurgy Spec. Vol.* **12:**159-169.

Hulsey, B. T. H., 1962, Philosophy Behind Trading Oil Properties, *Jour. Petroleum Technology* **14:**727-728.

Humphries, H. G., 1927, Payout Status of Oil and Gas Producing Properties, *Am. Petroleum Inst. Bull.,* Jan. 31, pp. 231-235.

Huntington, R. L., 1936, Estimation of Oil and Gas Reserves, *Oil and Gas Jour.* Oct. 15, **35:**124-127; *Oil Weekly,* Oct. 12, **83**(5):25-28.

Huntley, L. G., 1913, Possible Causes of the Decline of Oil Wells, *U.S. Bur. Mines Tech. Pap. 51,* 32p.

Huntley, L. G., and S. Huntley, 1921, Mexican Oil Field: Survey of Producing Areas, Known Reserves, and Geological Factors Which Point to Important Development and Cooperation of Producing Companies, *Mining and Metallurgy,* no. 177, Sept., pp. 27-32.

Huppler, J. D., 1974, Scheduling Gas Field Production for Maximum Profit, *Am. Inst. Mining and Metall. Engineers Trans.* Part 2, **257:**279-294.

Hurst, W., 1958, The Simplification of the Material Balance Formulas by the La Place Transformation, *Am. Inst. Mining and Metall. Engineers Trans.* **213:**292-303.

Hutton, G. H., 1921, Value of Placer Deposits, *Mining and Scientific Press* **123:**365-368.

Hymas, K. I., 1968, A Note on Sampling at the Questa Molybdenite Mine, *Canadian Inst. Mining and Metallurgy Spec. Vol.* **9:**319-320.

Ichisugi, N. et al., 1974, An Integrated Computer Application System for Mine Evaluation, *Mining Geology,* Part 2, **24**(124):137-148. (In Japanese.)

Ickes, E. L., 1936, Estimation of Probable Value of Wildcat Land, *Am. Assoc. Petroleum Geologists Bull.* **20:**1005-1018.

Imai, S., and S. Itho, 1971, Some Techniques for the Determination of Effective Drill Spacing, *Canadian Inst. Mining and Metallurgy Spec. Vol.* **12:**199-208.

Ingalls, W. R., 1922, Taxation on Mines by States, in *American Mining Congress: Proceedings of the 25th Annual Convention, Conference on Mine Taxation, Cleveland, Ohio, 1922,* Washington, D.C. pp. 29-44.

Ingham, W. I., 1929, Petroleum Valuation, *Colorado School Mines Mag.*

Institute of Internal Auditors, 1950, Coal, Iron and Other Mining, in *Internal Auditing in Industry,* pp. 9-24.

Internal Revenue Service, (issued annually), *Depreciation, Investment, Credit, Amortization, Depletion,* United States Treasure Department, Doc. No. 5050.

Ireton, G., 1960, How to Finance Oil and Gas Production, *Oil and Gas Jour.,* May 9, pp. 165-168.

Ishayev, U. G., and M. Ya. Rudkevich, 1978, Metodika otsenki perspektivnykh zapasov nefti (na primere Sredneobskoy neftegasonosnoy oblasti), *Geologiya Nefti i Gaza,* no. 8, pp. 28-31.

Ivanov, A. A., 1953, *Osnovy geologii i metodika poiskov, razvedki i otsenki mestorozhdenii mineral'nykh solei,* Gosgeolizdat, Moscow, 204p.

Ivanov, G. V., and L. K. Kuznetsova, 1968, Velichina progreshnostey podscheta zapasov, osnovnoy kriteriy razvedki, *Akad. Nauk SSSR Sibirsk. Otdeleniye Geologiya i Geofizika,* no. 7, pp. 67–71.

Jackson, C. F., and J. B. Knaebel, 1932, Sampling and Estimation of Ore Deposits, *U.S. Bur. Mines Bull. 356,* 155p.

Jacoby, R. H., and V. Berry, 1958, A Method for Predicting Depletion Performance of a Reservoir Producing Volatile Crude Oil, *Am. Inst. Mining and Metall. Engineers Trans.* **210:**27–33.

Jacoby, R. H., R. C. Koeller, and V. Berry, 1959, Effect of Composition and Temperature on Phase Behavior and Depletion Performance of Rich Gas-Condensate Systems, *Jour. Petroleum Technology,* July, **11:**58–63.

Janin, C., 1913, *Mining Engineers' Examination and Report Book,* 2 vols. in 1, Mining and Scientific Press, San Francisco, California.

Jarpa, S. G., 1977, Capital Investment and Operating Cost Estimation in Open Pit Mining, in *Application of Computer Methods in the Mineral Industry,* R. V. Ramani, ed., American Institute of Mining and Metallurgical Engineers, New York, pp. 920–931.

Jeal, E. F., 1956, Mining Accounting and Taxation in the Union of South Africa, *Accountant* **135:**4–5.

Jeffries, F., 1967, Reservoir Volume Calculation with a Well Data System, in *Transactions of the Society of Professional Well Log Analysts, 8th Annual Logging Symposium,* Denver, 17p.

Jeffries, F., and F. E. Reichl, 1968, Imperial Oil Using Wireline Logs for Computer Evaluation of Oil Reserves, *Canadian Petroleum* **9**(12):32–37.

Jerrett, H. D., 1938, Evaluation of Water, and of Timber and Mineral Land, in *Theory of Real Property Valuation,* H. D. Jerrett, Sacramento, California, 309p.

Jewett, G. A., 1956, Sampling Design and Grade Estimation of Mineral Deposits, *Canadian Inst. Mining and Metallurgy Bull.,* March, **49**(527):136–145.

Jirasek, J., 1970, The Objective of Economic Evaluation of Long-range Studies of Surface Lignite Mines on Computers, *Uhli* **12**(9):331–334 (In Czech.)

Johnson, B. H., 1954, *Maximum Tax Benefit from Percentage Depletion,* Tulane Tax Institute, Tulane, Louisiana, pp. 373–392.

Johnson, E. E., and H. J. Bennett, 1968, An Engineering and Economic Study of a Gold Mining Operation, *U.S. Bur. Mines Inf. Circ. 8374,* 53p.

Johnson, P. W., and F. A. Peters, 1969, A Computer Program for Calculating Capital and Operating Costs, *U.S. Bur. Mines Inf. Circ. 8426,* 110p.

Johnson, R. E., 1978, Preliminary Mine Planning at the Minnamax Project, in *Productivity in Lake Superior District Mining,* L. K. Graven, ed., University of Minnesota Mining Symposium No. 39, pp. 20.1–20.22.

Johnson, R. H., 1915–1916, Valuation of Oil Properties, *Petroleum Age* **2**(12):4–8 (1915); **3**(1):35–39; **3**(2):37–40 (1916).

Johnson, R. H., 1916, The Valuation of Oil Properties, in *The American Petroleum Industry,* R. F. Bacon and W. A. Hamor, eds., McGraw-Hill, New York, pp. 345–367.

Johnson, R. H., 1919, Decline Curve Methods, *Am. Assoc. Petroleum Geologists Bull.* **3:**421–426.

Johnson, R. H., 1921, Variation in Decline Curves of Various Pools, *Am. Inst. Mining and Metall. Engineers Trans.* **65:**365–373.

Johnson, R. H., 1922, Appraisal of Oil and Gas Properties, *Eng. Soc. West. Pennsylvania,* March, pp. 35–45.

Johnson, R. H., and A. L. Bollens, 1927, Crude Petroleum—Loss Ratio

Method of Extrapolating Oil Well Decline Curves, *Am. Inst. Mining and Metall. Engineers Trans.* **77**:771–778.

Johnson, R. H., and A. W. Foster, 1920, Barrel Costs Versus Well-Day Cost, *Am. Assoc. Petroleum Geologists Bull.* **4**:299–301.

Johnson, R. H., and L. C. Morgan, 1926, A Critical Examination of the Equal Pound Loss Method and of Estimating Gas Reserves, *Am. Assoc. Petroleum Geologists Bull.* **10**:901–904.

Johnson, R. H., and P. Ruedemann, 1924–1925, Appraisal of Oil and Gas Properties, *Natl. Petroleum News* Aug. 6, 13, 20, 27, Oct. 29, Nov. 5, 12, 19, Dec. 10, 17, 1924; Jan. 28, Feb. 18, Apr. 15, 1925. pp. 55–57, 68–72, 69–80, 63–69, 67–70, 81–82, 113–125, 91–93, 68–73, 89–96, 85–92, 75–76, 101–108.

Joint Economic Committee, 1964, Taxation of Income from Natural Resources. Federal Tax System, in *Facts and Problems 1964,* U.S. Government Printing Office, Washington, D.C., pp. 107–118.

Jones, C., 1968, Economic Analysis for Mining Ventures and Projects, in *Surface Mining,* E. P. Pfleider, ed., American Institute of Mining and Metallurgical Engineers, New York, pp. 997–1013.

Jones, E. A., and W. T. Pettijohn, 1973, Examinations, Valuation and Reports, in *Society of Mining Engineers Mining Engineering Handbook,* vol. 2, A. B. Cummins and I. A. Given, eds., American Institute of Mining and Metallurgical Engineers, New York, pp. 32-1-32-56.

Jones, P. C., 1942*a*, Development, Operation and Valuation of Oil and Gas Properties, Part 1, *Oil and Gas Jour.* May 28, **40**:45–47.

Jones, P. C., 1942*b*, Development, Operation and Valuation of Oil and Gas Properties, *Oil and Gas Jour.,* Aug. 13, **41**:48, 58–59, 62; Aug. 20, **41**:43–44.

Jones, W. A., 1943, Estimation of Average Value of Gold Ore, *Canadian Inst. Mining and Metallurgy Trans.* **46**:209–225.

Jones, W. R., 1954–1955, Ore Reserves, Their Definition and Classification, *Institution of Mining and Metallurgy,* B. no. 577, pp. 85–88 (1954); discussion, nos. 579–583, pp. 223–249, 342–348, 410–416, 472–474, 514–515 (1955).

Joralemon, I. B., 1925, Sampling and Estimating Disseminated Copper Deposits, *Am. Inst. Mining and Metall. Engineers Trans.* **72**:607–620.

Joralemon, I. B., 1928, The Weakest Link; Or, Saving Time in a Mine Examination, *Eng. and Mining Jour.* **125**:536–540.

Jorgensen, D., 1978, Some Factors Involved in Prospecting for Gypsum Deposits, A Brief Review, *Oklahoma Geol. Survey Circ. 79,* pp. 1–6.

Joughin, N. C., 1973, Technological Innovation and Its Potential Effect on the Opening of New Gold Mines in South Africa, in *Application of Computer Methods in the Mineral Industry,* M. D. G. Salamon and F. H. Lancaster, eds., South African Institute of Mining and Metallurgy, Johannesburg, pp. 115–121.

Journel, A. G., 1974, Geostatistics for Conditional Simulation of Ore Bodies, *Econ. Geology* **69**:673–687.

Journel, A. G., 1976*a*, Convex Analysis for Mine Scheduling, in *Advanced Geostatistics in the Mining Industry,* M. Guarascio, M. David, and C. Huijbregts, eds., Reidel, Dordrecht, Holland, pp. 185–194.

Journel, A. G., 1976*b*, Ore Grade Distributions and Conditional Simulations— Two Geostatistical Approaches, in *Advanced Geostatistics in the Mining Industry,* M. Guarascio, M. David, and C. Huijbregts, eds., Reidel, Dordrecht, Holland, pp. 195–202.

Journel, A. G., 1979, Geostatistical Simulation: Methods for Exploring and Mine Planning, *Eng. and Mining Jour.* **180**(12):86–91.

Journel, A. G., and C. Huijbregts, 1973, Estimation of Lateritic-type Orebodies, in *Application of Computer Methods in the Mineral Industry,* M. D. G., Salamon and F. H. Lancaster, eds., South African Institute of Mining and Metallurgy, Johannesburg, pp. 207–212.

Journel, A. G., and C. Huijbregts, 1978, *Mining Geostatistics,* Academic, New York, 600p.

Judson, S. A., H. D. Easton, and W. A. Schaeffer, 1935, Estimation of Petroleum Reserves in Prorated Fields, *Am. Inst. Mining and Metall. Engineers Trans.* **114**:1–24.

Juravsky, A., 1931, General Methods of Calculation of Ore Reserves, *USSR Geol. and Prosp. Serv. Trans.* vol. 116, 40p.

Kaas, M., 1968, Computers in the Mining Industry, *Canadian Inst. Mining and Metallurgy Spec. Vol.* **9**:103–108.

Kaas, M., 1969, Computer Graphics for Processing Geologic and Mining Data, in *A Decade in Digital Computing in the Mineral Industry,* A. Weiss, ed., American Institute of Mining and Metallurgical Engineers, New York, pp. 9–22.

Kaganovich, S. Y., 1974, Principles of Price Accounting of Explored Mineral Resources, *Razved. i Okhrana Nedr,* no. 6 pp. 32–36. (In Russian.)

Kal'chenko, V., M. P. Pedan, G. G. Grebenkin, and E. S. Olejnikov, 1973, *Economical and Mathematical Models for the Assessment and Exploitation of Mineral Deposits,* SOPS, Kiev, 35p. (In Russian.)

Kantor, M. Z., 1977, Mineralogical Criteria Used in the Economic Evaluation of Magnetite Occurrences: Examples from Siberia, the Turgaj and Other Regions (of the USSR), *Akad. Nauk Kazakh. SSR Izv. Ser. Geol.,* no. 2, pp. 40–45. (In Russian.)

Karpova, Ye. N., 1971, Otsenka oshibok podscheta zapasov pri zakonomernoy izmenchivosti orudeneniya, *Univ. Moscow Vestn., Ser. Geol.,* no. 2, pp. 72–80.

Karpushin, D. M., A. A. Glukhov, and S. A. Pozdnyakov, 1968, O geologoekonomicheskoy otsenke geofizicheskikh rabot. *Vyssh. Ucheb. Zavedeniy Izv., Geologiya i Razved.,* no. 7, pp. 145–149.

Katell, S., R. Stone, and P. Wellman, 1974, Oil Shale, a Clean Energy Source, *Colorado School Mines Quart.* **69**(2):1–19.

Katz, D. L., 1935, A Method of Estimating Oil and Gas Reserves, *Oil Weekly,* Oct. 14, **79**(5):37–43; *Am. Inst. Mining and Metall. Engineers Trans.* **118**:13–32 (1936).

Katz, D. L., 1953, Development of Techniques in Reserve Estimation, *Jour. Petroleum Technology* **5**(9):81–83.

Katz, D. L. et al., 1952, Sample Grading Method of Estimating Gas Reserves, *Jour. Petroleum Technology* **4**(8):207–212.

Kaufman, G. M., 1963, *Statistical Decision and Related Techniques in Oil and Gas Exploration,* Prentice-Hall, Englewood Cliffs, New Jersey, 307p.

Kaye, E., 1939, Recovery of Condensate from Distillate Wells, *Oil and Gas Jour.,* Mar. 23, pp. 86–89.

Kazak, V. M., 1965, Appraisal of Complex Nature of Ore Deposits in Krivoy Rog Basin. *Internat. Geology Rev.* **7**:696–701.

Kazhdan, A. B., and L. P. Kobakhidze, 1967a, Ob ekonomicheskoy prirode zatrat na geologorazvedochnyye raboty (v poryadke obsuzhdeniya), *Vyssh. Ucheb. Zavedeniy Izv., Geologiya; Razved.,* no. 4, pp. 131–134.

Kazhdan, A. B., and L. P. Kobakhidze, 1967b, Geologo-ekonomicheskaya otsenka mestorozhdeniy poleznykh iskopayemykh na raznykh stadiyakh geologorazvedochnykh rabot, Vyssh. Ucheb. Zavedeniy Izv., Geologiya i Razved., no. 12, pp. 147–152.

Kazhdan, A. B., and L. P. Kobakhidze, 1969, Osobennosti geologo-ekonomicheskoy ostenki mestorozhdeniy poleznykh iskopayemykh v usloviyakh deystvuyushchikh gornykh predpriyatiy, Vyssh. Ucheb. Zavedeniy Izv., Geologiya i Razved., no. 7, pp. 124–130.

Kazhdan, A. B., M. V. Shumilin, and V. A. Vikent'yev, 1974, Metodicheskiye osnovy kolichestvennoy otsenki razvedannosti zapasov tverdykh poleznykh iskopayemykh, Sovetskaya Geologiya, no. 11, pp. 7–19.

Keefer, C. M., and L. Borgman, 1979, A Geostatistical Evaluation for Underground Uranium Mining, Contr. Geology 18:19–31.

Keighin, C., 1975, Resource Appraisal of Oil Shale in the Green River Formation, Piceance Creek Basin, Colorado, Colorado School Mines Quart. 70(3):57–68.

Keil, K., 1942, Grundzüge der praktischen Durchführung von Erzvorratsberechnungen, Metall und Erz 39:62–70, 84–87.

Kennedy, T. F., 1926, Sinking fund formula, Eng. and Mining Jour. Press 121(12):491–494. (Discussion of paper by J. A. Church, Eng. and Mining Jour. Press 121(1):8–12.)

Keplinger, C. H., 1967, Case Histories of Actual Performance vs Appraised Prognostication—Petroleum Reservoirs, Exploration and Economics of the Petroleum Industry 5:281–307.

Keplinger, H. F., 1977a, Reserve Calculation Review Shows Why Engineering Estimates Differ, Oil and Gas Jour., Jan. 17, 75(3):60–70.

Keplinger, H. F., 1977b, Reserves—A Better Understanding, in An Analysis of the Production Performance of the Windalia Sand Reservoir of the Barrow Island Oil Field, by C. T. Williams, Oil Gas 23(6):19–23.

Kermeen, J. S., 1968, Ore Reserve Estimation and Grade Control at the Phoenix Mine, Canadian Inst. Mining and Metallurgy Spec. Vol. 9:273–380.

Khokhryakov, V. S., 1975, Proyektirovaniye optimal'nykh variantov otkrytoy razrabotki solzhnostrukturnykh mestorozhdeniy s ispol'zovaniyem EVM, in 13th International Symposium on the Application of Computers and Mathematics for Decision Making in the Mineral Industries, G. Dorstewitz, ed., Verlag Glueckauf GmbH, Essen, pp. C.II.1–C.II.16.

Kinard, C. H., 1950, Famed "Loophole"—Percentage Depletion, Arkansas Law Rev. and Bar Assoc. Jour. 4:333–337.

Kindl, S., 1973, Methodo per la valutazione delle riserve dei giacimenti minerari; La soluzione del problema della campionatura, Industria Mineraria 24:387–405.

King, H. F., 1955 Classification and Nomenclature of Ore Reserves, Australasian Inst. Mining and Metallurgy Proc. No. 174, pp. 5–23.

King, H. F., 1965, Estimation of Ore Reserves, in Exploration and Mining Geology, L. T. Lawrence, ed., 8th Commonwealth Mining and Metallurgical Congress vol. 2, Australia and New Zealand, Melbourne, pp. 296–300.

King, K. R., 1974, Petroleum Exploration and Bayesian Decision Theory, Internat. Symp. Appl. Comput. Math. Miner. Ind. Proc. No. 12, vol. 1, pp. D77–D117.

Kingston, G. A., M. David, R. F. Meyer, A. T. Ovenshine, S. Slamet, and J. J. Schanz, 1978, Workshop on Volumetric Estimation, Internat. Assoc. Mathematical Geology Jour. 10:495–499.

Kirby, E. B., 1894, The Sampling and Measurement of Ore Bodies in Mine Examination, *Colorado Scientific Soc. Proc.* **5**:81–105 (1894-1896).

Kircaldie, W., 1923, Retrospective Appraisals; Their Use in Determining Invested Capital, Depreciation and Depletion, *American Mining Congress: Proceedings of the 26th Annual Convention, Conference on Mine Taxation, Milwaukee, Wisconsin, 1923,* Washington, D.C., pp. 59–75.

Kiselev, V. M., 1974, Promyshlennaya otsenka i obosnovaniye minimal'no dopustimogo koeffitsiyenta rudonosnosti, *Razved. i Okhrana Nedr,* no. 12, pp. 10–13.

Klinge, U., 1971, Anwendungsmöglichkeiten für die Lagerstättenbewertung, insbesondere bei Uranerzen, *Erzmetall* **24**:220–226.

Klinge, U., 1972, Possibilities and Limits for the Application of Data Processing to the Valuation of Deposits and Mining Plans, *Erzmetall* **25**:486–489.

Knapp, A., 1921, Modified Oil-well Depletion Curves, *Am. Inst. Mining and Metall. Engineers Trans.* **65**:405–411.

Knowles, A., 1966, A Geological Approach to Determine the Pattern of Gold Distribution in a Reef, in *Symposium on Mathematical Statistics and Computer Applications in Ore Evaluation, Mar. 7–8, Johannesburg,* South African Institute of Mining and Metallurgy, pp. 157–172.

Knox, H. A., 1908, On Certain Errors in Computing Ore Values, *Eng. and Mining Jour.,* April 18, **85**(16):806.

Knudsen, H. P., Y. C. Kim, and E. A. Mueller, 1975, Comparative Study of the Geostatistical Ore Reserve Estimation Method Over the Conventional Methods, in *13th International Symposium on the Application of Computers and Mathematics for Decision Making in the Mineral Industries,* G. Dorstewitz, ed., Verlag Glueckauf GmbH, Essen, pp. M.VI.1–M.VI.19.

Koch, G. S., 1969, Computer Applications in Mining Geology, in *Computer Applications in the Earth Sciences,* D. F. Merriam, ed., Plenum Press, New York, pp. 121–140.

Koch, G. S., and M. Gomez, 1966, Delineation of Texas Lignite Beds by Statistical Techniques, *U.S. Bur. Mines Rept. Inv. 6833,* 38p.

Koch, G. S., and R. F. Link, 1964, Accuracy in Estimating Metal Content and Tonnage of Ore Body from Diamond-Drill-Hole Data, *U.S. Bur. Mines Rept. Inv. 6380,* 24p.

Koch, G. S., and R. F. Link, 1966, Some Comments on the Distribution of Gold in a Part of the City Deep Mine, Central Witwatersrand, South Africa, in *Symposium on Mathematical Statistics and Computer Applications in Ore Evaluation,* South African Institution of Mining and Metallurgy, Johannesburg, pp. 173–189.

Koch, G. S., and R. F. Link, 1967, Procedures and Precision of Ore Estimation from Assays of Vein Samples, in *Computer Short Course and Symposium on Mathematical Techniques and Computer Applications in Mining and Exploration, 1962, Colorado Mines,* vol. 1, University of Arizona, Tucson, pp. L1-1–L1-30.

Koch, G. S., and R. F. Link, 1970, A Statistical Interpretation of Sample Assay Data from the Getchell Mine, Humboldt County, Nevada, *U.S. Bur. Mines Rept. Inv., 7383,* 28p.

Koch, G. S., and R. F. Link, 1970–1971, Statistical Analysis of Geological Data, vols. 1 and 2, Wiley, New York, 375p., 417p.

Koch, G. S., and R. F. Link, 1971, The Coefficient of Variation, A Guide to the Sampling of Ore Deposits, *Econ. Geology* **66**:293–301.

Koch, G. S., R. F. Link, and S. W. Hazen, 1964, Statistical Interpretation of Sample Assay Data from Mi Vida Uranium Mine, Big Indian District, San Juan County, Utah, *U.S. Bur. Mines Rept. Inv. 6550*, 40p.

Kochetov, M. N., 1962, Problem of Simplifying the Techniques of Calculating Oil and Gas Reserves, *Petroleum Geology* **3**(12A):699–702.

Koepf, E. H., and R. J. Granberry, 1961, The Use of Sidewall Core Analysis in Formation Evaluation, *Jour. Petroleum Technology* **13**:419–424.

Kogan, I., 1961, Osnovnye trebovaniya k geologicheskim otchetam pri utverzhdenii zapasov v GKZ, *Sovetskaya Geologiya*, no. 5, pp. 121–133.

Kogan, I., 1974, *Podschet zapasov i geologopromyshlennaya otsenka rudnykh mestorzhdeniy*, 2nd ed., Izd. Nedra, Moscow, 303p.

Kol'berg, A. V., 1969, Statistical Processing and Evaluation of Geological Exploration for Future Planning of Mining, with Regard to Information Theory, in *Sbornik: Bestransportn sistemy razrabotki mestorozhdenii*, Chelyabinsk, pp. 80–84. (In Russian.)

Kolm, K. E., 1974, ERTS MSS Imagery Applied to Mapping and Economic Evaluation of Sand Dunes in Wyoming, *Contr. Geology* **12**:69–76.

Kontorovich, A. E., and V. I. Demin, 1979, A Method of Assessing the Amount and Distribution of Oil and Gas Reserves in Large Oil and Gas Basins, *Internat. Geology Rev.* **21**:361–367.

Kopp, R. S., 1965, Evaluation of Beryl-bearing Pegmatites of the Copper Mountain Mining District, Fremont County, Wyoming, *Compass* **43**:21–29.

Koranyi, G., 1970, Classification and Estimation of Natural Gas Resources and Reserves: International Comparison of Methods, *Földtani Kutatas* **13**(2):1–8. (In Hungarian.)

Kortsenshteyn, V. N., 1975, New Data on Gases in Ground Water of Large Artesian Basins and Their Effect on Estimates of Gas and Oil Reserves, *Acad. Sci. USSR Doklady, Earth Sci. Sec.* **215**(1–6):227–229.

Kosygin, M. K. et al., 1970, Assessing Accuracy in Calculating Reserves and Evaluating Basic Exploration Parameters, in *Sbornik matematicheski metody v poiskovo-razvedochn*, praktike Irkutsk, pp. 236–253. (In Russian.)

Kotyakhov, F. L. et al., 1968, Zur Beurteilung der Speichereigenschaften eines Erdölhorizonts nach dem Kernmaterial, *Zeitschr. Angew. Geologie* **14**(3):119–124.

Koval'chuk, N. R., and N. Predtechenskaya, 1973, Integral'nyy variant ob"yemnogo metoda podscheta zapasov nefti i gaza, *Geologiya Nefti i Gaza*, no. 8, pp. 32–37.

Koval'chuk, N. R., and N. Predtechenskaya, 1975, Kriterii otsenki promyshlennogo znacheniya razvedannykh zalezhey nefti, *Geologiya Nefti i Gaza*, no. 2, pp. 51–55.

Kraft, J. E., and D. F. H. Whitford, 1972, A Comparison of Computer and Realized Grades at Brenda Mines, in *Decision-making IV, Cooperative Decision-making by the Geologist and Mining Engineer, February 10–12, 1971*, Center of Continuing Education, University of British Columbia, Vancouver, 386p. (16p.)

Krakover, A. S., 1955, Tax Aspects of Uranium Mining, *Wyoming Law Jour.* **9**:189–198.

Krasil'shchikov, Ya. S., 1967, Fotodokumentatsiya gornykh vyrabotok (na primere flogopitovykh mestorozhdeniy), Vyssh. Ucheb. Zavedeniy Izv., *Geologiya; Razved.*, no. 8, pp. 79–84.

Krause, C., 1968, Ore Reserve Estimation and Grade Control at Campbell Chibougamau, *Canadian Inst. Mining and Metallurgy Spec. Vol.* **9**:147–159.

Kreiter, V. M., 1968, *Geological Prospecting and Exploration,* A. Gurevich, trans., Mir Publishers, Moscow, 383p.

Krige, D. G., 1951-1952, A Statistical Approach to Some Basic Mine Valuation Problems on the Witwatersrand, *Chem. Metall. and Mining Soc. South Africa Jour.* **52:**119-139 (Dec., 1951); **53:**201-215 (Mar., 1952); **53:**264-266 (May, 1952); **53:**25-26 (July, 1952); **53:**43-44 (Aug., 1952).

Krige, D. G., 1960, *On the Departure of Ore Value Distribution from the Lognormal Model in South African Gold Mines,* South African Institute of Mining and Metallurgy, Nov., pp. 231-241.

Krige, D. G., 1962, Significance of Limited Number of Borehole Results in Exploration with Particular Reference to Exploration of New South African Gold Fields, in *Computer Short Course and Symposium on Mathematical Techniques and Computer Applications in Mining and Exploration, College of Mines,* vol. 1, University of Arizona, Tucson, pp. Q1-Q11.

Krige, D. G., 1964a, A Brief Review of the Developments in the Application of Mathematical Statistics to Ore Valuation in the South African Gold Mining Industry, *Colorado School Mines Quart.* **59**(4):785-794.

Krige, D. G., 1964b, Recent Developments in South Africa in the Application of Trend Surface and Multiple Regression Techniques to Gold Ore Valuation, *Colorado School Mines Quart.* **59**(4):795-809.

Krige, D. G., 1966a, Two-dimensional Weighted Moving Average Trend Surfaces for Ore Valuation, in *Symposium on Mathematical Statistics and Computer Applications in Ore Evaluation, Mar. 7-8, Johannesburg,* South African Institute of Mining and Metallurgy, pp. 13-79.

Krige, D. G., 1966b, Ore Value Trend Surfaces for the South African Gold Mines Based on a Weighted Moving Average, in *Proceedings of the Symposium and Short Course on Computers and Operations Research in Mineral Industries, Pennsylvania State University, Apr. 17-23,* Min. Indus. Exper. Stat., Spec. Pub., vol. 1, G, 29p.

Krige, D. G., 1966c, A Study of Gold and Uranium Distribution Patterns in the Klerksdorp Gold Field, *Geoexploration* **4:**43-53.

Krige, D. G., 1970a, The Role of Mathematical Statistics in Improved Ore Valuation Techniques in South African Gold Mines, in *Topics in Mathematical Geology,* M. A. Romanova and O. V. Sarmanov, eds., Plenum Publishing Corporation, New York, pp. 243-261.

Krige, D. G., 1970b, The Development of Statistical Models for Gold Ore Valuation in South Africa, *6th Internat. Mining Congress, Madrid,* **6:**1-4, Paper 1-A6.

Krige, D. G., 1976, A Review of the Development of Geostatistics in South Africa, in *Advanced Geostatistics in the Mining Industry,* M. Guarascio, M. David, and C. Huijbregts, eds., Reidel, Dordrecht, Holland, pp. 279-293.

Krige, D. G., and A. H. Munro, 1968, A Review of Some Conceptual and Practical Implications of the Use of Valuation Surfaces for Gold Ore Reserve Estimations, *Canadian Inst. Mining and Metallurgy Spec. Vol.* **9:**33-40.

Krige, D. G., and J. M. Rendu, 1975, The Fitting of Contour Surfaces to Hanging and Foot-wall Data for an Irregular Ore Body, in *13th International Symposium on the Application of Computers and Mathematics for Decision Making in Mineral Industries* G. Dorstewitz, ed., Verlag Glueckauf GmbH, Essen, pp. C.V.1-C.V.12.

Krige, D. G. et al., 1969, The Use of Contour Surfaces as Predictive

Models for Ore Values, in *A Decade in Digital Computing in the Mineral Industry,* A. Weiss, ed., American Institute of Mining and Metallurgical Engineers, New York, pp. 127–162.

Krumlauf, H. E., 1960, Exploration Costs of Small Mines, *Mineral Information Service (State of California, Div. Mines)* **13**(12):1–4.

Kuck, D. L., 1966, Valuation of Equal Square Blocks from Randomly Spaced Data, 6th Ann. International Symposium on Computers and Operations Research, *Pennsylvania State Univ., Min. Ind. Expt. Stat. Spec. Pub. 2-64,* vol. 2, pp. T1–T9.

Kuntz, E., 1962–1980, *Law of Oil and Gas,* Anderson Publishing Company, Cincinnati, 7 vols., 11 books.

Kurtz, W., 1920, Mine Accounting in Relation to Federal Taxes, *Jour. Accountancy,* Jan., pp. 30–42.

Kurtz, W., 1923, Profit in Mining Ventures, *American Mining Congress: Proceedings of the 26th Annual Convention, Conference on Mine Taxation, Milwaukee, Wisconsin, 1923,* pp. 113–124.

Kuscevic, B., T. L. Thomas, and J. Penberthy, 1974, Sampling Distributions in South Crofty Tin Mine, Cornwall, in *Geological, Mining and Metallurgical Sampling,* M. J. Jones, ed., Institution of Mining and Metallurgy, London, pp. 102–109.

Kuz'min, V. I., 1966, Usloviya primeneniya formul srednego arifmeticheskogo i srednego vzveshennogo pri podschete zapasov poleznogo iskopayemogo, *Razved. i Okhrana Nedr,* no. 2.

Kuz'min, V. I., 1972, Effektivnoye primeneniye srednevzveshennykh i srednearifmeticheskikh otsenok pri podschete zapasov i otsenke geologorazvedochnoy informatsii, *Akad. Nauk SSSR Sibirsk. Otdeleniye Geologiya i Geofizika,* no. 7, pp. 74–81.

Kužvart, M., and M. Böhmer, 1978, *Prospecting and Exploration of Mineral Deposits,* Elsevier, Amsterdam, 431p.

Lahee, F. H., 1941, This Matter of Estimating Oil Reserves, *Am. Assoc. Petroleum Geologists Bull.* **25**:164–166.

Lahee, F. H., 1950, Our Oil and Gas Reserves, Their Meaning and Limitations, *Am. Assoc. Petroleum Geologists Bull.* **34**:1283–1287.

Lahee, F. H., 1955, The Terminology of Petroleum Reserves, *4th World Petroleum Congress Proc.,* Rome, June 6–15, sect. 2, pp. 561–565.

Laing, G. J. S., 1977, Effects of State Taxation on Mining Industry in Rocky Mountain States, *Colorado School Mines Quart.* **72**(2):1–126.

Lallement, B., 1976, La geostatistique au B. R. G. M., in *Advanced Geo-Statistics in the Mining Industry,* M. Guarascio, M. David, and C. Huijbregts, eds., Reidel, Dordrecht, Holland, pp. 327–332.

Lallement, B., 1977, Use of Geostatistics at the B. R. G. M. to Determine the Best Way to Prove an Orebody, in *Application of Computer Methods in the Mineral Industry,* R. V. Ramani, ed., American Institute of Mining and Metallurgical Engineers, New York, pp. 1026–1032.

Lamb, L., 1950, Mine Accounting—Commentary Written by R. Goyne Miller, *Chartered Accountant in Australia,* **20**:724–735.

Lampietti, F. J., and L. F. Marcus, 1971, Computer Simulation of Ringarooma Bay Offshore Sampling for Tin, *Canadian Inst. Mining and Metallurgy Spec. Vol.* **12**(7):223–228.

Lampietti, F. J., and L. F. Marcus, 1974, Computer Model Predicts Acceptable Risk for Commercial Nodule Mining Projects, *Eng. and Mining Jour.* **175**:53–59.

Lane, A. C., 1911, Stock Value and Mine Value, *Canadian Mining Jour.* Nov. 1, 15; Dec. 1, pp. 691, 729, 775.

Lane, K. F., 1964, Choosing the Optimum Cut-off Grade, *Colorado School Mines Quart.* **59**(4):811–829.

Langton, J., 1911, Commentary on "A Method of Calculating Sinking-Funds" by J. B. Dilworth, *Am. Inst. Mining and Metall. Engineers Trans.* **42**:908–910, 1912.

Larkey, C. S., 1925, Mathematical Determination of Production Decline Curves, *Am. Inst. Mining and Metall. Engineers Trans.* **71**:1322–1328.

Larkey, C. S., and R. T. Bright, 1925, Group Appraisals of Oil Properties, *Oil and Gas Jour.,* Feb. 19, **71**:86, 100.

Lasky, S. G., 1945, The Concept of Ore-Reserves, Many Factors Enter into a Proper Definition of the Term, *Mining and Metallurgy* **26**(466):471–474.

Lasky, S. G., 1950, How Tonnage and Grade Relationships Help Predict Ore Reserves, *Eng. and Mining Jour.* **151**(4):81–85.

Latyshova, M. G., and T. F. D'Yakonova, 1974, The Statistical Analysis of Exploratory Geophysical Data for Reserve Estimation: General Principles, *Neftegazovaya Geologiya i Geofizika,* no. 4, pp. 38–41. (In Russian.)

Lautsch, H., 1970, On the Determination of Economically Viable Coal Reserves in the Context of Special Storage Conditions, *Bergbauwissenschaften* **17**(10):368–374. (In German.)

Lawn, J., 1897, *Mine Accountants and Mining Book-keeping,* C. Griffin, London, 147p.

Leach, B., 1970, A Short Note on Calculating the Confidence Limits on Average Grade of Ore When the Lengths of Assayed Core Sample are Unequal, *Amdel Bull.* **9**:51–54.

Leach, B., 1975, Evaluating Vein-type Deposits, in *13th International Symposium on the Application of Computers and Mathematics for Decision Making in Mineral Industries,* G. Dorstewitz, ed., Verlag Glueckauf GmbH, Essen, pp. M.I.1–M.I.12.

Leake, P. D., 1920, *Depreciation and Wasting Assets and Their Treatment in Computing Annual Profit and Loss,* 3rd rev. ed., Pitman & Sons, Ltd, London.

Lees, G. M., 1950, Calculating Petroleum Reserves, *Inst. Petroleum Rev.* **4**(38):33–37; **4**(45):295–296.

Lefkovits, H. C., and C. S. Mathews, 1958, Application of Decline Curves to Gravity-Drainage Reservoirs in the Stripper Stage, *Jour. Petroleum Technology* **10**(11):275–280.

Leggett, T. H., and F. H. Hatch, 1904, An Estimate of the Gold Production and Life of the Main Reef Series, Witwatersrand, Down to 6000 Feet, *Inst. Mining and Metallurgy Trans.* **12**:39–46.

Leigh, R. W., and R. L. Blake, 1971, An Iterative Approach to the Optimal Design of Open Pits, *Canadian Inst. Mining and Metallurgy Spec. Vol.* **12**:254–260.

Leith, C. K., 1920, Geologists as Witnesses in Mining Litigation, *Econ. Geology* **15**:674–680.

Leith, C. K., 1921, *Economic Aspects of Geology,* Henry Holt, New York, 475p.

Leith, C. K., 1938, *Mineral Valuations of the Future,* American Institute of Mining and Metallurgical Engineers, New York, 116p.

Lentz, O. H., 1960, Mineral Economics and the Problem of Equitable

Taxation, a Study in the Legislative Rationale of Percentage Depletion Allowances, *Colorado School Mines Quart.,* vol. 5, no. 2, 111p.

Leon, V., A. T., 1973, The New General Mining Law in Peru and the Creation of the "Peruvian Mining Community," *Colorado School Mines Quart.* **68:**81-114.

Leviatov, G. O., 1937, Estimation and Registration of the Reserves in the Gold and Platinum Mining, *All-Union Gold and Platinum Prosp. Trust Trans.,* no. 5, pp. 100-170. (In Russian.)

Lewis, A. S., 1946, Risks of Mining Ventures Should be Evaluated, *Eng. and Mining Jour.,* Sept., **147:**61-63.

Lewis, C. K., 1969, An "Economic Life" for Property Evaluation, *Eng. and Mining Jour.* **170**(10):78-79.

Lewis, E. M., 1931, Mine Sampling and the Commercial Value of Ores, *Utah Eng. Expt. Sta. Bull. 10,* pp. 13-27.

Lewis, F. M., and R. B. Bhappu, 1975, Evaluating Mining Ventures Via Feasibility Studies, *Mining Eng.,* Oct., **27:**48-54.

Lewis, J. A., 1957, Interpretation of Core Analysis in Predicting Oil Recovery, *Producers Monthly* **21**(10):31-34, 36-37.

Lewis, J. O., 1917, Methods for Increasing the Recovery from Oil Sands, *U.S. Bur. Mines Bull. 148,* 128p.

Lewis, J. O., and C. H. Beal, 1918, Some New Methods of Estimating the Future Production of Oil Wells, *Am. Inst. Mining and Metall. Engineers Trans.* **59:**492-520; *Am. Inst. Mining and Metall. Engineers Bull.* **134:**477-504.

Lewis, J. O., and W. W. Cutler, 1920, A Numerical Expression for Production-Decline Curves, *Eng. and Mining Jour.* Sept., **110:**479-480.

Lichtblau, J. H., and D. P. Spriggs, 1959, *The Oil Depletion Issue,* Petroleum Industry Research Foundation, Inc., New York, 158p.

Liddy, J. C., 1971, The Appraisal and Evaluation of Mineral Prospects, *Australian Mining* **63**(7):50-53.

Lindley, A. H. et al., 1976, Mineral Financing, in *Economics of the Mineral Industries,* W. A. Vogley, ed., American Institute of Mining and Metallurgical Engineers, New York, pp. 420-432.

Lindley, C. H., 1914, A Treatise on the American Law Relating to Mines and Mineral Lands, Etc., 3 vols., 3rd ed., Bancroft-Whitney, San Francisco.

Link, R. F., and G. S. Koch, 1967, Linear Discriminant Analysis of Multivariate Assay and Other Mineral Data, *U.S. Bur. Mines Rept. Inv. 6898,* 25p.

Link, R. F., G. S. Koch, and G. Gladfelter, 1964, Computer Methods of Fitting Surfaces to Assay and Other Data by Regressive Analysis, *U.S. Bur. Mines Rept. Inv. 6508,* 69p.

Link, R. F., N. N. Yabe, and G. S. Koch, 1966, A Computer Method of Fitting Surfaces to Assay and Other Data in Three Dimensions by Quadratic-Regression Analysis, *U.S. Bur. Mines Rept. Inv. 6876,* 42p.

Lintern, W., 1872, *The Mineral Surveyor and Valuer's Complete Guide,* Lockwood & Co., London, 192p.; 146p.

Lisle, G., 1900, Colliery Sinking or Redemption Funds, in *Accounting Theory and Practice,* William Green & Sons, Edinburgh, 426p.

Lobach, V. I., and V. N. Moshkin, 1974, Dostovernost' opredeleniya koeffitsiyenta rudonosnosti na osnovanii geologicheskoy dokumentatsii gornykh vyrabotok zolotorudnogo mestorozhdeniya, *Vyssh. Ucheb. Zavedeniy Izv., Geologiya i Razved.,* no. 11, pp. 81-83.

Lombardi, M., 1915, The Valuation of Oil Lands and Properties, *Western Engineer* **6:**153–159.

London, E. E. et al., 1964, Principles of Evaluating Gas Prospects According to the Gases Dissolved in Subsurface Waters, *Petroleum Geology* 5(3):149–154.

Louis, H., 1923, *Mineral Valuation,* Charles Griffin & Co., London, 291p.

Lourie, G. B., 1951, Wasting Assets—The Treatment of and a Proposal For, *Tax Law Rev.* **6:**409–433.

Lovejoy, W. F., and P. T. Homan, 1965, Methods of Estimating Reserves of Crude Oil, Natural Gas, and Natural Gas Liquids, in *Resources for the Future,* Johns Hopkins Press, Baltimore, 163p.

Lovering, T. S., 1969, Mineral Resources from the Land, in *Resources and Man, a Study and Recommendations by the Committee on Resources and Man of the Division of Earth Sciences,* W. H. Freeman, San Francisco, pp. 109–134.

Lucero, H. N., 1971, Applications of Geostatistical Criteria to Uranium Ore Reserve Estimation, *Canadian Inst. Mining and Metallurgy Spec. Vol.* **12:**192–198.

Lukman, R., 1969, Some Problems Associated with the Valuation, Mining and Processing of Alluvial Cassiterite on the Plateau Tinfield of Nigeria, in *A Second Technical Conference on Tin, Bangkok, 1969,* W. Fox, ed., International Tin Council, London, pp. 299–324.

Lynch, E. J., 1962, *Formation Evaluation,* Harper & Row, New York, 422p.

Lyon, J. R., 1975, Using Multiple Cutoff Rates for Capital Investment, *Jour. Petroleum Technology* **27:**822–826.

Mabra, D. A., 1956, A Method for Evaluating Drillable Oil and Gas Prospects, *Gulf Coast Assoc. Geol. Socs. Trans.* **6:**241–246.

McCarthy, C. F., 1951, Tax Accounting for Quarries, in *Handbook of Tax Accounting Methods,* J. K. Lasser, pp. 629–636.

McCormack, C. P., 1924, Should Mines be Subjected to a Yearly Valuation?, *Coal Age,* July 31, pp. 145–148.

McCray, A. W., 1975, *Petroleum Evaluations and Economic Decisions,* Prentice-Hall, Englewood Cliffs, N. J., 544p.

McCrossan, R. G., N. L. Ball, and L. R. Snowdon, 1972, An Evaluation of Surface Geochemical Prospecting for Petroleum, Olds-Caroline Area, *Canadian Geol. Survey Pap. 71-31,* 101p.

McDaniel, G. A., 1970, How to Quickly Evaluate Exploration Projects, *World Oil,* July, pp. 131–137.

McDermott, W., 1894–1895, Mine Reports and Mine Salting, *Inst. Mining and Metallurgy Trans.* **3:**108–149.

MacDonnell, L. J., 1976, Public Policy for Hard-Rock Minerals Access on Federal Lands: A Legal-Economic Analysis, *Colorado School Mines Quart.,* vol. 71, no. 2, 109p.

McElvaney, E., 1961, Financing Property Aquisitions, in *Economics of Petroleum Exploration, Development, and Property Evaluation,* Prentice-Hall, Englewood Cliffs, N. J., pp. 174–189.

McGarraugh, R., 1920, *Mine Bookkeeping,* McGraw-Hill, New York, 126p.

McGiddy, T. C., and D. B. Whitfield, 1974, A Computer System for the Evaluation and Planning of Coal Mines (CMEPS), *Internat. Symp. Appl. Comput. Math. Miner. Ind. No. 12,* 2:H1–H24.

McGowen, R. L., 1952, Allowance for Depreciation and the Federal Income Tax Treatment of Depreciable Properties in the Oil and Gas Industry,

Proceedings of the 3rd Annual Institute on Oil and Gas Law, Southwest Legal Foundation, pp. 303–340.

McGrath, T. O., 1921, *Mine Accounting and Cost Principles,* McGraw-Hill, New York, 271.

Mach, B., 1975, Die geologische Aufgabenstellung und das geologisch-ökonomische Ergebnis im Mittelpunkt der Leitung. Planung und ökonomischen Stimulierung geoligischer Untersuchungasarbeiten, *Zeitschr. Angew Geologie* **21**(3):105–109.

McKechnie, D., 1930*a*, Valuation of a Prospect, *Mining Mag.,* Dec., **43**:371–373.

McKechnie, D., 1930*b*, The Valuation of Ore in a Prospect, *Canadian Inst. Mining and Metallurgy Bull.,* Nov., no. 223, pp. 1453–1460.

McKelvey, V. E., 1966, Reply to J. M. Ryan, Limitations of Statistical Methods for Predicting Petroleum and Natural Gas Reserves and Availability, *Jour. Petroleum Technology* **18**(3):287.

Mackenzie, B. W., 1973, Corporate Exploration Strategies, in Application of Computer Methods in the Mineral Industry, M. D. G. Salamon and F. H. Lancaster, eds., South African Institute of Mining and Metallurgy, Johannesburg, pp. 109–114.

MacKenzie, J., 1966, Evaluation, Sampling and Forecasting Grades of Ore at Nchanga (Zambia), in *Symposium on Mathematical Statistics and Computer Applications in Ore Evaluation, March 7–8, 1966,* South African Institute of Mining and Metallurgy, Johannesburg, pp. 358–371.

McKinstry, H. E., 1948, *Mining Geology,* Prentice-Hall, New York, 699p.

McLaughlin, D. H., 1939, Geologic Factors in the Valuation of Mines, *Econ. Geology* **34**:589–621.

McLaughlin, R. P., 1921, *Oil Land Development and Valuation,* McGraw-Hill, New York, 196p.

McMurry College, 1957–1958, *Petroleum Conference on Oil and Gas Taxation,* McMurry College, School of Business Administration, Abilene, Tex., 1957, 116p.; 1958, var. pages.

McNaughton, D. A., and F. A. Garb, 1975, Finding and Evaluating Petroleum Accumulations in Fractured Reservoir Rocks, *Exploration and Economics of the Petroleum Industry* **13**:23–49.

McQuillin, K. B., 1977*a*, A Computer Based Data Storage and Retrieval System for the Evaluation of Coalfield Data, in *Application of Computer Methods in the Mineral Industry,* R. V. Ramani, ed., American Institute of Mining and Metallurgical Engineers, New York, pp. 355–367.

McQuillin, K. B., 1977*b*, A Computer Based System for the Geological Evaluation of a Surface Coal Mine, in *Application of Computer Methods in the Mineral Industry,* R. V. Ramani, ed., American Institute of Mining and Metallurgical Engineers, New York, pp. 368–378.

Magnou, E., 1975, Consecuencia de la teoria de los factores de la producción de J. B. Say, en la valuación objetiva de yacimientos, in *Primer simposio nacional de geología económica,* vol. 2, V. Angelelli, Chairperson, Ed. Científicas Argent. Librart, Buenos Aires, pp. 465–470.

Magnus, B., 1916, Metallurgical Calculations as Affecting Mine Valuation, *Eng. and Mining Jour.* **102**:669–670.

Makarov, M. S., L. N. Dudenko, G. P. Erastov, and D. P. Bobritskij, 1973, The Optimal Grid of Observations for Ore Prospecting, *Tr. Vses. Ordena Lenina Nauchno Issled. Geol. Inst. (Leningrad)* **180**(2):109–115. (In Russian.)

Maksimov, M. I., 1960, Method of Calculation of Recoverable Oil Reserves in the Final Stage of Exploitation of Oil Strata Under Conditions of Displacement of Oil by Water *Petroleum Geology* **3**(3B):181–187.

Mallory, W. W., 1963, Analysis of Petroleum Potential Through Regional Geologic Synthesis, *Am. Assoc. Petroleum Geologists Bull.* **47**:756–776.

Mallory, W. W., 1975, Synopsis of Procedure; Accelerated National Oil and Gas Resource Evaluation (ANOGRE), in *Mineral Resources and the Environment,* appendix to sec. 2, rept. of panel on estimation of mineral reserves and resources, National Academy of Sciences, Washington, D.C., 5p.

Manefield, A. K., 1974, Mini-Computers in the Australian Petroleum Industry, *Australasian Oil and Gas Rev.* **20**(8):24–29.

Manhenke, V., 1969, Zur Spezifik und ökonomischen Bewertung geologischer Gebietsressourcen, *Zeitschr. Angew. Geologie* **15**(119):649–655.

Maréchal, A., 1972, An Example of the Application of Geostatistics to the Problem of Estimation in Iron Mines, *Minerales* **27**:17–21. (In Spanish.)

Maréchal, A., 1976a, Selected Minable Blocks: Experimental Results Observed on a Simulated Orebody, in *Advanced Geostatistics in the Mining Industry,* M. Guarascio, M. David, and C. Huijbregts, eds., Reidel, Dordrecht, Holland, pp. 137–161.

Maréchal, A., 1976b, The Practice of Transfer Functions: Numerical Methods and Their Application, in *Advanced Geostatistics in the Mining Industry,* M. Guarascio, M. David, and C. Huijbregts, eds., Reidel, Dordrecht, Holland, pp. 253–276.

Maréchal, A., and J. Serra, 1970, Random Kriging, in *Geostatistics — A Colloquium,* Plenum Press, New York, pp. 91–112.

Marino, J. M., and J. -P. Slama, 1973, Ore Reserve Evaluation and Open Pit Planning, in *Application of Computer Methods in the Mineral Industry,* M. D. G. Salamon and F. H. Lancaster, eds., South African Institute of Mining and Metallurgy, Johannesburg, pp. 139–144.

Markov, N. N., M. K. Lenskikh, and Yu. V. Shurubor, 1978, Construction of Maps of Oil-Saturated Thickness by Computer for Calculation of Oil and Gas Reserves, *Petroleum Geology* **15**(2):61–63.

Marriott, H. R., 1925, *Money and Mines; The Administration, Organization and Economics of Precious and Nonferrous Metal Mines,* Macmillan, New York, 270p.

Marshall, K. T., 1964, A Preliminary Model for Determining the Optimum Drilling Pattern in Locating and Evaluating an Ore Body, *Colorado School Mines Quart.* **59**(4):223–236.

Marston, A., and T. Agg, 1936, *Engineering Valuation,* McGraw-Hill, New York, 655p.

Marston, A., R. Winfrey, and J. C. Hempstead, 1953, Valuation of Mines, in *Engineering Valuation and Depreciation,* 2nd ed., A. Marston, R. Winfrey, and J. C. Hempstead, eds., McGraw-Hill, New York, pp. 364–383.

Martin, D. A., 1977, Canadian and U.S. Resource Tax Laws: The Significance to Computerized Economic Evaluation, in *Application of Computer Methods in the Mineral Industry,* R. V. Ramani, ed., American Institute of Mining and Metallurgy, New York, pp. 895–907.

Martin, R. H., 1972, The Money Tree, *Earth Sci. Bull.* **5**(2):11–22.

Martinez, A. R., 1966, Estimation of Petroleum Resources, *Am. Assoc. Petroleum Geologists Bull.* **50**:2001–2008.

Masters, K., 1975, Exploration und Bergbau unter Permafrost-Bedingungen auf der Black Angel-Grube, Marmirilik/Grönland, *Erzmetall* **28**(3):93–98.

Matheron, G., 1962–1963, Traité de géostatistique appliquée, tome 1–2, *Mém. Bur. Recher. Géol. Min.,* 1962, no. 14, 333p.; 1963, no. 24, 171p.

Matheron, G., 1965, *Les variables regionalisees et leur estimation,* Masson et Cie, Paris.

Matheron, G., 1967, Kriging, or Polynomial Interpolation Procedures?, *Canadian Inst. Mining and Metallurgy Trans.* **70:**240–244.

Matheron, G., 1969a, Course in Geostatistics, *Cahiers Centre Morph. Math. Fountainebleau,* no. 2, 82p. (In French.)

Matheron, G., 1969b, Le krigeage universel, *Cahiers Centre Morph. Math. Fountainebleau, Ecole d. Mines de Paris.*

Matheron, G., 1976a, A Simple Substitute for Conditional Expectation: The Distinctive Kriging, in *Advanced Geostatistics in the Mining Industry,* M. Guarascio, M. David, and C. Huijbregts, eds., Reidel, Dordrecht, Holland, pp. 221–236.

Matheron, G., 1976b, Forecasting Block Grade Distributions: The Transfer Functions, in *Advanced Geostatistics in the Mining Industry,* M. Guarascio, M. David, and C. Huijbregts, eds., Reidel, Dordrecht, Holland, pp. 237–251.

Mathews, P. S., 1936, Practical and Legal Aspects of Mine Financing, *Mining and Metallurgy* **17:**193–195.

Mathewson, E., 1910, The Depreciation of Factories, Mines, and Industrial Undertakings and Their Valuation, 4th ed., Spon & Chamberlain, New York.

Matthews, T. K., 1969, Tax Planning: A Guide to Financing in the Mining Industry, *Mining Eng.,* Jan., **21:**81–85.

Matveev, P. S., and A. V. Nikiforov, 1961, The extent of Exploration of a Mineral Deposit Preparatory to Its Commercial Development, *Internat. Geology Rev.* **3:**927–930.

Maxfield, P. C., 1973, *The Income Taxation of Mining Operations,* Rocky Mountain Mining Law Foundation, Boulder, Colo., 345p.

Maxwell, A. S., 1978, Underground Production Mining at Chingola, with Emphasis on a Computerized Ore-reserve Calculation, Depletion and Prediction System, *Commonwealth Mining and Metall. Congress Proc. No. 11.*

May, G. O., 1936, Valuation of Mines in *Twenty-five Years of Accounting Responsibility, 1911–1936,* vol. 1, B. C. Hunt, ed., American Institute Publishing Company, New York, pp. 247–262.

Mayer-Gürr, A., 1969, Erdgas-Vorratsberechnungen: ein Wort zur Klärung und ein Vorschlag zur Koordinierung, *Erdöl u. Kohle* **22:**129–132.

Mayer-Gürr, A., 1976, *Petroleum Engineering,* Wiley, New York, 208p.

Megill, R. E., 1971, *An Introduction to Exploration Economics,* Petroleum Publishing Company, Tulsa, Okla., 159p.

Melvill, L. V., 1929, On the Graphic Methods of Showing Variations in Values of Ore Bodies, *Geol. Soc. South Africa Trans.* **32:**27–30, 1930.

Mendes, H. C., and G. Melcher, 1975, Use of Simulation Model for Production Planning at the Jacupiranga Phosphate Mine, in *13th International Symposium on the Application of Computers and Mathematics for Decision Making in Mineral Industries,* G. Dorstewitz, ed., Verlag Glueckauf GmbH, Essen, pp. C.VI.1–C.VI.11.

Mensch, W. H., 1959, Calculation of Gas Reserves from Gas Density Data, *Petroleum Engineer* **31**(10):B49–B57.

Menzie, W. D., M. L. Labovitz, and J. C. Griffiths, 1977, Evaluation of Mineral Resources and the Unit Regional Value Concept, in *Application of Computer Methods in the Mineral Industry,* R. V. Ramani, ed.,

American Institute of Mining and Metallurgical Engineers, New York, pp. 322–339.

Merriam, D. F., 1972, Trend Analysis in Mining Exploration and Exploitation, in *Decision-making IV, Cooperative Decision-making by the Geologist and Mining Engineer, February 10-12, 1921,* Center of Continuing Education, University of British Columbia, Vancouver, 25p.

Messel, M. J., 1947, Examination and Valuation of Chrysotile Asbestos Deposits Occurring in Massive Serpentine, *Am. Inst. Mining and Metall. Engineers Trans.* **173:**79–84.

Meyer, R. F., 1978, The Volumetric Method for Petroleum Resource Estimation, *Internat. Assoc. for Mathematical Geology Jour.* **10:**501–518.

Meyer, W. G., 1958, *Use of Oil and Gas Property Appraisal Reports, Spec. Pap. No. 1* Willis G. Meyer & Associates, Dallas.

Michelson, R. W., and H. Polta, 1969, A Systematic Method for Evaluating a Mineral Deposit Using Cash Flow Analysis and Varying Cost Criteria, in *A Decade in Digital Computing in the Mineral Industry,* American Institute of Mining and Metallurgical Engineers, New York, pp. 211–240.

Michelson, R. W. et al., 1970, Evaluating the Economic Availability of Mesabi Range Taconite Iron Ores with Computerized Models, *U.S. Bur. Mines Inf. Circ. 8480,* 99p.

Mid-Continent Oil and Gas Association, 1968, *Percentage Depletion, Economic Progress and National Security,* Tulsa, Okla., 77p.

Mikolayevskiy, E. Yu, and R. Aliyev, 1968, K otsenke neftegazonosnosti glinistykh, *Razved. Geofizika,* no. 25, pp. 85–94.

Millan, J. R., 1975a, Casos de valuación en mineria, in *Primer Simposio Nacional de Geología Económica,* vol. 2, V. Angelelli, Chairperson, Ed. Cientificas Argent. Librart, Buenos Aires, pp. 471–476.

Millan, J. R., 1975b, Los metodos racionales de evaluación aplicados a proyectos mineros, in *Segundo Congreso Ibero-Americano de geología económica,* vol. 2, Dept. Argent. Sci. Publ., Buenos Aires, pp. 487–495.

Miller, G. W., 1907, *Mining Law in Practice,* 2nd ed., The Ores & Metal Publishing Company, Denver, Colo., 292p.

Miller, K. G., 1948, *Oil and Gas Federal Income Taxation,* Commerce Clearing House, Chicago, 251p. (See also subsequent editions.)

Miller, P. J., 1913, Assessment of Mines, *Eng. and Mining Jour.,* Nov. 22, pp. 969–971.

Miller, R. V., 1956, *Depletion Allowance, Property Aggregation and Other Current Problems,* New York University Institute on Federal Taxation, New York, pp. 535–543.

Mills, B., 1936, Over a Dozen Import Factors Must be Considered in Producing Property Evaluation, *Oil Weekly,* Aug. 17, **82**(10):21–24.

Mills, L., and J. C. Willingham, 1926, *The Law of Oil and Gas,* Callaghan & Co., Chicago, 791p.

Mills, L. J., 1972, The Utilization of Resources—Geology, *Mining Engineer* **131**(5):235–246.

Milovidov, K. M. et al., 1972, Voprosy ekonomiki, in *Matematicheskiye metody v gasoneftyanoy geologii i geofizike,* Izd. Nedra, Moscow, pp. 173–200.

Mining Congress Journal, 1959-1960, Economic evaluation of proposed mining ventures, Nov., Dec., and Jan. issues and as a separate pamphlet.

Mints, I. Ye., 1972, O metodike vybora ob"yektov poiskov i razvedki mestorozhdeniy nefti i gaza na osnove otsenki ikh narodnokhozyaystvennoy effektivnosti, *Geologiya Nefti i Gaza,* no. 8, pp. 26–29.

Miroshnikov, A., and V. Prokhorov, 1967, Kadastr mestorozhdeniy i rudoproyavleniy na perforirovannykh kartakh, *Akad. Nauk SSSR Sibirsk. Otdeleniye Geologiya i Geofizika,* no. 7, pp. 113–118.

Mishra, R. N., 1972, Estimation of Kumaraswamy Iron-ores, *Indian Minerals* **26**(3):24–27.

Missan, H., B. R. Cooper, S. M. el Raba'a, Griffiths, and C. Sweetwood, 1978, Workshop on Aeral Value Estimation, *Internat. Assoc. for Mathematical Geology Jour.* **10**:433–439.

Moberg, N., 1962, New Developments in the Exploration and Investigation of Iron Ore Properties, in Recent Advances in the Mining . . . iron ore, University of Minnesota Center of Continuing Study, Minneapolis, pp. 91–100.

Momdzhi, G. S., and I. I. Pastushenko, 1965, Preliminary Estimate of Iron Ore Reserves, *Internat. Geology Rev.* **7**:1747–1755.

Montague, W., 1948, Problems of Iron Ore Taxation, *Natl. Tax Assoc. Proc.,* pp. 243–249.

Montgomery, R. H., 1923a, Valuation of mines, in *Income Tax Procedure,* Ronald Press Co., New York, pp. 1118–1134.

Montgomery, R. H., 1923b, Valuation of Oil and Gas Wells, in *Income Tax Procedure,* Ronald Press Co., New York, pp. 1134–1148.

Moore, C. L., 1966, Reply to J. M. Ryan, Limitations of Statistical Methods for Predicting Petroleum and Natural Gas Reserves and Availability, *Jour. Petroleum Technology* **18**(3):286–287.

Moore, T. V., 1933, Open Flow Test is Practically Worthless in Gauging Reserve, *Oil Weekly,* May 22, **69**(10):16–18.

Moreau, G., 1906, *Étude sur l'état actuel des mines du Transvaal. Les gites — leur valeur, étude industrielle et financière,* C. Béranger, Paris, 222p.

Morkill, D. B., 1918, Formulas for Mine Values, *Mining and Scientific Press* **117**:276–277.

Morris, A. B., 1922-1923, Physical Valuation of Oil Properties, *Natl. Petroleum News,* Nov. 8–22, 1922; Jan. 3, 1923, pp. 49–53, 65–66, 69–70, 82–86.

Morris, A. B., 1923, Graphic Appraisal of Mining Property, *Eng. and Mining Jour.,* Aug. 11, **116**:241–242.

Morris, A. B., 1933, Probable Recovery Estimate of Field Requires Consideration of Much Data, *Oil Weekly,* Jan. 23, **68**(6):14–15; Feb. 6, **68**(8):38–39.

Morris, A. B., 1937, An Evaluation Table with Limiting Elements, *Oil Weekly,* Dec. 13, **88**:38–39.

Morrison, R., 1878, *Digest of the Law of Mines and Minerals . . .,* Bancroft, San Francisco, 448p.

Morse, C., 1964, Potential Hazards of Direct Investment in Raw Materials, in *Natural Resources and International Development,* M. Clawson, ed., Johns Hopkins Press, Baltimore, pp. 367–414.

Mortimer, N., 1980, Uranium Resource Economics, *Resources Policy* **6**(1):19-32.

Mosberg, L. G., 1979, Financing Oil and Gas Ventures, vols. 1 and 2, IED Press, Oklahoma City, 843p.

Moyer, W. I., 1923, Time to Pay Out as a Basis for Valuation of Oil Properties, *Am. Inst. Mining and Metall. Engineers Trans.* **68**:1121–1129.

Mukherjee, B., and S. P. Mukherjee, 1977, On the Estimation of Quality of Ore Reserve Through Statistical Model and Sequential Decision Process, *Indian Jour. Earth Sci.* **4**:137–140.

Mulligan, J., 1966, Tin-lode Investigations, Cape Mountain Area, Seward Peninsula, Alaska, *U.S. Bur. Mines Rept. Inv. 6737,* 43p.

Munro, A. H., 1966, A Review of the Statistical Techniques in Ore Valuation, in *Symposium on Mathematical Statistics and Computer Applications in Ore Evaluation, Mar. 7-8, 1966, Johannesburg,* South African Institute of Mining and Metallurgy, pp. 3-12.

Munroe, H. S., 1890, Examination of Mines, *School Mines Quart.* **11:**193-201; **12:**22-27, 117-127.

Murgu, M., and D. Sandu, 1972, Distributia componentilor utili si corelatii minereu-metal intr-un zacamint situat in scarn, *Bull. Inst. Pet., Gaze si Geol. Tech. (Bucharest)* **19:**19-28.

Muskat, M., and M. O. Taylor, 1946, Effect of Reservoir Fluid and Rock Characteristics on Production Histories of Gas-drive Reservoirs, *Am. Inst. Mining and Metall. Engineers Trans.* **165:**78-93.

Musgrove, P. A., 1976, Mathematical Aspects of the Grade-tonnage Distribution of Metals. Appendix to a Paper by D. B. Brooks, in *Economics of the Mineral Industries,* W. A. Vogely, ed., American Institute of Mining and Metallurgical Engineers, New York, pp. 192-205.

Nalivkin, V. D., M. D. Belonin, V. S. Lazarev, S. G. Neruchev, and G. R. Sverchkov, 1976, Criteria and Methods of Quantitative Assessment of Petroleum Prospects in Poorly Studied Large Territories, *Internat. Geology Rev.* **18:**1259-1268.

Napolskii, M. S., 1964, Quantitative Reserve Estimates, in *Printsipy otsenki perspektiv neftegazonosnosti krupnykh territorii,* Gos. geol. komitet SSSR, "Nedra" Press, Leningrad, no. 11, pp. 188-243. (In Russian.)

Nemec, V., 1965, Disadvantages of the Polygonal Method for Computations of Mineral Raw Material Reserves, in *Short Course and Symposium on Computer and Computer Applications in Mining and Exploration,* vol. 1, A, J. C. Dotson and W. C. Peters, eds., College of Mines, University of Arizona, Tucson, 13p.

Nemec, V., 1973, An Interscientific Basis for Computing Ore Reserves, *Mineralia Slovaca* **5**(4):449-452.

Nemec, V., 1974, Space Models of Ore Deposits; Further Practice Achievements, *International Symposium Appl. Comput. Math. Miner. Ind., Proc. No. 12,* vol. 2, pp. H40-H58.

Nemec, V., 1975, Multicomponent Models of Ore Deposits, in *13th International Symposium on the Application of Computers and Mathematics for Decision Making in Mineral Industries,* G. Dorstewitz, ed., Verlag Glueckauf GmbH, Essen, pp. C.IV.1-C.IV.11.

Neter, J., and W. Wasserman, 1966, *Fundamental Statistics for Business and Economics,* Allyn & Bacon, Boston, Mass., 738p.

Netzeband, F., 1929, Example of Prospecting and Valuing a Lead-Zinc Deposit in the Tri-state District, *Eng. and Mining Jour.* **127:**913-916.

Neubauer, W. H., 1971, Kosten der Lagerstättenerschliessung und Probleme ihrer Finanzierung; Versuch einer operationsanalytischen Fassung nach Wahrscheinlichkeitsmodellen, *Berg- u. Hüttenm. Monatsh.* **116:**279-293.

Newendorp, P. D., 1972, Bayesian Analysis—A Method for Updating Risk Estimates, *Jour. Petroleum Technology* **24:**193-198.

Newendorp, P. D., 1975, *Decision Analysis for Petroleum Exploration,* Petroleum Publishing Company, Tulsa, Okla., 668p.

Newton, M. J., and A. G. Royle, 1973, Mathematical Models of Orebodies, in *Application of Computer Methods in the Mineral Industry,* M. D. G. Salamon and F. H. Lancaster, eds., South African Institute of Mining and Metallurgy, Johannesburg, pp. 203–206.

Nichols, L., 1968, Field Techniques for the Economic and Geotechnical Evaluation of Mining Property for Opencase Mine Design, Knob Lake, Quebec, *Quart. Jour. Eng. Geology* **1**(3):169–180.

Nickerson, A. L., 1958, *Oil, Taxes, and Progress,* American Petroleum Institute, New York, 7p.

Nininger, R., 1954, *Minerals for Atomic Energy,* Van Nostrand, New York, 367p.

Niskanen, P., 1974, Sensitivity Maps and Tables for the Evaluation of Low-grade Ore Deposits, *Internat. Jour. Rock Mechanics and Mining Sci.* **11**(5):165–171.

Nordale, A. M., 1947, Valuation of Dredging Ground in the Sub-arctic, *Canadian Inst. Mining and Metallurgy Trans.* **50**:487–496.

Nordeng, S., C. Ensign, and M. Volin, 1964, Application of Trend Analysis to the White Pine Copper Deposit, *Stanford Univ. Pubs. Geol. Sci.* **9**(1):186–202.

Noren, N.-E., 1971, Mine Development—Some Decision Problems and Optimisation Models, *Canadian Inst. Mining and Metallurgy Spec. Vol.* **12**:240–253.

Norris, R. V., 1914, Taxation of Coal Lands: Methods of Assessment in the Anthracite Regions of Pennsylvania, *Colliery Engineer,* Feb., pp. 437–439.

Norris, R. V., 1915, Valuation of Anthracite Mines, *School Mines Quart.,* July, pp. 313–325.

Northern Miner Press Ltd, 1968, *Mining Explained,* Toronto, 264p.

Norton, J. J., and L. R. Page, 1957, Methods Used to Determine Grade and Reserves of pegmatites, *Mining Eng.* **8**(4):401–414.

Novik-Kachan, V. P., 1970, Some Peculiarities of Uranium Reserve Estimation for Underground Leaching in Sedimentary Rocks, *Atomnaya Energiya* **29**(7):3–6. (In Russian.)

Nowak, I. R., 1970, Gravity Evaluation of a Manganese Deposit at Woodie Woodie, Western Australia, *Western Australia Geol. Survey Ann. Rept. 1969,* pp. 59–62.

Nowak, T., 1953, Use of Volumetric Methods in Reserve Estimation, *Jour. Petroleum Technology* **5**(10):21–22.

Nyun, U, 1965, Testing and Evaluating Natural Gas Wells, *U.N. ECAFE Mineral Resources Devel. Ser. No. 25,* pp. 142–151.

Oblitas, J., 1969, Evaluación geológica de reservas de petróleo, *Instituto Boliviano del Petróleo* **9**(1):63–68.

O'Brian, D. T., 1969, Financial Analysis: A Tool for the Progressive Mining Man, *Mining Eng.* **21**(10):66–70.

O'Brian, D. T., and A. Weiss, 1968, Practical Aspects of Computer Methods in Ore Reserve Analysis, *Canadian Inst. Mining and Metallurgy Spec. Vol.* **9**:109–113.

O'Donahue, T. A., 1910, *Valuation of Mineral Property,* Crosby, Lockwood & Son, London, 158p.

O'Donahue, T. A., 1914, *New Interest Tables for the Valuation of Mineral Properties, Inst. Mining Eng. Trans.* **47**:100–104.

O'Donahue, T. A., 1921, Valuation of Mineral Properties, with Special Reference to Post-war Conditions, *Surveyors' Inst.,* May 9, pp. 309–344.

Oil and Gas Journal, 1953a, Calculation of Gas Reserves, in *Engineering Fundamentals,* Oil and Gas Journal, Tulsa, p. 53.

Oil and Gas Journal, 1953*b*, Calculation of Gas Reserves by Pressure-Volume Method, in *Engineering Fundamentals,* Oil and Gas Journal, Tulsa, p. 56.

O'Leary, J., 1979, Ore Reserve Estimation Methods and Grade Control at the Scully Mine, Canada—An Integrated Geological/Geostatistical Approach, *Mining Mag.,* April, pp. 300–315.

Oliver, E., 1920, Oil Property Appraisal Discussed by an Expert, *Oil and Gas Jour.,* Feb. 27, **18:**46, 48, 50.

Oliver, E., 1921, Appraisal of Oil Properties, *Am. Inst. Mining and Metall. Engineers Trans.* **65:**353–364.

Oliver, W. B., 1971, The Geologists' Role in Evaluation Economics, *Am. Assoc. Petroleum Geologists Bull.* **55:**796–800.

Olson, N. K., 1972, Rapid Technique for Estimating Tonnages of Mineral Resources, *South Carolina Div. Geology Geol. Notes* **16**(2):43–49.

Omaljev, V., 1974, Istrazenost i metodologija proracuna rezervi lezista urana Zirovski vrh, Inst. Geol.-Rud. *Istraživanja i Istpitivanja Nuklearnih i Drugih Mineral. Sci.* **9:**53–92.

Onoprienko, V. P., N. M. Svikhnushin, and V. V. Slasenkov, 1975, Ob ustanovlenii granits razlichnykh grupp zapasov nefti i gaza pri ob"yemnom metode podscheta, *Geologiya Nefti i Gaza,* no. 3, pp. 40–43.

Oosterveld, M. M., 1973, Ore Reserve Estimation and Depletion Planning for a Beach Diamond Deposit, in *Application of Computer Methods in the Mineral Industry,* M. D. G. Salamon and F. H. Lancaster, eds., South African Institute of Mining and Metallurgy, Johannesburg, pp. 65–71.

Owen, L., 1931, Empirical Formulae for the Production Curves of Oil from Wells, *Inst. Petroleum Technology Jour.* **17:**500–501.

Pack, F., 1917, The Estimation of Petroleum Reserves, *Am. Inst. Mining and Metall. Engineers Bull.* **128:**1121–1134.

Page, D., and B. Jaczkowski, 1977, A Dynamic Ore Reserve Calculation and Accounting System for a Stratiform Ore-body, *Internat. Symp. Appl. Comput. Oper. Res. Miner. Ind. No. 15,* pp. 79–87.

Paine, P., 1942, *Oil Property Valuation,* Wiley, New York, 203p.

Paione, J., 1975, Rules for the Estimation of Mineral Reserves, *Rev. Escola Minas* **32**(2):38–40. (In Portugese.)

Pallister, H. D., 1918, Formulas for Mine Valuation, *Mining and Scientific Press* **117:**682–684.

Panou, G., 1974, The Bukena (Tin) Deposit (Katanga): An Example of Reserves Estimation, *Acad. Royale Sci. Outre-Mer Cl. Sci. Tech.* **18:**1–91. (In French.)

Pantazis, T. M., 1978, Reserve Assessment of Cyprus Copper and Iron Pyrites and Chromites, *Internat. Assoc. for Mathematical Geology Jour.* **10:**555–564.

Panyity, L. S., 1926, Valuation of Properties in the Bradford District, in Petroleum Development and Technology, 1926, *Am. Inst. Mining and Metall. Engineers Trans. Spec. Vol.* **G-26:**235–240.

Panyity, L. S., 1931, Practical Interpretation of Core Analysis, in Petroleum Development and Technology, 1931, *Am. Inst. Mining and Metall. Engineers Trans.* **92:**320–328.

Paransevicius, J., 1968, Ore Reserve Estimation at the Pamour Mine, *Canadian Inst. Mining and Metallurgy Spec. Vol.* **9:**209–214.

Pardee, F., 1950, The Hoskold Formula is Justified, *Eng. and Mining Jour.,* Jan., **151:**76–77.

Pardee, F., 1952, Local Problems in the Assessment of Copper and Iron

Mines in Michigan, *Proceedings of the 45th Annual Conference on Taxation,* National Tax Assoc. pp. 600–608.

Pardee, F., 1957, The Michigan Mine Appraisal System, in *Examination and Valuation of Mineral Property,* 4th ed., C. H. Baxter and R. D. Parks, eds., Addison-Wesley, Reading, Mass., pp. 447–465.

Parfitt, P. O., 1968, Ore Reserve Estimation and Grade Control at McIntyre, *Canadian Inst. Mining and Metallurgy Spec. Vol.* **9:**197–202.

Park, J., 1905, *Examination and Valuation of Mines,* J. Mackay, gov. printer, Wellington, N. Z., 90p. (Reprinted from *New Zealand Mines Record*).

Park, J., 1907, A Text-book of Mining Geology, for the Use of Mining Students and Miners, 2nd ed., C. Griffin, 228p.

Parker, H., and P. Switzer, 1975, Use of Conditional Probability Distributions in Ore Reserve-Estimation, in *13th International Symposium on the Application of Computers and Mathematics for Decision Making in Mineral Industries,* G. Dorstewitz, ed., Verlag Glueckauf GmbH, Essen, pp. M.II.1–M.II.16.

Parkinson, E. A., and A. L. Mular, 1972, Mineral Processing Equipment Costs and Prelininary Capital Cost Estimations, *Canadian Inst. Mining and Metallurgy Spec. Vol.* 13, 142p.

Parks, R. D., and J. N. Galbraith, 1967, Computer Programming in Evaluation of Mineral Property, in *Computer Short Course and Symposium on Mathematical Techniques and Computer Applications in Mining and Exploration, Colorado Mines,* vol. 2, University of Arizona, Tucson, pp. F2-1–F2-11.

Parrish, I. S., 1968, Grade Control at the Copper Rand Mines Division, Patino Mining Corporation, *Canadian Inst. Mining and Metallurgy Spec. Vol.* **9:**160–167.

Parsons, A. B., 1933, *The Prophyry Coppers,* American Institute of Mining and Metallurgical Engineers, New York, 581p.

Parsons, C. P., 1928, Accurate Estimates of Gas Reserves, *Oil and Gas Jour.,* May 10, **26:**86, 156, 159.

Parsons, F. W., and R. D. Hall, 1917, Valuation of Federal Coal Lands, *Pan American Sci. Congress 15th Proc.* **3:**539–546.

Pasternak, V. S., 1969, Criteria for the Categorization of Reserves of Natural Resources Using a Minsk 22 Computer, in *Sbornik. Tezisy dokl. Resp. Nauchno- Techn. Konferentsii po Probl. Ugol'noi Pro-mysh,* CH-Z, Donetsk, pp. 118–189. (In Russian.)

Patterson, J., 1959, Estimation of Ore Reserves, *Mines Mag.* **49**(3):38–44.

Patterson, J., P. C. DeVergie, and R. J., Meehan, 1964, Application of Automatic Data Processing Techniques to Uranium Ore Reserve Estimation and Analysis, *Colorado School Mines Quart.* **59:**859–886.

Pavlishin, V. I., and P. K. Vovk, 1970, Ispol'zovaniye K/Rb-otnosheniya dlya otsenki produktivnosti pegmatitovykh tel, *Geologicheskiy Zhurnal* **30**(6):84–88.

Pederson, J. A., 1974, Economic Index for Increased Recovery from Stripper Wells, *Colorado School Mines Mineral Industries Bull. 17,* no. 2, 8p.

Peebles, G. A., and D. M. Ward, 1968, Ore Reserve Estimation and Grade Control at Eldorado's Beaverlodge Operation, *Canadian Inst. Mining and Metallurgy Spec. Vol.* **9:**239–250.

Peiker, E. W., and L. Forsythe, 1969, Financial Analysis—Henderson Project, *Colorado School Mines Quart.* **64**(3):159–174.

Peloubet, M. E., 1937, Natural Resource Assets; Their Treatment in Accounts and Valuation, *Harvard Business Rev.* **16**(1):74–92.

Peloubet, M. E., 1959, Accounting for the Extractive Industries, in *Economics of the Mineral Industries,* E. H. Robie, ed., American Institute of Mining and Metallurgical Engineers, New York, pp. 393–433.

Peltola, E., 1962, Malmin arviointi Outokummun kaivoksella, *Geologi* **14**(5):62–65.

Peltola, E., K. Paavo, and P. Voutilainen, 1971, How a Computer was Used in Evaluating the Vuonos Nickel Deposit in Finland, *World Mining* **24**(11):58–61.

Percy, R. F., 1933, Valuation Tables for Mineral Rents, Royalties, and Terminable Annuities, *Inst. Mining and Metallurgy Trans.,* part 3, Dec., **86**:102–105.

Perets, V., 1969, Evaluation of Ore Reserves by Computers and Optical Coincidence cards, *Trudy Tsentr. n-i. Tornorazved in-ta* **86**(1):104–117. (In Russian.)

Perišič, M., D. Bratičević, and D. Vitorović, 1977, Application of a Programming Language in Computer Processing of Mineral Deposits, in *Application of Computer Methods in the Mineral Industry,* R. V. Ramani, ed., American Institute of Mining and Metallurgical Engineers, New York, pp. 465–478.

Perry, J., D. Phillips, and L. Shearer, 1974, Mineralization Models for Mine Planning and Evaluation, *Internat. Symp. Appl. Comput. Math. Miner. Ind. Proc. No. 12,* vol. 2, pp. H72–H79.

Pester, L., and H. L. Planert, 1969, The Rapid and Comprehensive Evaluation of Some Important Data Resulting from the Geological Exploration of Brown-Coal Deposits by Means of Electronic Computers, *Bergbautechnik* **19**:11.

Peters, W. C., 1969, The Economics of Mineral Exploration, *Geophysics* **34**:633–644.

Peters, W. C., 1978, *Exploration and Mining Geology,* Wiley, New York, 719p.

Peterson, A. F., and A. R. Eshbach, 1962, Mining Cost Control, One Answer to Diminishing Profit Margins, *Mining Eng.,* Oct., **14**:39–41.

Petrascheck, W. E., 1951, Berechnung und Schätzung von Lagerstätten-vorräten, *Zeitschr. Erzbergbau u. Metallhüttenwesen* **4**(6):209–211.

Petrov, V. A., 1965, O primenenii sposobov srednego arifmeticheskogo i srednego vzveshennogo dlya rascheta srednikh parametrov pri podschete zapasov poleznogo iskopayenogo, *Sovetskaya Geologiya,* no. 2.

Petrov, V. A., 1977, O methodike vychisleniya plotnosti rud pri opredelenii yeye sposobom vyyemki tselikov, *Sovetskaya Geologiya,* no. 12, pp. 123–127.

Pfleider, E. P., and J. Scofield, 1967, A Preliminary Economic Analysis of the Underground Mining of Minnesota Taconite, *Mining Eng.,* Sept., **19**:107–112.

Phelps, E. R., 1968, Correlation of Development Data and Preliminary Evaluation, in *Surface Mining,* E. P. Pfleider, ed., American Institute of Mining and Metallurgical Engineers, New York, pp. 122–137.

Phillips, R., 1971, A Method for Estimating the Grade of Diamond Deposits, *Inst. Mining and Metallurgy Trans.* **80**(780):B357–B362.

Phillips, R. J., 1968, Trend Surface Analysis in Mining, *Western Miner* **41**(4):29–30, 32, 34, 36, 38.

Phillips, W. G. B., 1977, Statistical Estimation of Global Mineral Resources, *Resources Policy* **3**:268–280.

Phillips, W. G. B., and D. Edwards, 1976, Metal Prices as a Function of Ore Grade, *Resources Policy* **2**:167–178.

Pickering, J. C., 1917, *Engineering Analysis of a Mining Share*, McGraw-Hill, New York, 103p.

Pickering, J. C., 1919, Cost Keeping for Small Metal Mines, *U.S. Bur. Mines Tech. Pap. 223*, 46p.

Pierce, J. H., and T. F. Kennedy, 1960, *Mine Examination, Reports, Valuation*, Pierce Management Corporation, Scranton, Pennsylvania, 255p.

Pierre, M., 1974, The Sampling of Broken Ores; A Review of Principle and Practice, in *Geological Mining and Metallurgical Sampling*, Institute of Mining and Metallurgy, London, pp. 194–205.

Piloski, M. J., 1968, Some Comments on the "Application of Trend Analysis in the Evaluation of the Whalesback Mine, Newfoundland," by F. P. Agterberg, *Canadian Inst. Mining and Metallurgy Spec. Vol.* **9:**89–94.

Pirow, P. et al., 1966, Computer Programs for the Estimation of Two-dimensional Trend Surface Using a Weighted Moving Average, in *Symposium on Mathematical Statistics and Computer Applications in Ore Evaluation*, South African Institute of Mining and Metallurgy, Johannesburg, pp. 80–105.

Pirson, S. J., 1935, Production Decline Curve of Oil Well May be Extrapolated by Loss Ratio, *Oil and Gas Jour.*, no. 14, pp. 34–35.

Pirson, S. J., 1941, Probability Theory Applied to Oil Exploration Ventures, *Petroleum Engineer*, Feb., pp. 27–28; Mar., pp. 177–182; Apr., pp. 46–48; May, pp. 53–56.

Pirson, S. J., 1944, Equivalence of Material Balance Equations for Calculating Original Residual Oil Reservoir in Place, *Oil Weekly*, Apr. 3, **13:**28–30, 32, 36.

Pirson, S. J., 1963, *Handbook of Well Log Analysis for Oil and Gas Formation Evaluation*, Prentice-Hall, Englewood Cliffs, N.J., 326p.

Pitcher, R., 1947, Depletion and Depreciation in Oil Production Accounting, *Oil Weekly*, June 26, **126:**49–50, 51–52.

Plank, F. B., 1951, *Percentage Depletion*, Tulane University Tax Institute, New Orleans, La., pp. 227–239.

Pogrebitskiy, Ye. O., and V. I. Ternovoy, 1967, Nekotoryye voprosy konditsiy dlya mestorozhdeniy poleznykh iskopayemykh, *Geologiya i razvedka mestorozhdeniy poleznykh iskopayemykh*, Gorn. Inst., Zap. (Leningrad), **52**(2):111–121.

Polster, L. A., 1964, Qualitative Estimates of Reserves Magnitudes, in *Printsipy otsenki perspektiv neftegazonosnosti krupnyky territorii, Gos. geol. komitet SSSR*, "Nedra" Press, Leningrad, NILNEFTEGAZ, vyp. 11, pp. 7–187. (In Russian.)

Popoff, C. C., 1966, Computing Reserves of Mineral Deposits—Principles and Conventional Methods, *U.S. Bur. Mines Inf. Circ. 8283*, 133p.

Porter, S. P., 1965, *Petroleum Accounting Practices*, McGraw-Hill, New York, 557p.

Potapoff, P., 1968, Ore Reserve Estimation and Grade Control Methods for Nickel-Copper-Iron Sulphide Deposits at Sudbury, Ontario, *Canadian Inst. Mining and Metallurgy Spec. Vol.* **9:**130–138.

Pow, J. R. et al., 1963, Descriptions and Reserve Estimates of the Oil Sands of Alberta, in *The K. A. Clark Volume*, M. A. Carrigy, ed., Res. Counc. Alberta, Inf. Ser. No. 45, pp. 1–14.

Powell, D. H., 1960, Evaluation of Oil Properties in After-Tax Dollars, *Jour. Petroleum Technology*, Jan., pp. 28–36.

Powell, J. D., 1974, Evaluation of Phosphate Resources in Southeastern Idaho, *Idaho Bur. Mines Geol. Inf. Circ. 25*, 33p.

Powers, J. C., 1967, Geology and the Law: A Paradox in Land Evaluation, in *Economic Geology in Massachusetts,* O. C. Farquhar, ed., University of Massachusetts, Amherst, pp. 285–288.

Pozharitsky, K. L., 1963, Calculation and Practical Significance of Category C_2 Geologic Reserves of Ore Deposits, *Internat. Geology Rev.* **5:**1450–1456.

Prazsky, J., 1972, Evaluation of the Degree of Investigation of Deposits, Carried Out Using the Method of Relative Errors, *Geol. Pruzkum* **14**(6):173–174. (In Czech.)

Prescott, B., 1925, Sampling and Estimating Cordilleran Lead-Silver Limestone Replacement Deposits, *Am. Inst. Mining and Metall. Engineers Trans.* **72:**666–676.

Preston, F., and J. Van Scoyoc, 1964, Use of Asymmetry Frequency Distribution Curves of Core Analysis in Calculating Oil Reserves, in *Computers in the Mineral Industries,* G. A. Parks, ed., Stanford Univ. Pubs. Geol. Sci. **9**(2):694–720.

Preston, L. E., 1960, *Exploration for Non-ferrous Metals, An Economic Analysis,* Resources for the Future, Inc., Washington, D.C., 198p.

Pretorius, D. A., 1966, Conceptual Geological Models in Exploration for Gold Mineralization in the Witwatersrand Basin (South Africa), in *Symposium on Mathematical Statistics and Computer Applications in Ore Evaluation, Mar. 7–8,* South African Institute of Mining and Metallurgy, Johannesburg, pp. 225–275.

Probert, F. H., 1915, Valuation of Metal Mines, *Mining and Scientific Press* **111:**657–659.

Probert, F. H., and R. B. Earling, 1911, The Estimation of Ore Reserves, *Eng. and Mining Jour.* **92:**1179–1180.

Prokof'yev, A., 1967, Vyyavleniye i uchet uragannykh prob pri podschete zapasov rudnykh mestorozhdeniy, *Vyssh. Ucheb. Zavedeniy Izv., Geologiya i Razved.,* no. 4, pp. 48–57.

Prokof'yev, A., 1968, Recognition and Consideration of "Bonanzas" in Calculation of Ore Deposit Reserves, *Internat. Geology Rev.* **10:**697–703.

Pronk Van Hoogeveen, L. A. J., J. K. Cutland, and M. Weir, 1973, An Open Pit Design System for Stratiform Orebodies, *Internat. Symp. Appl. Comput. Methods Min. Ind.* (1972), no. 10, pp. 149–154.

Prouty, R. W., and R. T. Green, 1925, Methods of Sampling and Estimating Ore in Underground and Steam Shovel Mines of Copper Queen Branch, Phelps Dodge Corporation, *Am. Inst. Mining and Metall. Engineers Trans.* **72:**628–639.

Pryor, E. J., 1958, *Economics for the Mineral Engineer,* Pergamon, New York, 254p.

Pryor, R. N. et al., 1972, Sampling of Cerro Colorado, Rio Tinto, Spain, *Inst. Mining and Metallurgy Trans.* **81**(788):A143–A159.

Pundary, N., 1966, Applying Statistics to Estimate Coal Reserves, *Coal Age,* Mar., 96–100.

Purdie, J. J., 1968, Mineral Reserve Calculations at Lake Dufault, *Canadian Inst. Mining and Metallurgy Spec. Vol.* **9:**183–192.

Quinlan, T., and N. R. Crosby, 1977, Mining Reserves Using a Mini Computer, *Internat. Symp. Appl. Comput. Oper. Res. Miner. Ind. No. 15,* pp. 255–261.

Qureshy, M. N. et al., 1969, A Study and Practical Applications of Statistical Value Distribution for the Auriferous Deposits of the Kolar Gold Mines, India, in *A Decade of Digital Computing in the Mineral Industry,* A. Weiss, ed., American Institute of Mining and Metallurgical Engineers, New York, pp. 421–432.

Ramey, H. J., and E. T. Guerrero, 1970, The Ability of Rate-Time Decline Curves to Predict Production Rates, *Am. Inst. Mining and Metall. Engineers Petroleum Trans. Reprint Series No. 3,* pp. 103–105.

Rancourt, G., and J. H. Evans, 1968, Grade Control at East Malartic Mines, Limited, *Canadian Inst. Mining and Metallurgy Spec. Vol.* **9:**227–232.

Randolph, R. S., 1950, *Problems of the Oil and Gas Industry; Depletion Problems Including Those Arising from the Hudson and Abercrombie Decision,* New York University Institute on Federal Taxation, New York, pp. 491–504.

Rao, S. V. L. N., 1968, A Method for Grading Homogeneous Economic Mineral Deposits, *Geol. Soc. India Jour.* **9**(1):1–12.

Rau, W. I., 1980, Project Financing for International Mining Ventures, *Mining Eng.* **32:**1262–1264.

Rawlins, E. L., and M. A. Schellhardt, 1936, Extent and Availability of Natural Gas Reserve in Michigan "Stray" Sandstone Horizon of Central Michigan, *U.S. Bur. Mines Rept. Inv. 3313,* 139p.

Raymond, G. F., 1979, Ore Estimation Problems in an Eratically Mineralized Orebody, *Canadian Inst. Mining and Metallurgy Bull.* **72**(806):90–98.

Razumovsky, N. K., 1940, Distribution of Metal Values in Ore Deposits, *Acad. Sci. URSS, Comp. Rend. (Doklady)* **28**(9):814–816.

Read, T. T., 1943, Estimating Ore by the Polygon Method, *Eng. and Mining Jour.* **144**(8):84–85.

Redmayne, R. A. S., and G. Stone, 1920, *The Ownership and Valuation of Mineral Property in the United Kingdom,* Longmans, Green & Co., London, 263p.

Reed, W., 1920, How to Figure Leasehold Valuations for Taxation Purposes; Pointers that May Help You; Treasury Department's Views, *Coal Rev.,* Mar. 24, pp. 29–31.

Reeves, H. C., 1950, Assessment of Coal Producing Properties in Kentucky, *Natl. Tax Jour.,* June, pp. 173–178.

Reid, J. A., and C. Huston, 1945, The Practical Examination of Mineral Prospects, with a Discussion of the Form of Mining Reports, *Canadian Inst. Mining and Metallurgy Trans.* **48:**270–283.

Reis, B. J., 1923, Depletion and Other Factors Bearing on Coal Costs, *Coal Trade Bull.* Feb. 1 and 16, 29p. (Reprint.)

Rendu, J.-M., 1970, Geostatistical Approach to Ore Reserve Calculation, *Eng. and Mining Jour.* **171:**114–118.

Rendu, J.-M., 1971, Some Applications of Geostatistics to Decision-making in Exploration, *Canadian Inst. Mining and Metallurgy Spec. Vol.* **12**(6):175–184.

Requa, M., 1912, Present Conditions in the California Oil-fields, *Am. Inst. Mining and Metall. Engineers Trans.* **42:**837–846.

Requa, M., 1918, Methods of Valuing Oil Lands, *Am. Inst. Mining and Metall. Engineers Trans.* **59:**526–556.

Research Council of Alberta, 1977, Symposium on Coal Evaluation, Calgary, Alberta, October 31, 1974 to November 1, 1974, *Research Council Alberta Inf. Ser. No. 76,* 182p.

Reudelhuber, F. O., and R. F. Hinds, 1958, A Compositional Material Balance Method for Prediction of Recovery from Volatile Oil Depletion Drive Reservoirs, *Am. Inst. Mining and Metall. Engineers Trans.* **210:**19–26.

Rice, H., 1961, Discussion of J. B. Evans, "The Evaluation of Mining Properties—A Graphical Approach," *Canadian Inst. Mining and Metallurgy Trans.* **64:**120–121.

Richards, R., 1973, A Three-dimensional Model for Predicting Natural Gas Reserves and Availability, *Canadian Jour. Petroleum Technology* **12**(1):41–46.

Rickard, T. A., 1903, *The Sampling and Estimation of Ore in a Mine,* McGraw-Hill, New York, 222p. (2nd ed., 1907.)

Rickard, T. A., 1913, Valuation of Mines, *Mining and Scientific Press,* May 24, pp. 766–771.

Rickard, T. A., 1915, Valuation of Metal Mines, *Mining and Scientific Press,* Oct. 9, **111:**548–553.

Rickard, T. A., 1941, Salting, *Eng. and Mining Jour.,* Mar., **142:**42–45; May, pp. 52–54; June, pp. 50–51.

Rickard, T. A. et al., 1905, *The Economics of Mining,* Engineering and Mining Journal, New York, 421p. (2nd ed., 1907.)

Ricketts, A. H., 1911, *A Manual of American Mining Law,* Scientific Book Publishing Company, San Francisco, 486p.

Riddell, G. C., 1940, Fundamentals of Mineral Property Valuation, *Eng. and Mining Jour.* **141**(11):37–40.

Riddell, J. M., 1949, Enhancement and Hazard Factors as Related to Mine Valuation, *Am. Inst. Mining and Metall. Engineers Trans.* **184:**324–325.

Riesz, E. J., 1978, Can Rank-size "Laws" be Used for Undiscovered Petroleum and Mineral Assessments? *Jour. Australian Geology and Geography* **3**(3):253–256.

Ritchie, K. S., 1931, A Study of the Production Curves for the 8500 Foot Horizon in the Big Lake Oil Field, Reagan County, Texas, *Mining and Metallurgy,* June, pp. 266–269.

Roberts, D. V. et al., 1966, The Urban Threat to the Sand and Gravel Industry, *Nevada Bur. Mines Rept. 13,* pp. 51–64.

Roberts, J. E., 1972, Using Models to Cope with Uncertainty in Mine Investment Decisions, in *Decision-making IV, Cooperative Decision-making by the Geologist and Mining Engineer, February 10-12, 1971,* Center of Continuing Education, University of British Columbia, Vancouver, 3p.

Roberts, T. G., 1959, A Permeability Block Method of Calculating a Water Drive Recovery Factor, *Petroleum Engineer,* Sept., pp. B45–B48.

Roberts, W. A., 1944, *State Taxation of Metallic Deposits,* Harvard University Press, Cambridge, Mass.

Robertson, C. E., 1971, Evaluation of Missouri's Coal Resources, *Missouri Div. Geol. Survey and Water Resources Rept. Inv. No. 48,* 92p.

Rocky Mountain Law Review, 1955, *Symposium on Uranium Law,* vol. 27, pp. 375–537.

Rocky Mountain Mineral Law Foundation, 1955, *Proceedings of the Rocky Mountain Mineral Law Institute Annual, 1955,* Mathew Bender & Co., New York. (See also subsequent proceedings.)

Rodgers, T. A., 1965, The Use of Cash Flow Techniques in the Evaluation of Mining Investment, in *Exploration and Mining Geology,* L. J. Lawrence, ed., vol. 2, Australasian Institute of Mining and Metallurgy, Melbourne, pp. 289–295.

Rodionov, G. G., and B. M. Ronenson, 1972, *Assessment of Deposits When Prospecting and Exploring Mica,* Nedra, Moscow, 216p. (In Russian.)

Roe, D. W., 1955, Introduction to Tax Problems of Uranium, *Univ. Denver 5th Ann. Tax Inst.,* pp. 118–127.

Roebuck, F., 1979, *Economic Analysis of Petroleum Ventures,* Institutes for Energy Development, Tulsa, Okla., 191p.

Roeser, H. M., 1925, Determining the Constants of Oil Production Decline Curves, *Am. Inst. Mining and Metall. Engineers Trans.* **71:**1315–1321.

Rogers, G. S., and C. E. Lesher, 1914, Use of Thickness Contours in the Valuation of Lenticular Coal Beds, *Econ. Geology* **9:**707–729.

Roginskiy, F. N., 1968, Operdeleniye ekonomicheskoy effektivnosti geologorazvedochnykh rabot, *Razved. i Okhrana Nedr,* no. 7, pp. 35–38.

Rose, W., 1960, Material Balance Forecasts with Initial Production Data Lacking, *Producers Monthly,* Sept., pp. 24–29.

Rossen, R. H., 1975, A Regression Approach to Estimating Gas in Place for Gas Fields, *Am. Inst. Mining and Metall. Engineers Trans.,* part 1, **259:**1283–1289.

Roufaiel, G. S. S., 1966, On the Application of Geostatistics to Stratiform Ore Deposits, *Jour. Geology United Arab Republic,* **6:**121–131.

Rowe, R. B., 1953, Evaluation of Pegmatitic Mineral Deposits, *Canadian Inst. Mining and Metallurgy Bull. No. 499,* pp. 700–705.

Royle, A. G., 1978, A Probabilistic Basis for Ore Reserve Statements, *Mining Mag.* **139**(2):142–143.

Royle, A. G., 1979a, Why Geostatistics, *Eng. and Mining Jour.* **180**(5):92–94, 97, 98, 101.

Royle, A. G., 1979b, Estimating Small Blocks or Ore; How to Do it with Confidence, *World Mining* **32**(4):55–57.

Royle, A. G., and E. Hosgit, 1974, Local Estimation of Sand and Gravel Reserves by Geostatistical Methods, *Inst. Mining and Metallurgy Trans.* **83:**A53–A62.

Royle, A. G., and E. Hosgit, 1975, Local Estimation of Sand and Gravel Reserves by Geostatistical Methods, *Inst. Mining and Metallurgy Trans.* **84:**A73–A79.

Royle, A. G., and M. J. Newton, 1972, Mathematical Models, Sample Sets and Ore-Reserve Estimation, *Inst. Mining and Metallurgy Trans.* **81:**A121–A128.

Royle, A. G., M. J. Newton, and V. K. Sarin, 1974, Geostatistical Factors in Design of Mine Sampling Programmes, in *Geological, Mining and Metallurgical Sampling,* Institution of Mining and Metallurgy, London, pp. 71–77 (Reprint from *Inst. Mining and Metallurgy Trans.,* 1972.)

Rudenno, V., and E. C. Thomas, 1977, Development of a Simplified Economic Model to Allow More Flexible and Direct DCF Analysis in Mining Feasibility Studies, in *Application of Computer Methods in the Mineral Industry,* R. V. Ramani, ed., American Institute of Mining and Metallurgical Engineers, New York, pp. 908–919.

Rudkevich, M., 1972, *Methods for Resource and Reserve Estimation and the Basis of Calculating Parameters,* SBOR, Tyumen 179p. (In Russian.)

Ruedemann, P., 1922a, Some Graphical Methods for Appraising Oil Wells, *Am. Assoc. Petroleum Geologists Bull.* **6:**533–544.

Ruedemann, P., 1922b, Comparative Barrel Day Values for Different Sized Wells, *Natl. Petroleum News,* June 21, pp. 73–75.

Ruedemann, P., and I. Gardescu, 1922, Estimation of Reserves of Natural Gas Wells by Relationship of Production to Closed Pressure, *Am. Assoc. Petroleum Geologists Bull.* **6:**444–463.

Rummerfield, B. F., and N. S. Morrisey, 1964a, How to Evaluate Exploration Prospects, *Geophysics* **29:**434–444.

Rummerfield, B. F., and N. S. Morrisey, 1964b, The Application of Computers in Analyzing and Evaluating Petroleum Exploration Prospects, *Stanford Univ. Pubs. Geol. Sci.* **9**(2):530–560.

Ruskin, V. W., 1967, Versatile Applications of Computers to Mining Exploration and Evaluation, *Canadian Inst. Mining and Metallurgy Bull.* **60:**1051–1059.

Rutledge, R. W., 1976, The Potential of Geostatistics in the Development of Mining, in *Advanced Geostatistics in the Mining Industry,* M. Guarascio, M. David, and C. Hujbregts, eds., Reidel, Dordrecht, Holland, pp. 295–311.

Ruttan, G. D., 1968, Ore Estimation and Grade Control at Sherritt Gordon's Lynn Lake Operation, *Canadian Inst. Mining and Metallurgy Spec. Vol.* **9:**139–143.

Ruzicka, V., 1976a, Uranium Resources Evaluation Model as an Exploration Tool, in *Exploration for Uranium Ore Deposits,* International Atomic Energy Agency, Vienna, pp. 673–682.

Ruzicka, V., 1976b, Evaluation of Uranium Resources in the Elliot Lake-Blind River Area, Ontario, *Canada Geol. Survey Paper 76-1B,* pp. 127–129.

Ruzicka, V., 1978, Evaluation of Selected Uranium-bearing Areas in Canada, *Canada Geol. Survey Paper 78-A,* pp. 269–274.

Ryan, J. M., 1966, Limitation of Statistical Methods for Predicting Petroleum and Natural Gas Reserves and Availability, *Jour. Petroleum Technology* **18:**281–284.

Sage, B. H., and W. N. Lacey, 1935, Formation Volume and Viscosity Studies for the Dominquez Field, *Oil and Gas Jour.* **34:**16–18.

Sage, B. H., and W. N. Lacey, 1937, Applying Phase-Equilibrium Data to Estimation of Oil and Gas Reserves, *Oil and Gas Jour.,* May 23, **36:**124, 126, 128, 129–131.

Salgado, J. A., 1968, Evaluación de muestreos en minas de oro mediante computadores digitales, *Geos,* no. 18, pp. 48–57.

Saliers, E. A., 1922, Mine Valuation, in *Depreciation Principles and Applications,* Ronald Press, New York, pp. 466–468.

Salimbayev, A. S. et al., 1968, Sopostavleniya razvedochnykh i ekspluatatsionnykh dannykh po Sololovskomu kar'yeru, *Akad. Nauk Kazakhskoy SSR, Izvestiya Seriya Geologicheskaya,* no. 1, pp. 1–10.

Salski, W., 1974, Znaczenie badan tektonicznych w kopalniach glebinowych, *Przeglad Geol.* **22**(2–3):72–77.

Sandefur, R. L., and D. Grant, 1976, Preliminary Evaluation of Uranium Deposits — A Geostatistical Study of Drilling Density in Wyoming Solution Fronts, in *Exploration for Uranium Ore Deposits,* International Atomic Energy Agency, Vienna, pp. 695–714.

Sanders, T. P., 1936, Greater Accuracy in Estimating Oil and Gas Reserves Now Demanded, *Oil and Gas Jour.,* Aug. 13, **35:**17, 20.

Sandier, J., 1962, *Mise en valeur des gisements métallifères,* Masson et Cie, Paris, 149p.

Sani, E., 1976, Mineral Investment Valuation and the Cost of Capital, *Resources Policy* **2**(4):285–296.

Sani, E., 1977, The Role of Weighted Average Cost of Capital in Evaluating a Mining Venture, *Mining Eng.* **29:**42–46.

Sara, W. A., 1963, Reserves of Bituminous Coal in the Liverpool and Paparoa areas, Greymouth Coafield, *New Zealand Jour. Geology and Geophysics* **6:**566–581.

Sarkisyan, B. M., 1961, Problem of the Method of Calculating Remaining Oil Reserves at Depth, *Petroleum Geology* **3:**395–400. (Translated from 1959 edition.)

Sarma, D., 1968, A Preliminary Statistical Study on the Basic Mine Valuation Problems at Kolar Goldfields, Mysore State (India), *Geoexploration* **7:**97–105.

Sarma, D., 1970, A Statistical Study in Mining Exploration as Related to the Kolar Gold Fields (India), *Geoexploration* **8:**19–35.

Sassano, G. P., 1970, Estimation of Mineral Reserves with Computers, *Geologia Tecnica* **17**(1):3–9.

Savinskiy, I. D., A. P. Grudev, and A. Petrova, 1975, Primeneniye matematicheskikh metodov i EVM dlya povysheniya tochnosti podscheta zaposov i postroyeniya kart v izoliniyakh, *Geologiya Rudnykh Mestorozhdcniy.* **17**(6):108–114.

Savosin, M. N., 1971, O tochnosti opredeleniya moshchnosti, soderzhaniya i zapasov metalla po ekspluatatsionnomu bloku, razvedannomu gornymi vyrabotkami, *Sovetskaya Geologiya*, no. 3, pp. 77–89.

Schenck, G. H. K., 1966, Selected Bibliography Serves as Guide to Modern Mine Valuation Methods, *Eng. and Mining Jour.* **167**(9):206–208, 260.

Schillinger, A., 1964, The Application of Induced Polarization Probing Techniques Underground—Michigan Native Copper District, *Am. Inst. Mining and Metall. Engineers Trans.* **229:**409–415.

Schilthuis, R. J., 1935, Reservoir Energy and Oil Production Data Used in Estimating Reserves, *Oil and Gas Jour.*, Oct. 17, **32**(22):34–39.

Schilthuis, R. J., 1936, Active Oil and Reservoir Energy, *Am. Inst. Mining and Metall. Engineers Trans.* **118:**33–52. (Reprint pp. 31–50.)

Schilthuis, R. J., 1938, Connate Water in Oil and Gas Sands, *Am. Inst. Mining and Metall. Engineers Trans.* **127:**199–214.

Schilthuis, R. J., and W. Hurst, 1935, Variations in Reservoir Pressure in the East Texas Field, *Am. Inst. Mining and Metall. Engineers Trans.* **114:**164–176.

Schmidt, F. S., 1916, The Present Value of a Mine, *Mining and Scientific Press* **112:**207–208.

Schmitt, H., 1929, Extension of Oreshoots with Comments on the Art of Ore Finding, American Institute of Mining and Metallurgical Engineers, New York, 9p. *(Am. Inst. Mining and Metall. Engineers Tech. Pub. No. 164, Cl. I, Min. Geol., No. 20.)*

Schmitt, H., 1938, Valuation of Mines and Prospects, *Eng. and Mining Jour.* **139**(3):43–46.

Schmutz, G. L., and E. M. Rams, 1963, *Condemnation Appraisal Handbook.* Prentice-Hall, Englewood Cliffs, N. J., 812p.

Schoemaker, R., 1957, Gas Appraisal, a Graphic Short-cut for Geologists, *Alberta Soc. Petroleum Geologists Jour.* **5**(9):200–209.

Schoemaker, R., 1963, A Graphic Short-cut for Rate of Return Determinations, *World Oil,* July 1, pp. 73–84; Aug. 1, pp. 69–73; Sept. 1, pp. 64–68.

Schwab, B., and H. D. Drechsler, 1978, Evaluation of New Mining Ventures: Average Cost Versus Net Present Value, *Canadian Inst. Mining and Metallurgy Trans.* **81:**1–5.

Schwade, I. T., 1967, Geologic Quantification—Description-Numbers-Success Ratio, *Am. Assoc. Petroleum Geologists Bull.* **51:**1225–1239.

Schwingle, C. J., 1962, Problems of Economic Valuation of Industrial Assets, *Eng. Economist,* Jan.–Feb., pp. 23–30.

Scott, J. H., 1963, The GRADE Computer Program for Calculating Uranium Ore Reserves, *U.S. Atomic Energy Comm. Rept. RME-145,* 25p.

Seale, W. E., 1951, Problems of Depletion in Oil and Gas Leases, in *2nd Annual Institute on Oil and Gas Law,* Southwestern Legal Foundations, pp. 351-366.

Seda-Reyda, C. S., 1971, La geoestadistica y su aplicación en la evaluación de yacimientos mineros, *Soc. Geol. Boliviana Bol. No. 16,* pp. 17-30.

Semenyuk, V. D., Yu. I. Matkovskiy, V. I. Testov, and A. G. Glebov, 1964, Sopostavleniya razvedannykh zapasov (v blokakh) s zapasami po dannym ekspluatatsii. *Sovetskaya Geologiya,* no. 7, pp. 114-130.

Sen, J., 1921, Method of Calculating Average Value of Placer Ground, *Mining and Scientific Press* **122:**704.

Sergeev, A. A., 1973, *Economics of Drilling and Mining Exploration Activities,* VIMS, SBOR, Moscow, 95p. (In Russian.)

Sergijko, Y. A., A. R. Rakhimbaev, and P. F. Ashaev, 1972, *Elementary Operations with Matrices: Estimation of Reserves by the Digital Model,* PZCHM (for BESM-4), Alma-Ata, USSR, 37p. (In Russian.)

Serra, J., 1971, Quanitification of Lateritic-type Structures—Application to Their Estimation, *Canadian Inst. Mining and Metallurgy Spec. Vol.* **12:**170-173.

Shakespeare, W., 1965, Taxation of New Mines, *Canadian Inst. Mining and Metallurgy Bull.* **58:**999-1001.

Shalmina, G. G., S. D. Tegicheva, and V. I. Botvinnikov, 1974, *Estimation of Multi Component Mineral Deposits: Recommended Methods,* SNIIGGIMS, Novosibirsk, 77p. (In Russian.)

Shamel, C. H., 1906, The American Law Relating to Minerals. (Reprinted from) *School Mines Quart.,* Nov., vol. 27, 1905, 27p.

Shamel, C. H., 1907, *Mining, Mineral and Geological Law,* Hill Publishing Company, London and New York, 627p.

Shamkova, V. B., and S. G. Saubanova, 1969, Vydeleniye kollektorov v karbonatom razreze i otsenka kharaktera ikh nasyshchennosti s pomoshch'yu EVM, *Akad. Nauk SSSR Sibirsk. Otdeleniye Geologiya i Geofizika,* no. 2, pp. 78-83.

Sharp, W. E., 1976, A Log-normal Distribution of Alluvial Diamonds with an Economic Cutoff, *Econ. Geology* **71:**648-655.

Shaw, C. T., 1966, Current Valuation Practice on Western Areas Gold Mining Company, Limited (South Africa) and Application of the Computer in this work, in *Symposium on Mathematical Statistics and Computer Applications in Ore Evaluation, Mar. 7-8,* South African Institute of Mining and Metallurgy Johannesburg, pp. 314-323.

Shaw, E., 1929, Rough Estimation of Tonnage of Stone and Gravel Deposits, *Rock Products,* Mar. 16, **32**(6):66-68.

Shaw, E. W., 1919, The Principles of Natural Gas Land-Valuation, *Am. Assoc. Petroleum Geologists Bull.* **3:**378-388.

Shaw, S. F., 1933, Ultimate Recovery and Its Relation to Ultimate Profits, *Oil and Gas Weekly,* Apr. 3, **69**(3):13.

Shaw, S. F. 1935, Reserves Represented by the Important Fields Operating Under Proration and Unitization, *Oil Weekly* **76:**18-23, 38.

Sherrod, G. E., 1962, Factors Beyond the Engineering Evaluation that Affect Fair Market Values, *Jour. Petroleum Technology* **14:**1307-1310.

Shestakov, V. A., 1976, Nauchnyye osnovy vybora i ekonomicheskoy otsenki sistem razrabotki rudnykh mestorozhdeniy, Izd. Nedra, Moskow, 263p.

Shrivastava, D. K., 1977, Ore Gradation and Reserve Estimation of Magnesite of Chalk Hills of Salem, Tamil Nadu, *Indian Acad. Sci. Proc.*, part A, **43**:323-331.

Shumin, N. M., 1974, Metodika ekonomicheskogo obosnovaniya maksimal'no dopustimoy moshchnosti porodnogo prosloya pri podschete zapasov uglya, *Razved. i Okhrana Nedr,* no. 6, pp. 20-23.

Shurtz, R. F., 1959, The Electronic Computer and Statistics for Predicting Ore Recovery, *Mining Eng.* **11**:1035-1044.

Shurtz, R. F., 1966a, The Mathematics of Mine Sampling, *Am. Inst. Mining and Metall. Engineers Trans.* **235**:75-82.

Shurtz, R. F., 1966b, Statistical Control in the Evaluation of an Asbestos Deposit, *Canadian Inst. Mining and Metallurgy Trans.* **69**:141-146.

Sichel, H. S., 1947, An Experimental and Theoretical Investigation of Bias Error in Mine Sampling with Special Reference to Narrow Gold Reefs, *Inst. Mining and Metallurgy Trans.*, Feb., **56**:403-443.

Sichel, H. S., 1951-1952, New Methods in the Statistical Evaluation of Mine Sampling Data, *Inst. Mining and Metallurgy Trans.* **61**(6):261-288.

Sichel, H. S., 1973, Statistical Valuation of Diamondiferous Deposits, in *Application of Computer Methods in the Mineral Industry,* M. D. G. Salamon and F. H. Lancaster, eds., South African Institute of Mining and Metallurgical Engineers, Johannesburg, pp. 17-26.

Sichel, H. S., and R. S. Rowland, 1961, Recent Advances in Mine Sampling and Underground Valuation Practice in South African Gold Fields, *7th Commonwealth Mining and Metallurgical Congress Transactions,* pp. 1-21.

Simakov, V. A., and Sh. Sh. Baybusinov, 1973, Izmeneniye kharakteristik zapasov polimetallicheskikh mestorozhdeniy v zavisimosti ot bortovogo soderzhaniya, *Vyssh. Ucheb. Zavedeniy Izv., Geologiya i Razved.,* no. 3, pp. 146-151.

Simic, R., I. Trajkovic, and R. Gavric, 1972, Computer Calculation of Mass Volumes in Open Pit Mining with Special Reference to the Kosovo Coal Basin, *Zb. Rad. Rud. Geol. Fak. Univ. Beogr.* **15**:209-220. (In Serbo-Croatian.)

Sinclair, A. J., 1974, Probability Graphs of Ore Tonnages in Mining Camps— A Guide to Exploration, *Canadian Inst. Mining and Metallurgy Bull.* **67**(750):71-75.

Sinclair, A. J., 1979, Preliminary Evaluation of Summary Production Statistics and Location Data for Vein Deposits, Slocan, Ainsworth and Slocan City Camps, Southern British Columbia, *Canada Geol. Survey Paper 79-1B,* pp. 173-178.

Sinclair, A. J., and J. R. Deraisme, 1976, A 2-dimensional Geostatistical Study of a Skarn Deposit, Yukon Territory, Canada, in *Advanced Geostatistics in the Mining Industry,* M. Guarascio, M. David, and C. Huijbregts, eds., Dordrecht, Holland, pp. 369-379.

Sinclair, A. J., and G. Woodsworth, 1970, Multiple Regression as a Method of Estimating Exploration Potential in an Area near Terrace, B.C., *Econ. Geology* **65**:998-1003.

Singer, D. A., D. P. Cox, and L. J. Drew, 1975, Grade and Tonnage Relationships Among Copper Deposits, *U. S. Geol. Survey Prof. Paper 907A,* pp. A1-A11.

Skinner, E. N., and H. R. Plate, 1915, *Mining Costs of the World; A Compilation of Cost and Other Important Data on the World's Principal Mines,* McGraw-Hill, New York, 414p.

Skipsey, E., 1970, Role of the Geologist in Modern Coal Mining, with Special Reference to the East Midlands Coalfield of Britain, *Mining and petroleum geology, 9th Commonw. Min. Metall. Cong. Publ. Proc.* vol. 2, pp. 369–383.

Skopos, M. J., and M. D. Lawton, 1968, Method of Calculating Ore Reserves at the Anglo-Rouyn Property, Saskatchewan, *Canadian Inst. Mining and Metallurgy Spec. Vol.* **9:**193–196.

Slamet, S., 1977, Mineral Property Valuation Under the Indonesian Mining Law, *Internat. Assoc. for Mathematical Geology Jour.* **3:**333–337.

Slider, H. C., 1976, *Practical Petroleum Reservoir Engineering Methods,* Petroleum Publishing Company, Tulsa, Okla., 559p.

Smirnov, V. I., 1960, Geological Bases for the Exploration and Prospecting of Ore Deposits (pt. 2, sec. 3–4) *Internat. Geology Rev.* **2:**739–762.

Smirnov, V. I. et al., 1960, *Podschet zapasov mestorozhdenii poleznykh iskopaemykh,* Gosudar, Nauch.-Tekh. Izd. Lit. Geol. i Okhrane Nedr, Moscow, 672p.

Smith, J. W., L. G. Trudell, and K. E. Stanfield, 1963, Comparison of Oil Yield from Core and Drill-cutting Sampling of Green River Oil Shales, *U.S. Bur. Mines Rept. Inv. 6299,* 35p.

Smith, M. B., 1970, Probability Models for Petroleum Investment Decisions, *Jour. Petroleum Technology* **22:**543–550.

Smith, S., 1975, The Effect of Economic and Technological Factors on Estimates of Proved Reserves and Productive Capacity, in *Seminar on Reserves and Productive Capacity,* American Petroleum Institute, Washington, D.C., pp. 43–54.

Smith, W. H., 1978, Computer Evaluation and Classification of Coal Reserves, in *Coal Exploration,* W. L. G. Muir, ed., Miller Freeman Publishing, Inc., San Francisco, pp. 450–462.

Smith, W. H., and D. J. Berggren, 1963, Strippable Coal Reserves of Illinois, *Illinois Geol. Surv. Circ. 348,* 59p.

Smock, J. C., 1882, Valuation of Iron Mines in New York and New Jersey, *Am. Inst. Mining and Metall. Engineers Trans.* **10:**288–293.

Sobolevskiy, T. F., 1976, *Ekonokicheskaya otsenka rudnykh mestorozhdeniy,* Izd. Nedra, Moscow, 141p.

Society of Petroleum Engineers, 1962, *Symposium on Petroleum Economics and Valuation,* American Institute of Mining and Metallurgical Engineers, Dallas, 138p.

Sohnge, G., 1966, Outline of Ore Estimation Methods at Tsumeb, South Africa (Namibia), in *Symposium on Mathematical Statistics and Computer Applications in Ore Evaluation, Mar. 7–8,* South African Institute of Mining and Metallurgy, Johannesburg, pp. 342–352.

Sokolovskiy, Yu. A., 1966, Ekonomicheskaya sushchnost' i otsenka razvedannykh zapasov poleznykh isokopayemykh, *Sovetskaya Geologiya,* no. 1, pp. 109–118.

Soloman, E., 1970, Return on Investment: The Relation of Book Yield to True Yield, *Am. Inst. Mining and Metall. Engineers Petroleum Trans. Reprint Ser. No. 3,* 2nd ed., pp. 154–161.

Solovev, V. G., 1939, Methods of Variation Statistics Applied to the Prospecting and Calculations of the Reserves of Useful Mineral Deposits, *Cent. Geol. Prosp. Inst. Trans. Fasc. 115,* 55p. (In Russian.)

Sonosky, J. M., 1961, Engineering the Oil Loan, *Jour. Petroleum Technology,* Jan., **13:**19–23.

Sorrentino, S., 1948, Criteri ed osservazioni sulla valutazione dei giacimenti e dei permessi minerari, *Soc. Geol. Italiana Boll.* **66:**59–66.

Soukup, B., 1963, Mineral Deposit Evaluation by the Aid of a Computer, in *Short Course and Symposium on Computer and Computer Applications in Mining and Exploration,* vol. 3, part HH J. C. Dotson and W. C. Peters, eds., University of Arizona, Tucson, 17p.

South Texas Geological Society, 1939, *Notes on Valuation of Oil and Gas Properties by the Corpus Christi Geologists' Study Group,* Corpus Christi, Tex., 47p.

Spinks, A. F., and S. P. Nicholls, 1973, A Proposed Model of Dynamic Ore Reserve Assessment for a Caving System in Mining, in *Application of Computer Methods in the Mineral Industry,* M. D. G. Salamon and F. H. Lancaster, eds., South African Institute of Mining and Metallurgy, Johannesburg, pp. 247–249.

Sproule, J. C., 1955, Classifications of Oil and Gas Reserves Estimates, *Canadian Oil and Gas Industries* **8**(8):51–55.

Stalker, W. H., 1937, Process Costs, *Costs and Management,* pp. 34–46.

Stammberger, F., 1974, Zur Bestimmung der ökonomischen Effektivität geologischer Erkundungsarbeiten, *Zeitshr. Angew. Geologie* **20**(5):225–232.

Standard Oil Company of New Jersey, 1958a, *An Analysis of the Depletion Provision as it Applies to the Petroleum Industry,* Standard Oil Company of New York, 24p.

Standard Oil Company of New Jersey, 1958b, *The Depletion Provision in Taxing Natural Resources: How and Why it was Created,* Standard Oil Company of New York, 15p.

Stanley, L., 1946, Taxation of Oil and Gas Properties, *Jour. Accountancy* **82:**333–334.

Stanley, L., 1957, *Percentage Depletion,* American Petroleum Institute, New York, 11p.

Stearn, E. W., 1966, Wake Up to Travesty of Zoning, *Rock Products,* July, pp. 88–93.

Steckley, R. C., 1978, Recently Developed Bureau of Mines Programs to Aid Mineral Deposit Exploration and Evaluation, in *Exploration,* American Min. Cong., vol. 16, 13p.

Steele, H., 1914, Mine Taxation, *Eng. and Mining Jour.* **98:**381–384.

Stephenson, E. A., 1933, Valuation of Natural Gas Properties, in *Geology of Natural Gas,* H. A. Ley, ed., American Institute of Petroleum Geologists, Tulsa, Okla., pp. 1011–1033.

Stephenson, E. A., 1968, Estimation of Natural Gas Reserves, *Am. Assoc. Petroleum Geologists Mem. 9,* vol. 2, pp. 2046–2103.

Stephenson, E. A., and I. G. Grettum, 1930, Valuation of Flood Oil Properties, *Am. Inst. Mining and Metall. Engineers Tech. Pub. 323,* 20p.

Stermole, F. J., 1977, *Economic Evaluation and Investment Decision Methods (and Problem Solutions Manual),* 2nd ed., vols. 1 and 2, Investment Evaluation Corporation, Golden, Colo.

Stevens, W. F., and G. Thodos, 1961, Estimating Oil Reserves from Production Data, *Petroleum Engineer* **33**(2):B46–B48, B50–B51.

Stoddard, S., 1955, Uranium—A New Industry and Its Tax Problems, in *Proceedings of the 6th Annual Institute on Oil and Gas Law,* Southwestern Legal Foundation, Dallas, pp. 539–589.

Stol, H., and G. Drever, 1977, Evaluation of Techniques for Assessing the Uranium Potential of Lake Covered Areas. Part 1: Underwater Radiometry, *Summ. Rept. Field Invest. Sask. (Regina),* pp. 72–77.

Storrar, C. D., 1966, Ore Valuation Procedures in the Gold Fields Group (with discussion), in *Symposium on Mathematical Statistics and Computer Applications in Ore Evaluation, Mar. 7-8,* South African Institute of Mining and Metallurgy, Johannesburg, pp. 276-298.

Strain, M. M., 1947, Some Specialized Phases of Accounting Practice, in *Accounting for the Extractive Industries,* pp. 35-79.

Stretch, R. H., 1904, Prospecting, Locating and Valuing Mines, 4th ed., McGraw-Hill, N. Y. (Reprinted from *Eng. and Mining Jour.*)

Stucke, H. J., 1970, Some General Formulae for the Geometry of Opencast Mines, in *Planning Open Pit Mines,* A. A. Balkema, Cape Town, South Africa, pp. 133-142.

Sugiyama, R., 1955, A Device for Prediction About the Variation of Thickness of the Ore-body, *Mining Geology* 5(17):162-172. (In Japanese.)

Sugiyama, R., 1958, L-S Fabric Analysis, *Japanese Jour. Geology and Geography* 29(1-3):75-98.

Sullivan, R., 1955, *Handbook of Oil and Gas Law,* Prentice-Hall, Englewood Cliffs, N. J., 556p.

Sultanov, S. A., V. I. Azamatov, and Y. S. Porman, 1973, Using Computers for Assessing Oil Supplies in Major Deposits, *Neftegazovaya Geologiya i Geofizika,* no. 11, pp. 25-27. (In Russian.)

Summers, W. L., 1927, *A Treatise on the Law of Oil and Gas,* Vernon Law Book Company, Kansas City, Mo., 863p.

Sundeen, S. W., 1968, Preliminary Evaluations of Surface Mine Prospects, in *Surface Mining,* E. P. Pfleider, ed., American Institute of Mining and Metallurgical Engineers, New York, pp. 54-65.

Surkan, A. J., J. R. Denny, and J. Batcha, 1964, Computer Contouring: New Tool for Evaluation and Analysis of Mines, *Eng. and Mining Jour.* 165(12):72-76.

Svenson, D., 1979, Efficient Australian Practice in Exploration and Coal Prospects, in *Coal Exploration,* vol., 2, G. O. Argall, Jr., ed., Miller Freeman, San Francisco, pp. 372-411.

Sverzhevskiy, V. L., 1975, Inzhenerno-geologicheskiye karty dlya tseley prognoza inzhenerno-geologicheskikh usloviy razrabotki ugol'nykh mestorozhdeniy Donetskogo basseyna, in *Problemy inzhenerno-geologicheskogo kartirovaniya; trudy Vsesoyuznogo simpoziuma,* G. A. Golodkovskaya et al., ed. et al., (30 Jan.-1 Feb., 1974), Izd. Mosk. Univ., Moscow, pp. 146-152.

Swanson, C. O., 1945, Probabilities in Estimating the Grade of Gold Deposits, *Canadian Inst. Mining and Metallurgy Trans.* 48:323-350.

Swanson, C. O., 1956, Evaluation of Ore Deposits, *Western Miner* 29(6):44-47.

Swanson, C. O., and W. T. Irvine, 1968, Ore Estimation in Cominco Mines, *Canadian Inst. Mining and Metallurgy Spec. Vol.* 9:69-70.

Swanson, E. A., 1968, Ore Reserve Estimation and Grade Control at Buchans, *Canadian Inst. Mining and Metallurgy Spec. Vol.* 9:119-122.

Switzer, P., and H. M. Parker, 1976, The Problem of Ore Versus Waste Discrimination for Individual Blocks; the Lognormal Model, in *Advanced Geostatistics in the Mining Industry,* M. Guarascio, M. David, and C. Huijbregts, eds., Reidel, Dordrecht, Holland, pp. 203-218.

Tapi, R. D., and P. Satyanarayana, 1973, Specific Gravity in Ore Reserves Estimation, *Indian Mining Eng. Jour.* 12(4):23-24.

Taylor, H., 1966a, Ore Valuation and Underground Metal Accounting at Bancroft and Other Zambian Copper Mines, *Inst. Mining and Metallurgy Trans.* 75:A109-A136.

Taylor, H., 1966*b*, Ore Valuation and Underground Metal Accounting in a Zambian Copper Mine, in *Symposium on Mathematical Statistics and Computer Applications in Ore Evaluation, Mar. 7–8,* South African Inst. Mining and Metallurgy, Johannesburg, pp. 353–357.

Taylor, H., 1972, General Background Theory of Cutoff Grades, *Inst. Mining and Metallurgy Trans.* **81**(788):A160–A179.

Taylor, R. G., 1979, *Geology of Tin Deposits,* Elsevier, Amsterdam, 543p.

Tegmeier, W., 1972, Calculation of Deposits by Means of Electronic Data Processing as a First Step for Production Planning, *Erzmetall* **25**(10):481–486. (In German.)

Terchroew, D. et al., 1965, An Analysis of Criteria for Investment and Financing Under Certainty, *Mining Science,* Nov., pp. 151–179.

Terry, F. T., 1946, Provisions of Income Tax Law and Regulations Affecting Owners of Oil and Gas Properties, *Jour. Accountancy* **81**:217–226.

Terry, F. T., and K. E. Hill, 1953, Valuation of Producing Properties for Loan Purposes, *Jour. Petroleum Technology,* July, sec. 1, **5**:23–26.

Thomas, J. C., 1954, Gold Mining Taxation in South Africa, *Canadian Tax Jour.,* Sept.–Oct., **2**:290–300.

Thomas, L. J., 1973, *An Introduction to Mining; Exploration, Feasibility, Extraction, Rock Mechanics,* Wiley, New York, 445p.

Thomas, L. K., and L. Hellums, 1972, A Nonlinear Automatic History Matching Technique for Reservoir Simulation Models, *Am. Inst. Mining and Metall. Engineers Trans.,* part 2, **253**:508–514.

Thorne, W. E., 1926, Testing and Estimating Alluvials for Gold, Platinum, Diamonds, or Tin, Mining Publishing Ltd, London, 60p.

Thurlow, E., 1939, The Role of Geology in the Valuation of Mines, *Glück Auf* **5**:6, 20.

Tixier, M. P., and R. P. Alger, 1970, Log Evaluation of Nonmetallic Mineral Deposits, *Geophysics* **35**:124–142.

Tixier, M. P., and M. R. Curtis, 1967, Oil Shale Yield Predicted from Well Logs, in *Drilling and Production,* Proceedings of the 7th World Petroleum Congress, Mexico, vol. 3, pp. 713–715.

Todd, J. D., 1950, Valuation and Subsurface Geology, in *Subsurface Geologic Methods,* 2nd ed., L. W. LeRoy, ed., Colorado School of Mines, Golden, Colo., pp. 792–809.

Tracy, G. W., 1955, Simplified Form of Material Balance Equation, *Jour. Petroleum Technology,* Jan., **5**:23–26.

Tracy, T. G., 1948, *Valuation of Illinois Oil Producing Properties for Tax Assessment,* State of Illinois Department of Revenue, 75p.

Trafton, B. W., and M. Sheinkin, 1969, Computer Applications in Financial Analysis, in *A Decade in Digital Computing in the Mineral Industry,* A. Weiss, ed., American Institute of Mining and Metallurgical Engineers, New York, pp. 241–254.

Troly, G., 1979, Procédure d'estimation des dépenses de recherche minière, *Chronique Mines et Recherche Minière,* no. 447, pp. 27–34.

Truscott, S., 1930, Valuation of Ore Reserves, *Mining Mag.,* May, **42**:313–318.

Truscott, S., 1947, Mine Economics; Sampling, Valuation, Organization, Administration, 2nd ed., Mining Publishing Ltd, London, 374p.

Trushkov, N. I., 1935, *Ekspertiza rudnykh mestorozhdenii,* ONTI NXTP SSSR, Glav. red. gorno-toplivnoi lit-ry, Leningrad-Moska.

Tucker, J. I., 1923, *Oil Valuation and Taxation,* Gulf Publishing Company, Houston, 332p.

Tulane University, 1952, *Tulane University Annual Tax Institute,* College of Law and Company Business Administration, New Orleans. (See also subsequent editions.)

Tulcanaza Navarro, E., 1970, Los procedimientos geoestadisticos en el análisis y estimación de yacimientos, *Santander Univ. Indus. Bol. Geología* **11**(25):61–79.

Tyler, P. M., 1959, Cost of Acquiring and Operating Mineral Properties, in *Economics of the Mineral Industries,* E. H. Robie, ed., American Institute of Mining and Metallurgical Engineers, New York, pp. 163–219.

Uchida, K., N. Ichisugi, N. Aratake, and S. Shigeno, 1970, Application of Two-Dimensional Trend Surface Analysis for the Ore Reserve Estimation at Lornex Mine, B.C., Canada, *Mining Geology* **20**(2):92–105.

Uglow, W. L., 1914, A Study of Methods of Mine Valuation and Assessment, with Special Reference to the Zinc Mines of Southwestern Wisconsin, *Wisconsin Geol. and Nat. History Survey Bull. 41* (Econ. Ser. no. 18), 73p.

Ulyanov, A. V., 1954, K voprosu o nauchnykh osnovakh perspektivnoi otsenki neftegazonosnykh oblastei, *Akad. Nauk SSSR, Institut Nefti, Tr.* **3:**3–13.

United States Treasury Department, 1919, *Manual for the Oil and Gas Industry Under the Revenue Act of 1918,* Bureau of Internal Revenue, Government Printing Office, Washington, D.C., 136p.

United States Treasury Department, 1922, *Memorandum 2039* in re *revaluation of Copper and Silver Properties,* 3p.

University of Arizona, 1969, *Symposium on Mine Taxation,* Society of Mining Engineers and College of Mines, University of Arizona, Tucson, var. pag.

Upadhyay, R. P., W. N. Hoskins, and R. L. Bolmer, 1977, Economic Evaluation of Oil Shale Mining in Colorado Using Sensitivity and Risk Analysis, in *Application of Computer Methods in the Mineral Industry,* R. V. Ramani, ed., American Institute of Mining and Metallurgical Engineers, New York, pp. 950–963.

Urasin, M. A., and I. L. Shamanskii, 1970, Computerizing Calculations for the Estimation of Some Types of Mineral Reserves, in *Sbornik matematicheskii metody v poiskovo-razvedochn praktike Irkutsk,* pp. 333–350. (In Russian.)

Vallée, M., J. Belisle, and M. David, 1977, Kriging as a Tool to Avoid Overestimation of Grade in Sulphide Orebodies, in *Application of Computer Methods in the Mineral Industry,* R. V. Ramani, ed., American Institute of Mining and Metallurgical Engineers, New York, pp. 1013–1025.

Vance, H., 1959, *Oil Field Evaluation,* privately published, Houston, Tex., 80p.

Vance, H., 1968, Requirements for Making Gas and/or Oil Loans and Summary of Current Factors in Estimating Reserves in Natural Gases of North America, B. W. Beebe and B. F. Curtis, eds., *Am. Assoc. Petroleum Geologists Mem. 9,* vol. 2, pp. 2104–2113.

Van Everdingen, A. F., E. H. Timmerman, and J. J. McMahon, 1953, Application of the Material Balance Equation to a Partial Water Drive Reservoir, *Jour. Petroleum Technology,* Feb., **5:**51–60.

Van Meurs, A. P. H., 1971, *Petroleum Economic and Offshore Mining Legislation: A Geological Evaluation,* Elsevier, Amsterdam, 208p.

Van Ostrand, C. E., 1926, Interpretation of Production Curves, *Oil and Gas Jour.,* Dec. 2, pp. 168, 306.

Van Rensburg, W. C. J., 1975, "Reserves" as a Leading Indicator to Future Mineral Production, *Resources Policy* **1:**343–356.

Varvill, W., 1935, The Examination of Abandoned Mines, *Inst. Mining and Metallurgy Bull. 371,* Aug., 16p.

Vasilev, P. A., 1968, Expedient Methods of Analyzing and Calculating the Reserves of Disseminated Gold, Silver and Cadmium in the Modzarovo Complex ore deposit, *Jubileen geologiceski sbornik, Bulg. Akad. Nauk.,* pp. 365-374. (In Bulgarian.)

Vasilev, P. A., 1972, On the Number of Hurricane Samples with Block Boundaries in which Reserves are Calculated, *Jour. Internat. Assoc. for Mathematical Geology* **4:**115-123.

Vasilev, P. A., 1973, Vurkhu predstavitelnostta na oprobvaneto pri prouchvane na tvurdi polezni izkopaemi, *Bulgarska Akademiya Naukite, Geologicheski Institut, Izvestiya Seriya Rudni i Nerudni Polezni Izkopoemi* **22:**197-204. (In Bulgarian.)

Vasilev, P. A., 1974, Otnosno opredelyaneto na obcmnoto teglo, *Bulg. Geol. Druzh., Spis.* **35**(2):214-217.

Vasil'yev, Yu. R., A. N. Dmitriyev, and V. V. Zolotukhin, 1973, Raspoznavaniye i otsenka nikelenosnykh differentsirovannykh trappovykh intruziy severa Sibirskoy platformy, *Akad. Nauk SSSR Sibirsk. Otdeleniye Geologiya i Geofizika,* no. 1, pp. 13-23.

Vassoevich, N. B., A. Y. Arkhipov, Y. K. Burlin, and Y. K. Sokolov, 1979, Improving the Methods of Evaluation of Oil Prospects of Sedimentary Basins, *Vestn. Mosk. Univ., Ser. 4, Geol., No. 6,* pp. 65-74.

Veatch, A. C., 1922, Estimation of Petroleum Reserves, *Econ. Geology* **17:**132-139.

Vegh, S., 1976, Asvanyvagyon-ertekitelet as kutatastervezes a nemzetkozi piacon, *Hung., Foeldt. Intez., Evi Jel.* **1974:**519-523. (In Hungarian.)

Velasquez, J. D., 1973, The Economics of Bauxite-Alumina-Aluminum in Western Canada, *Colorado School Mines Quart.* **68**(4):163-189.

Venter, R. H., 1976, A Statistical Approach to the Calculation of Coal Reserves for the Plains Region of Alberta, *Canadian Inst. Mining and Metallurgy Bull.* July, **69:**49-52.

Verma, B. K., 1972, Efficiency of Regression Techniques for Ore Valuation at Kolar Gold Fields, India, *Jour. Internat. Assoc. for Mathematical Geology* **4:**25-34.

Verner, W. J., and R. F. Shurtz, 1966, For Mine Evaluation—A Fresh Model, *Mining Eng.* **18**(11):65-71.

Versluys, J., 1928, An Investigation of the Problem of the Estimation of Gas Reserves, *Am. Assoc. Petroleum Geologists Bull.* **12:**1095-1105.

Vicente, C., 1966, Integração de dados de valorização de jazigos tabulares com vista à sua exploração a céu aberto, *Portugal Serviços Geol. Comun.* **50:**145-165.

Vickers, E. L., 1961, Marginal Analysis—Its Application in Determining Cut-off Grade, *Mining Eng.* **13:**579-582.

Vidyarthi, R. C., and P. P. Kala, 1975, Application of Probability Models and Sequential Decision-making in the Evaluation of the Balaria Zinc-Lead Deposit, Zawar, India, *Inst. Mining and Metallurgy Trans.* **84**(818):1-6.

Vikent'yev, V., O. Gus'kuv, and M. Shumilin, 1968, Connections Between Variation in Thickness and Values of an Ore Body, *Internat. Geology Rev.* **10:**648-652.

Vitorovic, D., D. Braticevic, and V. Lepojevic, 1975, Ein Programmsystem für die Lagerstättenerfassung und Darstellung mit komplexen berggeologischer Charakteristiken, in *13th International Symposium on the Application of Computers and Mathematics for Decision Making in Mineral Industries,* G. Dorstewitz, ed., Verlag Glueckauf GmbH, Essen, pp. E.II.1–E.II.16.

Vlasov, M. B. et al., 1970, Issledovaniye effektivnosti vzveshennoy i arifmeticheskoy otsenok srednego soderzhaniya pri podschete zapasov, *Vyssh. Ucheb. Zavedeniy Izv., Geologiya i Razved.,* no. 6, pp. 73–78.

Vogel, E., 1954, Analyse der in verschiedenen Ländern gebräuchlichen Vorratskategorien und eigene Vorschläge zur Klassifikation von Vorräten mit besonderer Berücksichtigung des Gangbergbaues, *Freiberger Forschungshefte* **10:**5–32.

Volodomonov, N. V., 1944, O methodike podscheta zapasov shil'nykh mestorozhdenii, *Gorny i Zhurnal,* no. 3–4.

Von Wahl, S., 1973, Criteria for the Economic Valuation of Non-Ferrous Ore Deposits, Erzmetall. **26**(1):26–32. (In Germany.)

Wahl, W. L., L. Mullins, and E. B. Elfrink, 1959, Estimation of Ultimate Recovery from Solution Gas-drive Reservoirs, *Am. Inst. Mining and Metall. Engineers Trans.* **213:**132–138.

Walker, B. D., 1968, The Log Analyst as an Expert Witness in Court, *Log Analyst* **9**(2):3–6.

Wallace, D., 1909, *Simple Mine Accounting,* 2nd ed., McGraw-Hill, New York, 78p.

Waller, R. E., 1956, *Oil Accounting; Principles of Oil Exploration and Production Accounting in Canada,* University of Toronto Press, 99p.

Walstrom, J. E., 1964, A Statistical Method for Evaluating Functions Containing Indeterminate Variables and Its Application to Recoverable Reserves Calculations and Water Saturation Determinations, *Stanford Univ. Pubs. Geol. Sci.* **9**(2):823–832.

Walstrom, J. E., T. D. Mueller, and R. C. McFarlane, 1967, Evaluating Uncertainty in Engineering Calculations, *Am. Inst. Mining and Metall. Engineers Trans.,* part 1, **240:**1595–1603.

Wanielista, K., 1978, Ekonomiczne kryteria bilansowosci zloz rud, *Archiwum Gornictwa* **23:**273–284.

Wansbrough, R. S., 1960, An Approach to the Evaluation of Oil-Production Capital Investment Risks, *Jour. Petroleum Technology,* Sept., **12:**25–29.

Warren, J. E., 1956, Considerations Concerning Bank Financing of Oil and Gas Properties, *Jour. Petroleum Technology,* May, **8:**11–14.

Washburne, C. W., 1915, The Estimation of Oil Reserves, *Am. Inst. Mining and Metall. Engineers Trans.* **51:**645–648.

Wasson, P. A. et al., 1962, Lake Superior Iron Resources—Metallurgical Evaluation and Classification of Nonmagnetic Taconite Drill Cores from the West Central Mesabi Range, *U.S. Bur. Mines Rept. Inv. 6081,* 62p.

Waterman, G. C., and S. Hazen, 1968, Development Drilling and Bulk Sampling, in *Surface Mining,* E. P. Pfleider, ed., American Institute of Mining and Metallurgical Engineers, New York, pp. 69–102.

Watermeyer, G. A., 1919, Application of Theory of Probability to Ore Reserves, *Chem., Metall. and Mining Soc. South Africa Jour.* **19:**97–108.

Watkins, P. B., 1959, Economic Evaluations, *Jour. Petroleum Technology* **11**(11):20–28.

Watson, D. M., 1977, Resources Development, Regional Planning and Ore Reserves, *Canadian Inst. Mining and Metallurgy Bull.* **70**(781):104–108.

Watson, J. W., 1966, The Estimation of Natural Gas Reserves, in *Natural Gas,* (Inst. Pet., Explor. and Prod. Group Symp. Proc., London), Elsevier, pp. 32–50, 111–112.

Weaton, G. F., 1965, Estimating Minnesota's Natural Iron Reserves, *Mining Eng.,* Jan., **17:**38–41.

Weaton, G. F., 1973, Valuation of Mineral Deposits, *Mining Eng.,* May, **25:**29–36.

Weaver, R., 1964, Relative Merits of Interpolation and Approximating Functions in the Grade Prediction Problem, *Stanford Univ. Pubs. Geol. Sci.* **9**(1):171–185.

Webber, C. E., 1961, Estimation of Petroleum Reserves Discovered, in *Petroleum Exploration Handbook,* G. B. Moody, ed., McGraw-Hill, New York, pp. 25-1-25-25.

Webber, M., 1921, Valuing Partly Exhausted Mines, *Mining and Scientific Press* **123:**385–389, Mar. 19; 489–494, Apr. 9.

Webster, J. S., 1936, Valuation of Mining Investments, *Chem. Eng. and Mining Rev.* **28:**188–193, 215–218.

Weeks, J. P., 1968, Ore Reserve Estimation and Grade Control at the Bralorne Mine, *Canadian Inst. Mining and Metallurgy Spec. Vol.* **9:**233–238.

Weeks, R. M., 1968, Ore Reserve Estimation and Grade Control at the Quemont Mine, *Canadian Inst. Mining and Metallurgy Spec. Vol.* **9:**123–129.

Weeks, W. G., 1932, Oil Well Production Curve Formulae, *Inst. Petroleum Technology Jour.* **18:**805–815.

Weinaug, C. F., 1964, Proposed Ad Valorem Tax Property Valuation, *Stanford Univ. Pubs. Geol. Sci.* **9**(2):721–728.

Weis, P. L., 1978, Estimating Reserves of Amorphous Graphite in Sonora, Mexico, *Eng. and Mining Jour.* **179**(10):123–128.

Weisner, R. C., J. F. Lemons, and L. Coppa, 1980, Valuation of Potash Occurrences Within the Nuclear Waste Isolation Pilot Plant Site in Southeastern New Mexico, *U.S. Bur. Mines Inf. Circ. 8814,* 94p.

Weller, C., and O. Lipscomb, 1955, Some Problems of Depreciation, *Texas Law Rev.* **33:**886–913.

Wellmer, F.-W. and P. Podufal, 1977, A Statistical Model for the Exploration of the Ramsbeck Pb/Zn Mine (F. R. G.). in *Application of Computer Methods in the Mineral Industry,* R. V. Ramani, ed., American Institute of Mining and Metallurgical Engineers, New York, pp. 431–440.

Wells, J. H., 1973, *Placer Examination—Principles and Practice,* U.S. Department of the Interior and Bureau of Land Management, Washington, D.C., 209p.

Welwood, R. J. R., 1968, Ore Reserve Estimation and Grade Control at Brunswick Mining and Smelting, *Canadian Inst. Mining and Metallurgy Spec. Vol.* **9:**168–171.

Westervelt, W., 1941, Mine Examinations, Valuations and Reports, in *Mining Engineers' Handbook,* vol. 2, sec. 25, R. Peele and J. Church, eds., Wiley, New York, 33p.

Wharton, H. M., 1972, Barite Ore Potential of Four Tailings Ponds in the *Washington County Barite District, Missouri, Missouri Div. Geol. Survey and Water Resources Rept. Inv. No. 53,* 91p.

White, E. E., 1914a, Commentary on J. R. Finlay, "Valuation of Iron Mines," *Am. Inst. Mining and Metall. Engineers Trans.* **45:**304–321.

White, E. E., 1914b, Further Discussion on J. R. Finlay's "Valuation of Iron Mines," *Am. Inst. Mining and Metall. Engineers Trans.* **50:**188–196.

White, G. H., and C. E. Gee, 1977, Computerized Geological and Mining Ore Reserve Systems at Western Mining's Kambalda Nickel Operations, *Internat. Symp. Appl. Comput. Oper. Res. Miner. Ind.,* no. 15, pp. 263-274.

White, P. C., 1940, Estimating Availability of Natural Gas Reserves, *Petroleum Engineer* **12**(3):23-28.

White, V. C., and A. W. Brainerd, 1956, Percentage Depletion—A Costly Study in Definitions, *Taxes: The Tax Mag.* Feb., **34**:97-106.

Whitehead, W. L., 1957, Valuation of Oil Property, in *Examination and Valuation of Mineral Property,* C. H. Baxter and R. Parks, eds., Addison-Wesley, Reading, Mass., pp. 299-343.

Whitford, D. F. H., 1970, The Computer at Brenda Mines, in *Decision-making in mineral exploration III, computer assistance in the management of exploration programs,* Exension Department, Engineering Program, University of British Columbia, Vancouver, pp. 54-62.

Whitten, E. H. T., 1966a, The General Linear Equation in Prediction of Gold Content in Witwatsrand Rocks, South Africa, in *Symposium on Mathematical Statistics and Computer Applications in Ore Evaluation, Mar. 7-8,* South African International Association for Mathematical Geology, Johannesburg, pp. 124-156.

Whitten, E. H. T., 1966b, Quantitative Models in the Economic Evaluation of Rock Units, Illustrated with the Donegal Granite and the Gold-bearing Witwatsrand Conglomerates, *Inst. Mining and Metallurgy Trans.* **75**:B181-B198.

Whitton, W. W., 1918, Formulas for Valuation of Mines, *Mining and Scientific Press* **116**:691-694.

Wilke, F. L., 1973, Capacity Calculations, Investment Allocation and Long-range Production Scheduling in German Coal Mines, in *Application of Computer Methods in the Mineral Industry,* M. D. G. Salamon and F. H. Lancaster, eds., South African Institute of Mining and Metallurgy, Johannesburg, pp. 133-138.

Wilke, F. L., and E. Von der Linden, 1974, Bauwürdigkeitsbeurteilung von Erzvorräten mit dem Cut-off-grade, *Erzmetall.* **27**(10):465-469.

Willcox, F., 1938, *Metal-Mining Company Accounting and Administration,* Montreal, 128p.

Willcox, F., 1949, *Mine Accounting and Financial Administration,* Pitman, New York, 489p.

Williams, D. W., 1951, Engineering Aspects of Depreciation, *2nd Annual Institute on Oil and Gas Law,* Southwestern Legal Foundation, Dallas, pp. 415-423.

Williams, D. W., 1952, Valuation of Natural Resource Properties, Tulane University Tax Institute, New Orleans, La., pp. 218-229.

Williams, H. R., and C. J. Meyers, 1975, *Oil and Gas Law,* Bender, New York, 678p.

Williamson, D. R., and E. R. Mueller, 1977, Ore Estimation at Cyprus Pima Mine, *Am. Inst. Mining and Metall. Engineers Trans.* **262**:17-29.

Williamson, D. R., and T. Thomas, 1974, Trend Analysis of Alluvial Deposits by Use of Rolling Mean Techniques, in *Geological Mining and Metallurgical Sampling,* M. J. Jones, ed., Institution of Mining and Metallurgy, London, pp. 94-101.

Willis, C., 1917-1918, Selling prospects, *Univ. Arizona Bull. 62, Econ. Ser. No. 13,* 18p.

Wilshusen, R. C., 1962, The Combined Use of Core Analysis and Log

Data for Better Productivity Evaluation, *South Texas Geol. Soc. Bull.* **2**(9):13–20.

Wilson, R. G., 1976, Estimation of the Potential of a Coal Basin, in *Coal Exploration,* W. L. G. Muir, ed., Miller Freeman, San Francisco, pp. 374–400.

Wilson, W. W., 1961, Oil and Gas Property Acquisition, *Jour. Petroleum Technology* **13**(6):527–530.

Wilson, W. W., and A. J. Pearson, 1962, How to Determine the Market Value of Secondary Recovery Reserves, *Jour. Petroleum Technology,* Aug., pp. 829–833.

Wimpfen, S. P., and H. J. Bennett, 1975, Copper Resource Appraisal, *Resources Policy* **1**:126–141.

Winckworth, K. C., 1968, Methods of Grade Control at the Rio Algom Uranium Properties in the Elliot Lake Area, *Canadian Inst. Mining and Metallurgy Spec. Vol.* **9**:261–264.

Winfrey, R., and J. C. Hempstead, 1952, *Engineering Valuation and Depreciation,* Iowa State University Press, Ames, 508p. (Reprinted 1965.)

Wintermute, T. J., 1971, Gulf Coast "Rule of Thumb" Economics, *Gulf Coast Assoc. Geol. Socs. Trans.* **21**:353–362.

Wisconsin Department of Natural Resources, 1976, Zoning and Financial Incentives for Reservation of Mineral Lands in Wisconsin, *Univ. Wisconsin Geol. Nat. History Survey Spec. Rept. No. 6,* 151p.

Wittman, C. W., 1955, Aggregation of Gas and Oil Properties, in *Budgeting, Forecasting, Return on Investment, and Related Papers,* Controllers Institute of America, pp. 91–99.

Wolf, H. J., 1958, How to Estimate Ore Reserves, *Eng. and Mining Jour.* **159**(12):92–95.

Wolff, J., 1925, Estimating on the Gogebic Range, *Am. Inst. Mining and Metall. Engineers Trans.* **72**:653–656.

Wolff, J., E. L. Derby, and W. A. Cole, 1925, Sampling and Estimating Lake Superior Iron Ores, *Am. Inst. Mining and Metall. Engineers Trans.* **72**:641–658.

Wood, H. S., 1922, Establishing the Value of Oil and Gas Leases as of March 1, 1913, *Natl. Petroleum News,* Oct. 18, pp. 43–47.

Wooddy, L. D., and T. D. Capshaw, 1960, Investment Evaluation by PresentValue Profile, *Jour. Petroleum Technology,* June, **12**:15–18.

Woodtli, R., 1971, L'enseignement de la prospection miniere par simulation, *Schweizer. Mineralog. u. Petrog. Mitt.* **51**:544–550.

Wright, C. W., 1936, Essentials in Developing and Financing a Prospect into a Mine, *California Jour. Mines and Geology* **32**(2):167–188.

Wright, R. J., 1959, An Experimental Comparison of U_3O_8 Ore Estimates vs. Production, *Eng. and Mining Jour.* **160**(11):100–102.

Wyer, S., 1914, *Notes on Depreciation of Natural Gas Wells,* Columbus, Ohio, 31p.

Wyer, S., 1917, Principles of Natural Gas Leasehold Valuation, *Am. Inst. Mining and Metall. Engineers Trans.* **56**:782–798.

Xhacka, P., 1972, Mbi vleresimin sasior te rezervave prognoze te naftes dhe te gazit, *Permbledhje Stud.,* no. 3, pp. 71–75.

Yatselenko, V. S., 1973, Sravnitel'naya otsenka dobyvnykh vozmozhnostey zapasovosnovnoy faktor ratsional'nogo razmeshcheniya razvedochnykh rabot, *Geologiya Nefti i Gaza,* no. 4, pp. 9–12.

Yershov, V. V., 1975, Oprativnaya otsenka kachestva i zapasov mineral'nogo syr'ya EVM., in *13th International Symposium on the Application of Computers and Mathematics for Decision Making in the Mineral Industries*, G. Dorstewitz, ed., Verlag Glueckauf GmbH, Essen, pp. C.III.1-C.III.13.

Young, C. M., 1949, Estimation of Reserves, in *Natural Gas Operations*, Columbia Gas System, ed., Columbia Engineering Corporation, 11p.

Young, L. E., 1916, Mine Taxation in the United States, *Univ. Illinois Stud. Soc. Sci.*, vol. 5, no. 4, 275p.

Yui, S., 1959, Precision of the Estimates of the Ore Reserves in the Fully Developed Ore Blocks, *Mining Geology* **9**(38):319-333. (In Japanese.)

Yumatov, B. P., P. K. Beketov, V. I. Papichev, and A. V. Silakov, 1973, Modern Methods of Averaging Non-ferrous ores in pits, *Vyssh. Ucheb. Zavedeniy Izv., Geologiya i Razved.* **73**(4):78-84. (In Russian.)

Zambo, J., 1970, Die Rolle der Verzinsung bei der Wahl der Förderkapazität für Bergwerksbetriebe, *Acta Geod. Geophys. Montan.* **5**(1-2):143-153.

Zambo, J., 1971, Über die Abbauwürdigkeit, Acta *Geodaetica Geophysica et Montanistica* **6**(1-2):163-174.

Zambo, J., 1972, Eine Möglichkeit zur Verkürzung der Rücklaufszeit von Investitionskosten im Erzbergbau, *Acta Geodaetica Geophysica et Montanistica* **7**(3-4):433-439.

Zapffe, C., 1925, Estimating in the Cuyana Iron Ore District, Minnesota, *Am. Inst. Mining and Metall. Engineers Trans.* **72**:661-664.

Zeidler, W., 1969, Economic Feasibility Studies on Wadi Fatima Iron Ore, Saudi Arabia, *Dir. Gen. Min. Resour. Res.*, **1969**:64-66 (1970).

Zemlyanov, V. N., and Yu. M. Olonov, 1970, Application of the Relative Probability Formula for Quantitative Appraisal of Information Value of Prospecting Indications, *Sovetskaya Geologiya* **70**(5):119-127. (In Russian.)

Zenin, M. F., 1938, Produkty okisleniya i vyshchelachivaniya sulifidov kak kriterii otsenki mestorozhdenii medno-porfirovykh rud po vykhodam, *Sredneaz. Ind. Inst., Gor. Fak., Tr., No. 1(9)*, 87p.

Zenkov, D. A., 1937, An Attempt of Using the Coefficient of Variability of Properties of an Ore Body to the Estimation of Reserves and Some Sampling Problems, *All-Union Gold and Platinum Prosp. Trust, Tr.*, issue 4, pp. 102-106. (In Russian.)

Zenkov, D. A., 1963, Puti ratsionalizatsii oprobovaniya gornykh vyrabotok, *Sovetskaya Geologiya*, no. 5, pp. 98-103.

Zeschke, G., 1964, Prospektion und feldmässige Beurteilung von Lagerstätten, Springer-Verlag, Vienna, 307p.

Zezulka, J., 1971, On the Problem of the Value of the Deposit of Mineral Raw Materials, *Geologicky Pruzkum* **13**:335-337. (In Czech.)

Zharov, E., and A. Trofimov, 1975, Contouring Study of Ore Deposits by Spatial Modeling, *Vyssh. Ucheb. Zavedeniy Izv., Geologiya i Razved.*, no. 3, pp. 94-99. (In Russian.)

Zhdanov, M. A., 1960, Problem of Calculation of Casing-head Gas Reserves, *Petroleum Geology* **3**(3-B):177-180. (In translation.)

Zhdanov, M. A., and I. S. Gutman, 1969, Nekotoryye aspekty rascheta iskhodnykh parametrov pri primenenii ob"yemnoy formuly podscheta zapasov nefti i gaza, *Geologiya Nefti i Gaza* **13**(1):29-32.

Zhdanov, M. A., M. N. Soson, and Z. G. Borisenko, 1973, O metodike otsenki stepeni dostovernosti parametrov zaleshey nefti i gaza, *Geologiya Nefti i Gaza,* no. 10, pp. 10–12.

Zhdanov, M. A., M. G. Ovanesov, and M. A. Tokarev, 1974, Kompleksnyy uchet geologicheskoy neodnorodnosti pri analize vyrabotki nefti i prognoze konechnogo koeffitsiyenta nefteotdachi, *Geologiya Nefti i Gaza,* no. 3, pp. 19–23.

Ziesmer, H. M., 1926, Sampling and Estimating Ore at Sacramento Hill, *Arizona Mining Jour.* **10**(9):7–11, 19–21.

Zodrow, E., and D. P. Harris, 1967, An Ore Grading Model for the Smallwood Mine, *Mining Eng.,* Aug., **19:**70–73.

Zorileanu, D., 1970, A Model of Simulation by Monte Carlo Methods of the Daily Ore and Metal Output in a Mining Exploitation, *Rev. Minelor* **21**(7):270–275. (In Romanian.)

AUTHOR CITATION INDEX

SUBJECT INDEX

375

About the Editor

SAM L. VanLANDINGHAM has been an oil-field laborer, mining geologist, university professor, research associate, and currently is a consulting geologist and environmentalist at 3741 Woodsong Drive, Cincinnati, Ohio 45239. His broad interests are typified in the authorship of ten books and many papers on such varied subjects as theoretical crystallography, analysis of meteorites, petrology, stratigraphy, paleoecology, environmental biology, and taxonomy. Experience has taken him from the oil fields of west Texas to the coal fields of Kansas and Kentucky, and to the gold fields and diatomite deposits of California and the Pacific Northwest. Dr. VanLandingham has received numerous grants from funding agencies, the most recent from the National Science Foundation for studies at the California Academy of Sciences in San Francisco. His most current interests include feasibility studies and geological preevaluation of mining prospects.

Benchmark Papers
in Geology

Series Editor: Rhodes W. Fairbridge
Columbia University